Biochemistry of Storage Carbohydrates in Green Plants

Biochemistry of Storage Carbohydrates in Green Plants

Edited by

P. M. Dey

R. A. Dixon

Department of Biochemistry
Royal Holloway College
(University of London)
Egham Hill, Egham
Surrey, England

1985

ACADEMIC PRESS

(Harcourt Brace Jovanovich, Publishers)

London   Orlando   San Diego   New York
Toronto   Montreal   Sydney   Tokyo

ACADEMIC PRESS INC. (LONDON) LTD.
24–28 Oval Road
LONDON NW1 7DX

United States Edition published by
ACADEMIC PRESS, INC.
Orlando, Florida 32887

British Library Cataloguing in Publication Data

Biochemistry of storage carbohydrates in
    green plants.
    1. Botanical chemistry 2. Carbohydrates----
    Chemical analysis 3. Plants----Analysis
    I. Dey, P.M.   II. Dixon, R.A.
    581.19'248   QK861

Library of Congress Cataloging in Publication Data
Main entry under title:

Biochemistry of storage carbohydrates in green plants.

    Includes bibliographies and index.
    1. Carbohydrates. 2. Botanical chemistry. I. Dey,
P. M. II. Dixon, R. A. III. Title: Storage carbohydra-
tes in green plants.
QK898.C3856 1984   581.19'248   84-16778
ISBN 0-12-214680-8 (alk. paper)

PRINTED IN THE UNITED STATES OF AMERICA

85 86 87 88   9 8 7 6 5 4 3 2 1

Contents

Chapter 1
Sucrose

  J. S. HAWKER

Chapter 2
D-Galactose-Containing Oligosaccharides

  P. M. DEY

Chapter 3
Glycosides

  P. M. DEY AND R. A. DIXON

Chapter 4
Starch

D. J. MANNERS

Chapter 5
Fructans

HORACIO G. PONTIS AND ELENA DEL CAMPILLO

Chapter 6
β-(1→3)-Linked Glucans from Higher Plants

R. A. DIXON

Chapter 7
Galactomannans

J. S. GRANT REID

Chapter 8

Mannans and Glucomannans

J. DEREK BEWLEY AND J. S. GRANT REID

Chapter 9

Algal Polysaccharides

ELIZABETH PERCIVAL AND RICHARD H. McDOWELL

Chapter 10

Polysaccharides Containing Xylose, Arabinose, and Galactose
in Higher Plants

K. BRINSON AND P. M. DEY

Contributors

Numbers in parentheses indicate the pages on which the author's contributions begin.

J. DEREK BEWLEY (289), Plant Physiology Research Group, Department of Biology, University of Calgary, Alberta T2N 1N4, Canada

K. BRINSON (349), Department of Biochemistry, Royal Holloway College (University of London), Egham Hill, Egham, Surrey TW20 0EX, England

ELENA DEL CAMPILLO* (205), Centro de Investigaciones Biologicas, F.I.B.A., 7600 Mar del Plata, Casillo de Correo 1348, Argentina

P.M. DEY (53, 131, 349), Department of Biochemistry, Royal Holloway College (University of London), Egham Hill, Egham, Surrey TW20 0EX, England

R.A. DIXON (131, 229), Department of Biochemistry, Royal Holloway College (University of London), Egham Hill, Egham, Surrey TW20 0EX, England

J.S. HAWKER (1), CSIRO, Division of Horticultural Research, Adelaide, South Australia 5001, Australia

D.J. MANNERS (149), Department of Brewing and Biological Sciences, Heriot-Watt University, Edinburgh EH1 1HX, Scotland

RICHARD H. McDOWELL (305), Bourne Laboratory, Department of Chemistry, Royal Holloway College (University of London), Egham Hill, Egham, Surrey TW20 0EX, England

ELIZABETH PERCIVAL (305), Bourne Laboratory, Department of Chemistry, Royal Holloway College (University of London), Egham Hill, Egham, Surrey TW20 0EX, England

HORACIO G. PONTIS (205), Instituto de Investigaciones Biologicas, Facultad de Ciencias—Exactas, Naturales y Biologicas, Universidad Nacional de Mar del Plata, and Centro de Investigaciones Biologicas, F.I.B.A., Casillo de Correo 1348, Argentina

J.S. GRANT REID (265, 289), Department of Biological Science, University of Stirling, Stirling FK9 4LA, Scotland

*Present address: Department of Agronomy, Cornell University, Ithaca, New York 14853.

Preface

The past few years have witnessed an explosion in the research on molecular plant sciences. This mainly reflects the attention given to plant systems by molecular biologists and biotechnologists. Underlying these approaches, however, is the often stressed need for more research in basic plant metabolism. Plant carbohydrate biochemistry has never been regarded as one of the most glamorous areas of plant science. However, because the ultimate goals of plant molecular biology and biotechnology must include both an increase in the yield of food crop products and a full exploitation of natural plant products in industrial situations, an understanding of the molecular aspects of the synthesis and mobilisation of plant storage carbohydrates is an essential basis for future biotechnological developments. This book therefore aims to present a detailed account of the biochemistry of various classes of storage carbohydrates found in green plants. Because most information in this area is presently found in large, multivolume works, the present format should particularly appeal to undergraduate and postgraduate students. However, we have tried to ensure that detail has not been lost at the expense of space considerations, and thus we hope that the book will provide useful information for research workers in plant carbohydrate chemistry and biochemistry and in food and other technologies.

The first three chapters of the book deal with low molecular weight reserve materials, while subsequent chapters deal with polysaccharides. Sucrose, sucrose-based oligosaccharides, starch, fructans, and mannose-containing polysaccharides are widely recognised as constituting major storage carbohydrates; these are discussed in Chapters 1, 2, 3, 4, 5, 7, and 8. In addition, we have chosen to include sections on the biochemistry of carbohydrates having less clearly defined storage functions; these include certain oligosaccharides, glycosides, and cell-wall associated polysaccharides. A case is made for the inclusion of such "fringe" compounds in the individual chapters. We hope this broad approach increases, rather than dilutes, the usefulness of the book. Finally, we felt that a chapter on algal polysaccharides was justified in view of their commercial importance, even though some of these compounds may not serve a true storage function. Each chapter aims to present information on the structure(s), occurrence, biosynthesis, degradation, and physiological role of the compound(s) under discussion.

P. M. Dey
R. A. Dixon

Acknowledgements

The contributors of chapters are grateful to various publishers and authors for permission to use copyright materials. The senior editor (P. M.D.) expresses his gratitude to Professor J. B. Pridham, Department of Biochemistry, Royal Holloway College, for valuable advice and support at the stage of planning of this book.

Chapter 1

Sucrose

J. S. HAWKER

CSIRO Division of Horticultural Research
Adelaide, South Australia

I. OCCURRENCE AND PROPERTIES OF SUCROSE IN STORAGE TISSUES

Sucrose is the major form in which carbon is translocated in most plants and is usually the first product of photosynthesis to accumulate in photosynthetic cells. It occurs in nearly all organs of plants and accumulates to high concentrations in some organs of some species. Of the estimated 2×10^{10} tons of carbon fixed in photosynthesis by land plants annually, about 4×10^7 tons (0.2% of fixed carbon or 11% of world food) ends up in the sucrose that is

BIOCHEMISTRY OF STORAGE
CARBOHYDRATES IN GREEN PLANTS

produced commercially from sugar cane, sugar beet, and other minor crops (Anonymous, 1980). A large amount of sucrose is also consumed annually as food in fruits and vegetables, e.g., about 0.6×10^6 tons of carbon as sucrose in 3.5×10^7 tons of apples. By contrast, wheat, rice, maize, potato, yam, and cassava contribute about 2.8×10^8 tons of carbon as starch to the world's annual food supply (Jenner, 1982).

Being readily soluble, sucrose can be relatively easily extracted from plant crops and purified by crystallization. High yields, good keeping qualities, and ready acceptance as food ensures a place for sucrose in the world markets. In time it is also likely to partially replace oil as a feedstock for industrial processes and as a precursor of liquid fuel (power alcohol).

In this chapter the biochemistry of the accumulation of sucrose in storage organs as it is now understood is discussed and compared with the better known mechanisms occurring in leaves and conducting tissue. For detailed literature surveys and historical aspects of sucrose synthesis, metabolism, transport, and accumulation, readers could consult recent essays (Pontis, 1977; Whittingham *et al.*, 1979; Akazawa and Okamoto, 1980; Avigad, 1982; Komor, 1982; Preiss, 1982; Willenbrink, 1982).

Sugar cane is a C4 tropical plant, whereas sugar beet is a C3 temperate plant, and over 90% of the world's sucrose is produced by these two plants. The choice of crop depends on the climate, and there is about twice as much sucrose produced in the world from sugar cane as from sugar beet. The other 10% or so of sucrose is called noncentrifugal sugar and is produced by other than centrifugal processes in relatively few areas. Included are such kinds of sugar as pinoncillo, panela, papelon, chancaca, radura, jaggery, gur, muscovado, and panocha (Anonymous, 1979). In many countries the farming, harvesting, and processing of both sugar cane and sugar beet is highly mechanized, which makes sucrose a competitively priced food and which has allowed its use as a liquid fuel precursor in Brazil and possibly eventually in other countries in the next century as oil supplies decline.

Many ripe fruits contain sucrose, either as a consequence of starch hydrolysis during the climacteric (banana, apple) or accumulated during development (grape, orange). Changes occurring in Red Delicious apples are shown in Fig. 1. Apples, apricots, bananas, dates, melons, oranges, peaches, pineapples, and plums contain 3–8% sucrose on a fresh weight basis in the edible portion. Other fruits such as lemon and pears contain less than 1% sucrose, and most fruits contain in addition glucose and fructose (Whiting, 1970). In grapes of the most widely grown species of grapevine (*Vitis vinifera*), between 0.2 and 1.0% sucrose occurs, but there is evidence that the large amount of glucose and fructose accumulated (up to 15% of each) is a result of hydrolysis of sucrose within the vacuole (Hawker *et al.*, 1976). In some other species of grapevine, up to 5.6% sucrose is accumulated (Lott and Barrett,

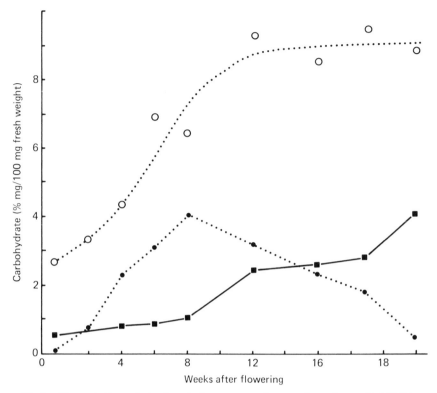

Fig. 1. Concentrations of starch, reducing sugars, and sucrose in apples, cv. Red Delicious, during growth in an orchard near Adelaide, South Australia. From W. J. S. Downton and J. S. Hawker (unpublished). ○, reducing sugars; ●, starch; ■, sucrose.

1967). Grapes contain no starch. The differences occurring in relative concentrations of glucose, fructose, and sucrose in various fruits is interesting biochemically because it may provide experimental material of possibly different enzyme constitutions for further research into pathways of synthesis and accumulation of sucrose.

The two major sources of commercial sucrose, sugar cane stalks and sugar beets, store up to 20% fresh weight of sucrose. In ripe sugar cane stalks there is little reducing sugar (<0.05%), but higher amounts (up to 5%) are found in young growing internodes and in mature stalks of actively growing plants. Starch levels in sugar cane stalks and sugar beets are usually very low in cultivated varieties (<0.01% in sugar cane). As will be discussed later, studies on the accumulation of sugars by these tissues have been and are being carried out in several laboratories. In addition, red beet tissue is popular experimental material, especially in studies where protoplasts and vacuoles are prepared, because of the anthocyamin present, which serves as an endogenous coloured marker in the vacuoles.

Sucrose is a nonreducing disaccharide of an α-D-glucopyranose and β-D-fructofuranose joined by a (1→2) linkage:

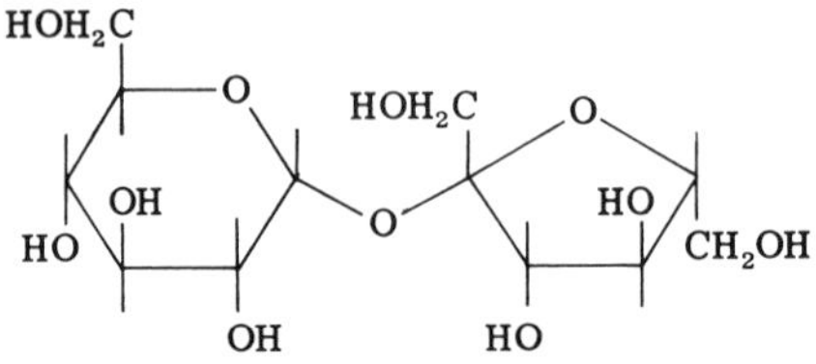

Sucrose

It is hydrolysed by even dilute acids, but it is reasonably stable in the presence of alkalis. Hydrolysis is also catalysed by the widely distributed enzyme β-fructofuranosidase (invertase) and by α-glucosidase, which is present in honey and animal intestine. The lability of sucrose is due to its glycosidic linkage between an α-glucopyranosyl moiety in the chair conformation and a β-D-fructofuranosyl moiety in the envelope conformation. The β-fructofuranoside nature of sucrose is quite uncommon and, although it does occur in other compounds, they can all be considered derivatives of sucrose. The high free energy of hydrolysis of sucrose (−7.0 kcal/mol) has been put forward as an explanation for the distinctive role of sucrose in plants, but this argument seems unlikely when the potential yield of ATP molecules from sucrose and a mixture of glucose and fructose are compared (about 69 compared to 68). In any case, if invertase hydrolyses sucrose the free energy of hydrolysis is lost to the organism, and it is only where sucrose synthase catalyses the synthesis of UDP-glucose from sucrose (to be discussed later), that the energy is conserved.

Few of the properties of sucrose considered seemed to suggest a reason for its role in translocation and storage, and only by comparison with trehalose, which is the only other nonreducing disaccharide found in nature and which in fungi and insects finds a role like sucrose in plants, did the nonreducing nature of sucrose appear significant. It has been proposed that sucrose, as a relatively unreactive derivative of glucose, protects the glucose from attacks by enzymes which catalyse its metabolism. Perhaps the significant feature of sucrose compared to glucose is that it is not phosphorylated directly, since sucrose kinases are not known in nature. It is of interest that mannitol is translocated in the lower plant group, brown algae, and a mannitol kinase is known, whereas sorbitol is translocated in some species of Prunoideae and sorbitol kinases are not known. Whereas in animals glycogen hydrolysis to release glucose shows a high degree of hormonal and enzymatic control, in plants the control of starch breakdown is not as sophisticated. However, in plants excess glucose can be converted to sucrose via sucrose phosphate and temporarily stored in the vacuole rather than enter a futile cycle of phosphorylation and hydrolysis by kinases and phosphatases as might otherwise occur. The possible role of a specific phosphatase in the synthesis and storage processes will be discussed later.

The major roles of sucrose in most plants are as a temporary storage product in leaves and as a translocation form of carbon. Sucrose is also important in many flowers as a pollinator attractant and/or as a germination medium for pollen on stigmas. Its role as a longer term storage compound is important more in the quantity stored worldwide in a few species than in its distribution in the plant kingdom. Starch as a store of carbon is of more widespread importance than sucrose, and the success of plants that store high concentrations of sucrose is mainly due to selection and protection by man. Sucrose is a highly soluble sugar having no apparent inhibitory effect on most biochemical reactions, properties that make it a useful compound in the control of osmotic pressure and flow of water between cellular compartments. However, at high concentrations in a particular compartment (e.g. vacuole), it might generate the requirement for a high concentration of other compounds in the cytoplasm to maintain osmotic balance. Insoluble starch does not generate a similar requirement. On the other hand, absolute compartmentation of sucrose probably does not occur. Regardless of the properties that might detract from its usefulness as a reserve material in plants, sucrose seems certain to remain an important world agricultural product because of its appeal and usefulness to man. Hopefully, man will survive to nurture sugar cane, sugar beet, and other sucrose storing crops.

II. SUCROSE SYNTHESIS

The enzymes of sucrose synthesis and sucrose breakdown have been described by Pontis (1977) and more recently and extensively by Avigad (1982). These two works were used as source material in the current description of the enzymes, and they should be consulted if more detailed references are required regarding some statements.

Three enzymes are considered to be associated with sucrose synthesis in green plants. They are

1. Sucrose phosphate synthase (UDP-D-glucose:D-fructose 6-phosphate 2-α-D-glucosyltransferase, EC 2.4.1.14)

UDP-D-glucose + fructose 6-phosphate $\longrightarrow$ sucrose 6^F-phosphate + UDP + H$^+$

Sucrose 6^F-phosphate is sucrose phosphorylated at the 6 position of the fructose moiety. Sucrose phosphorylated at the 6 position of the glucose moiety occurs in bacteria but is not known in green plants (see Section IV).

2. Sucrose phosphatase (Sucrose-6^F-phosphate phosphohydrolase, EC 3.1.3.24)

$$\text{Sucrose } 6^F\text{-phosphate} + H_2O \longrightarrow \text{sucrose} + P_i$$

3. Sucrose synthase (UDP-D-glucose:D-fructose 2-α-D-glucosyltransferase, EC 2.4.1.13)

$$\text{UDP-D-glucose} + \text{D-fructose} \rightleftharpoons \text{sucrose} + \text{UDP} + H^+$$

The three enzymes occur in the cytoplasm of cells and are not located within organelles. Current evidence suggests that sucrose synthesis usually proceeds via the pathway involving sucrose-P and catalysed by the first two enzymes. The third enzyme is more likely to catalyse the breakdown of sucrose *in vivo*, although some recent evidence discussed later suggests that different iso-enzymes of sucrose synthase may catalyse the reaction in different directions *in vivo*. Whether or not sucrose synthase and sucrose phosphatase are involved in sucrose loading into phloem and accumulation of sucrose into vacuoles, respectively, will also be discussed in the following sections. Sucrose phos-phorylase (EC 2.4.1.7) occurs in bacteria and fungi, but its presence in plants has not been confirmed and its properties make it unlikely to be involved in sucrose synthesis in green plants.

Several methods are used for the assay of sucrose-P synthase and sucrose synthase. These include determination of sucrose-P or sucrose or UDP by chemical or enzymatic means or the estimation of labelled sucrose-P or sucrose formed from labelled fructose 6-P, or fructose or UDP-glucose. With sucrose synthase, the back reaction can also be measured by estimation of fructose or UDP-glucose. Sucrose phosphatase activity has been determined by measuring the release of [^{14}C]sucrose from [^{14}C]sucrose-P. Other methods are feasible but have not been used, probably due to the unavail-ability of commercially produced sucrose-P. Invertases are assayed by measuring glucose, fructose, or glucose plus fructose released from either ^{14}C-labelled or unlabelled sucrose.

A.   Sucrose Phosphate Synthase

First discovered in and partially purified from wheat germ by Leloir and Cardini (1955), this enzyme has since been found in many plant tissues, especially those tissues which are synthesizing sucrose. The enzyme has recently been partially purified from spinach leaves (Harbron *et al.*, 1981; Amir and Preiss, 1982; Doehlert and Huber, 1983). Sucrose-P synthase has a high specific requirement for its two substrates, UDP-glucose and fructose 6-P, and the reaction towards sucrose synthesis is almost irreversible, especially when coupled to specific sucrose phosphatase. Whether or not cooperativity of substrates in effecting sucrose-P synthase activity occurs is not clear. Some reports indicate that saturation curves for UDP-glucose are sigmoidal, but most reports show hyperbolic curves. There are several examples of sigmoidal saturation curves for fructose 6-P, but a 50-fold purified enzyme from spinach leaves that was free of interfering enzymes (including glucose-P isomerase) showed hyperbolic saturation curves (Harbron *et al.*, 1981; Doehlert and Huber, 1983). These authors point out that the presence of glucose-P isomerase (EC 5.3.1.9) in assay media could result in changing fructose 6-P concentrations and apparent sigmoidal substrate saturation curves for fructose 6-P. Further work is required to determine whether the different saturation curves are due to interfering reactions or to species or varietal differences in sucrose-P synthase. The rate of sucrose synthesis by the spinach leaf enzyme was doubled by 5 mM glucose 6-P (Doehlert and Huber, 1983), although previous work had shown no effect (Harbron *et al.*, 1981).

Inorganic phosphate inhibits the enzyme, being competitive with UDP-glucose. Two reported values of K_i for P_i were 11 and 1.75 mM for the spinach leaf enzyme, and P_i slightly decreases the apparent affinity for fructose 6-P (Harbron *et al.*, 1981; Amir and Preiss, 1982). Fructose 1,6-bisphosphate inhibits the enzyme with a K_i of 0.8 mM, competitively with fructose 6-P and noncompetitively with UDP-glucose (Harbron *et al.*, 1981). Both the products of the reaction, UDP and sucrose-P, are inhibitory to the spinach leaf enzyme, with K_i values of 0.7 and 0.4 mM, respectively. The wheat germ enzyme is less sensitive to P_i inhibition and is not inhibited by sucrose-P (Salerno and Pontis, 1978). In the presence of Mg^{2+}, 50% inhibition occurs with 50 mM P_i, while greater than 100 mM P_i is required for 50% inhibition in the absence of Mg^{2+}. Sucrose inhibits the wheat germ enzyme in the presence of Mg^{2+} (Salerno and Pontis, 1980), but 100 mM sucrose does not inhibit partially purified spinach leaf enzyme (Harbron *et al.*, 1981; Amir and Preiss, 1982). Activity of sucrose-P synthase in crude leaf extracts of bean, peanuts, peas, and tobacco was significantly inhibited by 50 mM sucrose, but no effect was found in wheat, barley, or spinach (Huber, 1981). In addition, Mg^{2+} activated sucrose-P synthase activity in the crude extracts of these species (Huber, 1981; Harbron *et al.*, 1981), whereas Mg^{2+} did not stimulate the activity of purified spinach leaf enzyme (Harbron *et al.*, 1981; Amir and Preiss, 1982). Mg^{2+} relieved the inhibitory action of UDP with the same enzyme, and it was suggested that the apparent stimulation by Mg^{2+} in crude samples was due to chelation of inhibitory materials. Mn^{2+} can replace Mg^{2+} in many if not all of these effects. EDTA activated the enzyme in crude extracts (Hawker, 1967b), but it had no effect on purified enzyme (Amir and Preiss, 1982), adding support to the idea of chelation of inhibitory compounds. The enzyme does not require sulphhydryl compounds for stability. Much remains to be learnt about sucrose-P synthase in plants. Conflicting reports of its properties are almost certainly due to studies having been carried out on crude extracts, with enzyme purified to different degrees. The presence of several other enzymes, including hexose-P isomerase, sucrose phosphatase, UDPase, nonspecific phosphatases, and invertase, some of which require divalent cations for activity, could result in different activities under different conditions in which different products are measured. However, in view of the instability of the enzyme and the difficulties encountered in its purification (Pontis, 1977; Harbron et al., 1981; Amir and Preiss, 1982), workers can be excused for having used desalted crude extracts in attempts to determine changing activities under different environmental conditions in different organs.

The relative instability of the enzyme has hindered research into the properties of the enzyme protein. Molecular weights range from 2.8×10^5 to 4.5×10^5 for enzymes from spinach leaf wheat germ, and rice scutellum, and the wheat germ enzyme is built up of several subunits of unknown molecular weight.

Evidence suggests that the reaction mechanism of the wheat germ and spinach leaf enzymes is a sequential one in which UDP-glucose is first bound to the enzyme, followed by fructose 6-P. Further work is needed before it can be concluded whether an ordered bi–bi or Theorell–Chance mechanism operates (Harbron et al., 1981).

As more becomes known about sucrose-P synthase, it seems likely that the view held currently by several workers that the enzyme is a regulatory one important in the control of sucrose synthesis in plants (Harbron et al., 1981; Amir and Preiss, 1982; Huber, 1983) may become generally accepted. More work is needed on purified enzymes to determine whether sucrose-P synthases from different plant species have many properties in common and whether the enzyme is involved in regulation of sucrose synthesis throughout the plant kingdom. Further reference to control mechanisms are made later in this chapter.

B. Sucrose Phosphatase

After the discovery of sucrose-P synthase, the presence of a specific sucrose phosphate phosphatase was suspected in plants and evidence of higher percentage hydrolyses of sucrose-P than fructose 6-P at different concentrations of substrate was suggestive but not conclusive (Pontis, 1977). Work by Hatch (1964) indicated that sucrose-P is an intermediate in sucrose accumulation in sugar cane. He tested for a specific sucrose phosphatase at two pH values (5.4 and 8.2), both of which were obvious choices for phosphatases at that time. It was subsequently shown that a specific enzyme occurs in sugar cane, that it is relatively unstable, and that it has a pH optimum between 6.4 and 6.7 (Hawker and Hatch, 1966). The pH activity curve for the enzyme and the instability of the enzyme probably explain the failure to detect the enzyme in the earlier work.

The enzyme has been partially purified from only two sources, sugar cane stem and carrot roots. The enzyme requires Mg^{2+} for activity and is almost completely inhibited by EDTA (Hawker and Hatch, 1975). It is inhibited by sucrose, maltose, melezitose, 6-kestose and turanose but not by glucose or fructose. The K_m values for sucrose-P range from 45 to 170 μM. The partially purified enzyme from carrot hydrolysed P-enol pyruvate and fructose 6-P at 5 and 2% the rate observed for sucrose-P, respectively, part or all of which may have been due to interfering nonspecific phosphatase.

Crude enzyme extracts can be assayed for sucrose phosphatase activity under conditions that maximise activity of the specific enzyme but minimise activity of nonspecific phosphatases (Hawker and Hatch, 1975). These conditions include small amounts of enzyme and low concentrations of sucrose-P at pH 6.7 in the presence of bovine serum albumin. Specificity is checked by the addition of EDTA, sucrose, and maltose to some reaction mixtures, and it is usually found that very little interference is caused by other phosphatases. Using these techniques, evidence for sucrose phosphatase has been found in extracts from green algae, mosses, liverworts, ferns, gymnosperms, and angiosperms. It occurs in monocots and dicots, C3 plants, both mesophyll and bundle sheath cells of C4 plants, herbaceous annual and woody perennial plants, latex-containing plants, and sorbitol-translocating plants (Hawker and Smith, 1984a). Brown algae, red algae, and mushrooms do not contain sucrose and also do not contain sucrose phosphatase. The enzyme is found in roots, stems, leaves, fruits, and seeds at a level usually 10-fold greater than sucrose-P synthase. Both enzyme activities are lower in tissues that are not rapidly synthesizing sucrose.

Partially purified sucrose phosphatase from both sugar cane and carrot showed partially competitive inhibition by sucrose with K_i values of about 10 and 5 mM, respectively. With this type of inhibition the enzyme is never completely inhibited, even in sugar cane stem tissue where the concentration of sucrose can be higher than 600 mM. Melezitose inhibition was qualitatively similar to that of sucrose, with a K_i of 7 mM, but the inhibition by maltose was of mixed type with a K_i of 23 mM. The evidence suggests that sucrose and sucrose-P probably combine with different groups on the enzyme and melezitose combines at the sucrose site. Maltose also combines with the enzyme at the sucrose site (Hawker, 1967a). The possibility that sucrose inhibition of sucrose phosphatase is involved in the control of sucrose synthesis or that the enzyme, by virtue of its affinity for sucrose, is involved in sucrose transport is discussed in later sections, along with P_i effects on the enzyme.

Sucrose phosphatase has been detected in vacuole preparations from sugar beet roots, red beet roots, and immature sugar cane stem. It is not tightly bound to membranes but its exact location has not been proven. Whether the enzyme is associated with the tonoplast or whether it is located in the vacuole or in adhering cytoplasm needs to be determined. Ammonium molybdate at a concentration greater than 10^{-4} M was needed to cause 50% inhibition of the sucrose phosphatase, whereas sodium vanadate at 50 μM caused greater than 50% inhibition of the enzyme. (Hawker and Smith, 1984b,c and unpublished results).

C. Sucrose Synthase

Sucrose synthase was the first enzyme discovered in green plants that was capable of synthesizing sucrose (Cardini *et al.*, 1955). The enzyme seems to be ubiquitous in higher plants and occurs in most, if not all, tissues of plants. Knowledge of sucrose synthase is greater than that of sucrose-P synthase, probably due to its wider distribution, higher activity, and greater stability than that of the latter enzyme. In view of labelling patterns obtained during incorporation of $^{14}CO_2$ into sucrose, the properties of the enzyme, the distribution of the enzyme in plant tissues, and the presence of low levels of starch in maize mutants almost lacking sucrose synthase (see Preiss, 1982), it is now widely thought that sucrose synthase is involved in sucrose breakdown rather than in sucrose synthesis.

Sucrose synthase from maize kernels has been purified to homogeneity, and the enzyme from mung bean seedlings, rice grains, potato tubers, and bamboo shoots has been partially purified. The native enzyme is a tetramer of MW 4×10^5 with subunits of $0.9-1.0 \times 10^5$, and the enzyme can exist in higher aggregate forms.

The reaction catalyzed by the enzyme is reversible with a K'_{eq} [(sucrose)(UDP)/(UDP-glucose)(fructose)] of 1.3–6.7 in a range of temperatures and pH values between 25° and 37°C and pH 7.2–7.6. The optimum pH for sucrose cleavage is pH 6.5–7.0, whereas for sucrose synthesis it is around pH 7.5–8.0 The involvement of H^+ in the stoichiometry of the reaction probably accounts for the different pH optima and the effect of pH on the equilibrium constant.

Other compounds can act as substrates for sucrose synthase besides UDP-glucose, fructose, sucrose, and UDP, the most important being ADP to release ADP-glucose. However, the uridine nucleotide has a higher affinity for the enzyme, and V_{max} is higher. The K_m for fructose is about 5 mM, whereas the K_m for sucrose can vary between 10 and 400 mM. The enzyme from various plants has been reported to be inhibited by nucleotides, sugars, sugar phosphates and β-phenylglucoside and stimulated by Mg^{2+}. Sucrose synthase from maize leaves is more stable in the presence of Mg^{2+} (W. Claussen, personal communication). It is not possible to make general conclusions about the physiological significance of these effectors due to the lack of uniformity of the results. Current information suggests that the kinetic mechanism for sucrose synthase is sequential, as is the mechanism for sucrose-P synthase.

It has been suggested that sucrose synthesis and sucrose cleavage may be catalyzed by different isozymes of sucrose synthase, and recently renewed interest has been shown in the demonstration of multiple forms (isozymes) of the enzyme in plant tissues (Su, 1982; Gross and Pharr, 1982a,b). Two partially purified forms have been studied from cucumber fruit peduncle and fruit pericarp (Gross and Pharr, 1982a). One form (SS2) was more abundant in peduncle tissue than in pericarp tissue, and it had a pH optimum of 7.5 for sucrose synthesis compared to pH 9.0 for the other form. Cucumber peduncles do not contain sucrose-P synthase, but there is evidence that galactose moieties of transported stachyose are converted to sucrose by a pathway involving UDP-glucose (Gross and Pharr, 1982b). It has been postulated that SS2 may operate mainly to synthesize sucrose in cucumber peduncle tissue. Further purification of the enzymes to remove invertase and UDPase will allow determinations of K_m values for sucrose and UDP, which should provide more insight into the roles of the different forms of the enzymes. The apparent MW of 540,000 for both forms of the enzyme approximately corresponds to the minor hexameric form of the enzyme found in maize (Su and Preiss, 1978). The majority of the enzyme from maize kernels appears in a band corresponding to the tetrameric form of MW 365,000. Isozymes from pea seedlings and rice grains have been separated and found to have differences in kinetic properties and nucleotide specificities, but their physiological significance is not understood (Su, 1982).

All forms of the enzyme catalyse both synthesis and cleavage of sucrose, and previous attempts to separate the synthetic and cleavage activities of other sucrose synthases have not been successful. Mercaptoethanol was found to activate cleavage much more than synthesis with potato sucrose synthase (Pressey, 1969), and the activity of wheat germ enzyme that had been incubated with oxidized glutathione or oxidized thioredoxin was reduced in the cleavage direction but was unaffected in the synthesis direction (Pontis *et al.*, 1981). Sucrose cleavage activity was restored by dithiothreitol or reduced glutathione, and it was claimed that two forms of sucrose synthase exist, one reduced and one oxidized. The interconvertible forms catalyse sucrose cleavage and sucrose synthesis, respectively, and it was suggested that oxidation or reduction modified the active centre of the enzyme. Studies with thioredoxin-like protein from wheat germ and with sucrose synthase from other plants should prove interesting. The results with cucumber isozymes and oxidized and reduced forms of sucrose synthases suggest that the enzyme may have a role in the synthesis of sucrose in some instances. It has been suggested that it synthesizes sucrose in latex of *Hevea brasiliensis* (Tupý and Primot, 1982). Nevertheless, cleavage of sucrose must still be regarded as the major role of the enzyme.

Maize endosperm messenger RNA has recently been translated both in wheat germ extracts and rabbit reticulocyte lysates to produce a protein of molecular weight 88,000, which, it was shown, was the translation product *in vitro* of sucrose synthase mRNA (Wöstemeyer *et al.*, 1981). The length of the sucrose synthase mRNA was shown to be 2800 nucleotides, which is sufficient to code for a protein of molecular weight 88,000, the subunit size found for maize endosperm sucrose synthase (Preiss, 1982). This work has shown that it is now possible to assay for sucrose synthase mRNA by cell-free translation followed by gel electrophoresis and, in principle, to isolate cDNA clones derived from this message. The simple assay is made possible because sucrose synthase is one of the major endosperm proteins (Su, 1982). Sucrose synthase is encoded by the *shrunken* gene in maize endosperm, and transposable elements can decrease or increase gene expression (see Geiser *et al.*, 1982).

III. SUCROSE BREAKDOWN

Two enzymes are considered to be associated with sucrose breakdown in green plants. They are

1. Sucrose synthase. The equation in Section II can be reversed and simplified to

$$\text{Sucrose} + \text{UDP} \rightleftharpoons \text{UDP-glucose} + \text{fructose}$$

or

$$\text{Sucrose} + \text{ADP} \rightleftharpoons \text{ADP-glucose} + \text{fructose}$$

2. Invertase (β-D-fructofuranoside fructohydrolase, EC 3.2.1.26)

$$\text{Sucrose} \longrightarrow \text{glucose} + \text{fructose}$$

The enzyme hydrolyses terminal nonreducing β-D-fructofuranoside residues in β-D-fructofuranosides and also catalyses fructotransferase reactions.

Both acid and neutral invertases occur in plants, with pH optima of about 5 and 7.5, respectively. Acid invertase is found in the free space (which might include some of the cytoplasm) and vacuoles of plant cells, while neutral invertase probably resides in the cytoplasm, as does sucrose synthase. The concept that sucrose synthase is involved in polysaccharide synthesis from sucrose while sucrose storage and sucrose hydrolysis to provide hexose for respiration involves invertases is an oversimplification but is nevertheless a useful generalization of the roles of the enzymes in green plants (ap Rees, 1977).

A. Sucrose Synthase

The properties of the enzyme have been described in Section II,C, and it was pointed out that the role of sucrose synthase is most often the breakdown of sucrose. Several recently discovered facts lend support to this conclusion (Su, 1982). Firstly, the enzyme is one of the major soluble proteins in many growing plant tissues, constituting 1–2% of the soluble proteins of maize kernels and bamboo shoots. The enzyme exhibits high K_m values for sucrose (20–50 mM) and shows sigmoidal saturation kinetics with sucrose, which results in the enzyme being reactive with sucrose only at concentrations above a certain level. Thus the plant can accumulate some sucrose even in the presence of large amounts of sucrose synthase. However, the behavior of the enzyme towards sucrose varies between species and even depends on the buffer used. The relatively high K_m for sucrose has led to the suggestion that rates of cleavage of sucrose in plant tissues are likely to be much less than the potential V_{max} even in storage tissues containing 0.5 M sucrose, most of which is in the vacuole (Avigad, 1982). The concentration of sucrose in the cytoplasm is not known, but in photosynthetic leaf cells it was estimated to be 10–50 mM. However, Su (1982) has preliminary evidence that the monomeric form of the enzyme from asparagus could cleave 1 mM sucrose while the tetrameric form was unreactive. On the other hand, the corn kernel enzyme occurs in the tetrameric state *in vivo*. It is possible that the enzyme might exist in different forms in different tissues, which could result in different threshold values for sucrose cleavage and hence different levels of sucrose in different tissues. This concept is highly speculative, and much more research is required in this area.

Results using rice grain preparations show that there is sufficient sucrose synthase to cleave sucrose for starch synthesis and, *in vitro*, to provide sufficient ADP-glucose as a substrate for starch synthase (Su, 1982). Many workers have suggested that *in vivo* sucrose synthase also provides ADP-glucose for starch synthesis by the cleavage of sucrose in the presence of ADP. The ADP-glucose is assumed to act as a substrate for starch synthase. This scheme, however, does not take into account the fact that the two enzymes are in different cellular compartments and that the amyloplast membranes are unlikely to be permeable to ADP-glucose. Certainly there seems to be sufficient sucrose synthase in many plant tissues to provide ample UDP-glucose for membrane-bound transglycosylases responsible for the synthesis of glycoproteins, glycolipids, and reserve polysaccharides (except starch), all of which are able to use cytoplasmically produced sugar nucleotides. The sucrose synthase present can also cleave sucrose to provide glucose moieties eventually for starch synthesis, but the pathway is not known. One possibility is based on evidence that suggests that amyloplasts have similarities to chloroplasts (Preiss, 1982; Macdonald and ap Rees, 1983a,b). Glucose 1-P is formed from UDP-glucose by UDP-glucose pyrophosphorylase, or perhaps UDP-glucose phosphorylase (Gibson and Shine, 1983), and converted to triose phosphates, which traverse the amyloplast membranes to be reconverted to glucose 1-P and thus act as a substrate for ADP-glucose pyrophosphorylase to produce ADP-glucose. There are unknowns about this pathway (see also Chapter 4), and one of the outstanding questions is whether amyloplast membranes have permeability properties similar to those of chloroplasts.

The starch-producing colourless alga, *Polytoma uvella*, does not contain sucrose, sucrose-P synthase, or sucrose synthase (Mangat and Kerson, 1982), and ADP-glucose is presumably synthesized by ADP-glucose pyrophosphorylase as occurs in chloroplasts of green plants, because sucrose synthase is not present in the alga to provide ADP-glucose from sucrose.

The cleavage of sucrose by sucrose synthase to UDP-glucose depends on a supply of UDP. Plants are known to contain specific nucleoside diphosphatases with maximum activity towards UDP and none towards ADP (Huber and Pharr, 1981). If such an enzyme occurred in the cytoplasm, it could reduce the level of UDP and decrease the cleavage of sucrose. The role of the enzyme could become clearer if further studies were carried out on the distribution of the enzyme in various tissues and on its intracellular location in plant cells. Thus far, it has been proposed that its role is to reduce UDP levels and hence stimulate sucrose formation catalysed by either sucrose-P synthase or sucrose synthase.

The growth of fruit and accumulation of storage products in fruit usually requires the breakdown of sucrose to provide precurors. Sucrose synthase has been found in grapes, bananas, citrus fruits, apples, pears, and watermelons. The activity is particularly high in young grape berries and young eggplant fruits and probably provides nucleotide diphosphate sugars as precursors of polysaccharides in the rapidly growing tissues in which cell division and cell expansion are occurring (Hawker, 1969a; Claussen, 1983a). Further discussion of this is given in later sections.

B. Invertase

Although invertases are widespread in the plant kingdom, most of the information on molecular structure and mechanism of action comes from work done on fungi and yeast. Nevertheless, there is still a large body of knowledge regarding invertases in green plants, although this is sometimes glossed over by reviewers, probably partly because of its complexity and also partly because there is another enzyme in plants capable of cleaving sucrose, sucrose synthase. Other writers have dealt in detail with invertases (Glasziou and Gayler, 1972; ap Rees, 1974; Avigad, 1982), and the following description where unreferenced is taken from these authors.

There is sufficient evidence to show that at least in some plants, e.g., sugar cane stems, acid invertase occurs in the free space of the tissues, that part of tissue that comes into rapid diffusion equilibrium with a medium in which it is placed. Some of the invertase can be bound to cell walls, probably by salt linkages, while some of it is soluble and probably resides in the cytoplasm. Active accumulation of sucrose into some plants depends on hydrolysis by invertase in the free space, whereas in other plants sucrose can be accumulated without hydrolysis. In other tissues, e.g., mature sugar cane stem. alkaline invertase in the cytoplasm is involved in sucrose hydrolysis prior to uptake into the vacuole. In many species of plants, much, or the major part, of the acid invertase occurs in the vacuole, conclusive proof having been obtained for beet roots, maize scutellum, tobacco cells, and leaf tissues. Other hydrolases are situated in plant vacuoles much as in mammalian lysosomes, but unlike lysosomes plant vacuoles can contain quite high concentrations of sucrose. As tissues mature they often lose acid invertase activity and accumulate sucrose. On the other hand, the pathway of uptake of sugar in grape berries includes sucrose, but this is hydrolysed by high levels of acid invertase that persist in the vacuoles of these fruits.

Acid invertase activities are usually high in tissues that are rapidly growing and developing, such as root apex, callus cultures, and stem internodes. Much of the invertase is in vacuoles, where it hydrolyses stored or recently translocated sucrose to produce hexoses for respiration and synthesis of building blocks for cells.

Acid invertases have K_m values for sucrose between 2 and 13 mM and alkaline invertases between 9 and 25 mM. Evidence suggests that the acid invertases are glycoproteins, which would explain the vast variations found in molecular weights and their affinity for other macromolecular structures in plant extracts. Glucose is known to inhibit the production of invertase in yeast, and at least in one higher plant (sugar cane) there is strong evidence that glucose concentrations in the metabolic compartment (cytoplasm) control the synthesis of intracellular invertase. This is not a universal phenomenon, since incubation in glucose did not affect invertase activity in carrot or lentil tissues.

Alkaline invertase has received relatively little attention from biochemists. The enzyme was purified twofold from bean pods and appeared to be specific for sucrose. Bean leaf enzyme was purified 20-fold, but its specificity was not reported. Alkaline invertase from soybean nodules has been partially purified and shown to be a β-D-fructofuranosidase that is specific for sucrose (Morell and Copeland, 1984).

Tris buffer inhibits alkaline invertases quite markedly but inhibits acid invertases much less, a property that has been used in physiological experiments.

Naturally occurring invertase inhibitors have been found in several tissues, but more detailed studies have only been carried out with the potato tuber inhibitor. It is a small protein, of MW $\sim 1.8 \times 10^4$, and binds to the enzyme. The physiological significance and binding properties of the invertase inhibitors are not known. Further work using immunochemical probes might prove useful.

Although sucrose synthase is mainly responsible for sucrose cleavage in most developing starch-storing cereal grains, there is evidence that invertase is active in the early stages of some cereal grain development (Duffus, 1979) and that sucrose is hydrolysed by invertase during the unloading of phloem in the pedicel of maize kernels (Felkner and Shannon, 1980).

Research into invertases and their roles continues. Activation of invertase in *Hevea brasiliensis* latex, invertases in soybean root nodules, invertase in the free space of maize scutellum, cell wall invertases of sugar cane, and the possible role of free space invertase in phloem unloading are some examples of recent work (Eschrich, 1980; Humphreys and Echeverria, 1980; Jacob *et al.*, 1982; Prado *et al.*, 1982; Streeter, 1982).

IV. PATHWAYS OF SUCROSE ACCUMULATION INTO STORAGE COMPARTMENTS

A. General Transport of Sugar in Plants

Transport of compounds across membranes of cells of living organisms is of fundamental importance to the maintenance of life. Proteins with specific binding sites are involved; for this reason, transport is a biochemical reaction, even though the transported compound often suffers no permanent alteration of its chemical structure. There are only a few exceptions, but the transport of sucrose as sucrose 6^G-P by some bacteria and the probable transport of sucrose as sucrose 6^F-P by some plants such as sugar cane and grape berries are important and are discussed here together with proton cotransport of sucrose in plants such as sugar beet and red beet.

Knowledge of sugar transport processes in plants is relatively meagre compared to that in animals and bacteria, and it is too early to judge whether there are mechanisms specific to plants. Komor (1982) has described sugar transport in plants, a subject that is sometimes neglected in favour of ion uptake. A brief summary is provided here as a background for later discussion of sucrose accumulation by specific tissues. Transport of sugar across membranes can be classified as follows.

1. Passive Transport (Diffusion)

This process does not depend on energy, is proportional to sugar concentration, and in practice is usually slow. It probably becomes significant when large gradients exist, e.g., between vacuoles and apoplast of sucrose storing tissues.

2. Mediated Diffusion

Proteins combine with the transported sugar by chemical interaction to make its permeation through the membrane more rapid. The rate of permeation in this system is saturated at high substrate concentrations, and the process is usually only found in yeasts, moulds, and erythrocytes. Mediated diffusion is not known in higher plants.

3. Active Transport

Active transport of sugar occurs if sugar is either accumulated or excreted against a chemical gradient, in which case metabolic energy-yielding systems must be involved. A membrane component (usually protein) catalyses the transport, sugar can be accumulated against a concentration gradient, there is a stoichiometric use of metabolic energy, and V_{max} and K_m values can be calculated. Sometimes in higher plants two phases of uptake are found, an active uptake with a K_m of about 10 mM and a nonactive uptake with a high K_m or nonsaturable properties (Thorne, 1982; Maynard and Lucas, 1982a). Evidence has more recently been obtained that the nonsaturable phase in *Beta vulgaris* leaf tissues is also active (Maynard and Lucas, 1982b).

Active transport of sugars can theoretically be divided into three major categories.

a. *Conversion of permeating sugar to an impermeable derivative after passage through the membrane.* Although known in cells, this mechanism is not common. Nevertheless, sucrose can be taken up by *Streptococcus lactis* by a phosphoenolpyruvate-dependent phosphotransferase system that involves transport of sucrose 6^G-phosphate (Thompson and Chassy, 1981). Sucrose 6^F-phosphate hydrolysis has been proposed as a transport mechanism in the accumulation of sucrose in the vacuole of sugar cane (Glasziou and Gayler, 1972), and this mechanism fits into this first category of active transport.

b. *Primary active transport.* Here the coupling of metabolism to transport is direct, with an enzyme performing uptake and energy use together. H^+-ATPase of chloroplasts is a well documented example, and there is increasing evidence that plasmamembrane-located and tonoplast-located H^+-ATPases are active in plants. So far, a direct ATPase-sugar transport mechanism has not been discovered.

c. *Secondary active transport.* Here an ion gradient produced by a primary active transport mechanism is used to drive an uphill sugar transport by a downhill flow of the ion. Sugar transport in cells is driven by either sodium or hydrogen ion gradients. In plant cells it is usually a proton gradient maintained by ATPase that is responsible for the transport of sugar. Research in this field is currently quite active, and it has been speculated that sugar transport in all plants involves the proton cotransport system (Komor, 1982).

Before sugar can be accumulated in sugar cane stems, sugar beet roots, red beet roots, fruit, or other sinks, sucrose must be moved from the source (leaves) via the phloem by a much studied but not well understood process called translocation. Recent essays (Avigad, 1982; Giaquinta, 1983) provide detailed descriptions and discussion of the possible processes and mechanisms involved. Briefer reviews also describe recent concepts of phloem loading, translocation, and phloem unloading (Ho and Baker, 1982; Evert, 1982). In summary, evidence from many experiments with many species suggests that movement of sucrose from mesophyll cells of leaves to vacuoles of storage organs involves either symplastic or apoplastic transport of sucrose from the leaf cells to the phloem, and transport into the sieve tubes either through the symplasm (plasmodesmata) or from the apoplast, with or without hydrolysis by invertase or sucrose synthase. Several mechanisms proposed for the long-distance movement of sucrose through the phloem have included an osmotically driven pressure-flow mechanism, electroosmotic mechanisms, movement through transcellular strands, and mechanisms involving P-protein. Unloading of the phloem in the region of the sink can take place either via symplastic or apoplastic routes and, if apoplastic, energy-dependent transport must be involved.

Evert (1982) has suggested that of the many pathways and mechanisms described, the two steps that are gaining greatest acceptance and are most likely to be operative in the majority of plants involve active transport from the apoplast into the phloem across the plasmalemma by a sucrose proton cotransport system with the proton gradient maintained by H^+-ATPase activity (Fig. 2). Sucrose is moved through the phloem by the osmotically generated pressure flow mechanism, in which sucrose is moved to the sink from the source along a gradient of hydrostatic (turgor) pressure developed osmotically. Figure 3 shows that a high sucrose concentration near the source

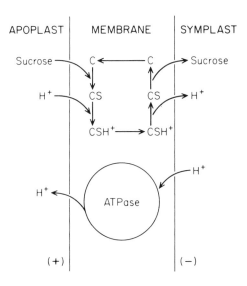

Fig. 2. A model for sucrose-proton cotransport across phloem membranes. In this model, a carrier protein (C) located at the external surface of the membrane binds sequentially with a sucrose molecule and a proton. The carrier–sucrose–H^+ complex (CSH^+) is then driven across the membrane in response to the electrochemical proton gradient created by the membrane ATPase, which pumps protons out of the cell. After the proton and the sucrose molecule dissociate, the carrier returns to the external membrane surface. From Evert (1982) © 1982 by the American Institute of Biological Sciences.

promotes water uptake from the xylem, with the reverse occurring at the sink.

There is no general agreement on the movement of sucrose from mesophyll cells to the apoplast of the phloem, nor on the mechanism of phloem unloading. Huber and Moreland (1981) showed a proportional release of sucrose and K^+ from wheat and tobacco mesophyll protoplasts and suggested that sucrose is cotransported with K^+ across the plasma membrane. Anderson (1983) found that K^+ stimulated sucrose release from *Vicia faba* leaf disks. Ho and Baker (1982) commented that perhaps phloem unloading is also controlled by K^+ levels. It should be noted that K^+-sucrose cotransport is not a mechanism accepted by all researchers. Phloem unloading is closely linked with sink filling, and mechanisms of accumulation of sucrose in storage organs (sinks) are discussed in the following sections.

Some species of trees contain high concentrations of sucrose in their vessels during spring, and there is evidence that in *Populus* reabsorption of sucrose into the parenchyma occurs by a pathway involving hydrolysis of sucrose by invertase followed by transport of hexoses across the plasmalemma (Sauter, 1981).

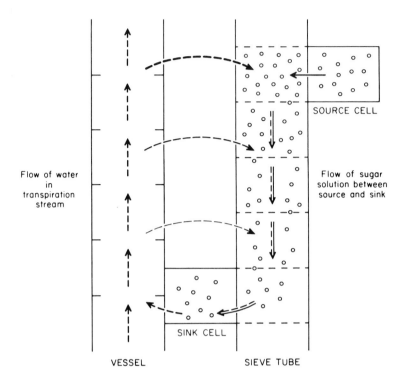

Flow of water in transpiration stream

Flow of sugar solution between source and sink

SOURCE CELL

SINK CELL

VESSEL   SIEVE TUBE

Fig. 3. Diagram of osmotically generated pressure-flow mechanism. Circles represent sucrose molecules. From Evert (1982).

B. Sugar Cane Stem

In 1972 Glasziou and Gayler published a review of work carried out mainly in the early and mid-1960s on the accumulation of sucrose in sugar cane stalk tissue. The scheme proposed by workers on sugar cane has been used as a model system, and it has been suggested that a similar pathway operates in grape berries (Hawker, 1969a; Brown and Coombe, 1982). The scheme has its critics, and recently alternative pathways involving sucrose-ion cotransport in other tissues have been proposed (Komor *et al.*, 1982b; Willenbrink, 1982). One of the major contentious steps is the involvement of sucrose-P in the transport of sucrose across the tonoplast from the cytoplasm to the vacuole where sucrose is stored. Even if sucrose-ion cotransport is operative at the tonoplast of some plants, it does not preclude the operation of a sucrose-P transport system in other plants. The sugar cane pathway of sucrose accumulation remains the most studied of any in plants and deserves description. Time and the use of modern techniques using isolated organelles should determine whether the scheme is applicable to some, a few or all plants.

Sugar cane stalk tissue was chosen to study sugar storage mechanisms because of its relative simplicity and its commercial importance. By removing the rind and using mid-internodal regions, typical monocot stem tissue can be obtained, which, in mature tissue, is made up of 92% storage parenchyma cells, 7.5% xylem and sclerenchyma, and 0.5% phloem. The vacuoles of the parenchyma cells constitute about 80% of the total tissue volume, the cytoplasm about 2.5%, and the free space of the tissue *in vivo* about 17% (Hawker, 1965). The sucrose concentration in the vacuole can be as high as 23% (w/v) and can be maintained at this level for many months. Young expanding internodes from the top of the stalk contain glucose and fructose at about 5% fresh weight and much lower levels of sucrose.

The pathway of sugar accumulation is summarised diagrammatically in Fig. 4. Sucrose moves into the apoplast (free space) from the phloem, it is hydrolysed by invertase, and glucose and fructose are transported by separate carriers into the symplasm (cytoplasm, metabolic compartment). Sucrose-P is synthesized and the hydrolysis of sucrose-P is an integral part of the transfer into the storage compartment (vacuole). In addition, all three sugars diffuse between the free space and storage compartment in the direction of the gradient. The evidence for the pathway has been reviewed in detail by Glasziou and Gayler (1972), but some of the salient features are described here.

Assymmetrically ^{14}C-labelled sucrose moves through phloem tissue without randomisation of label, indicating that sucrose is translocated without breakdown and resynthesis, which is the usual mechanism in plants. It was suggested that most of the sucrose moves from the phloem to the free space of the storage parenchyma cells via the cell walls rather than via plasmodesmata, since there is a correlation between free space invertase and dry matter input in immature tissue and since it has been shown that there are high concentrations of sucrose in the free space of mature tissue. It has subsequently been shown that radial photosynthate transfer occurs in the free space of stems of other species (see Patrick and Turvey, 1981). Sucrose in the free space of immature tissue is hydrolysed by a soluble acid invertase with a pH optimum of about 5, while in mature tissue a bound enzyme is present on cell walls with a pH optimum of 3.8.

Uptake studies with glucose, 3-*O*-methylglucose, and fructose at various temperatures, pH values, and under N_2 indicated that transport could be due to two specific carriers, one for glucose and another for fructose, situated at the plasmalemma. Uphill transport occurs, but not by counterflow. Energy is involved, but the sugars are not phosphorylated, nor is coupling to an Na^+

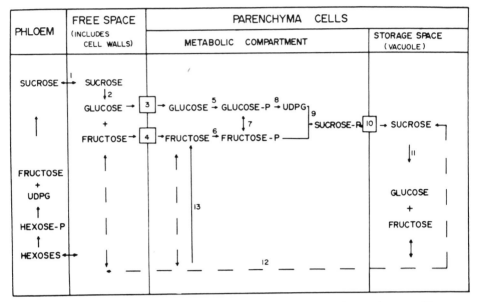

Fig. 4. The sugar cycle in sugar cane storage tissue. 1. Movement of sugars between phloem and free space (which includes cell walls). 2. Hydrolysis of sucrose in free space by acid invertase. 3. Carrier-mediated transfer of glucose into metabolic compartment. 4. Carrier-mediated transfer of fructose into metabolic compartment. 5, 6, and 7. Hexose phosphorylation and interconversion. 8. Synthesis of uridine diphosphate-glucose. 9. Synthesis of sucrose phosphate. 10. Transfer of sucrose moiety of sucrose phosphate into storage. 11. Hydrolysis of stored sucrose by acid invertase (immature tissue only). 12. Diffusional movement of sugars in the direction of the prevailing gradient. 13. Hydrolysis of sucrose by neutral invertase. From Glasziou and Gayler (1972).

pump likely. Perhaps a glucose proton cotransport system operates (see the work on sugar cane suspension cultures described later). In addition, a slow diffusion of glucose and fructose occurs, but this may be only one-fiftieth of the carrier-mediated process. Sucrose is not transported as such across the plasmalemma except at low rates by diffusion.

Once inside the metabolic compartment (cytoplasm), the glucose and fructose are either used as substrates for respiration or cell growth, or are converted to sucrose 6^F-P prior to or during transport across the tonoplast into the vacuole. All of the enzymes required for this conversion of monosaccharides to sucrose-P have been demonstrated in sugar cane stalk tissue, as has a specific sucrose phosphatase. This enzyme could act as a sucrosyl transferase, releasing free sucrose into the vacuole. Sucrose phosphatase seems unlikely to be located in the vacuole due to its pH optimum of 6.7, and it is not tightly bound to the tonoplast as evidenced by its lack of sedimentation at 100,000 g. It may be loosely associated with the tonoplast *in vivo*.

In immature tissue, some of the sucrose in the vacuole is hydrolysed by a soluble acid invertase, the synthesis of which is under regulatory control by its end product, glucose. The enzyme has a function in the regulation of growth, its activity is high in expanding cells above the intercalary meristem, and it disappears soon after cell expansion ceases. Varieties of sugar cane that retain this acid invertase do not store high levels of sucrose. Sucrose at high concentrations can be stored in the vacuoles of mature, fully expanded internodes of sugar cane, with losses being due to diffusion of the sucrose across the membranes.

The concentration of sucrose in the vacuole depends on the rate of active uptake of sucrose from the free space and the gradient between the free space and the vacuole. Active transport in some mature tissue could only maintain a 5% gradient, even though the concentration of sucrose *in vivo* was up to 23%. This anomaly was removed by the experimental finding that the concentration of sucrose in the aqueous phase of the intercellular spaces and cell walls of the tissue approached the concentration of sucrose in the vacuoles. Hence, the maintenance of a high gradient is not required. Prior to this somewhat unexpected result, it had been assumed that free-space concentrations of metabolites were very low. In some quarters the idea of relatively high concentrations of organic molecules in the free space is still unpopular. Nevertheless, the results obtained by Hawker (1965), using many techniques with sugar cane grown under different environmental conditions, have been verified by Glasziou and Gayler (1972) and have not been repudiated by evidence from other workers. Patrick and Turvey (1981) have suggested that radial transport of photosynthate through the apoplasm may occur in many stems. The concentration of sucrose in the xylem was found not to be in equilibrium with the intercellular and cell wall aqueous phase of the parenchyma, an expected result. How the sucrose moves from the phloem to the free space of the parenchyma and vice versa is not known. When sugar cane resumes growth during changes of environment, e.g., from 18 to 30°C, free space concentrations of sucrose decrease followed by decreases in vacuolar concentrations of sucrose. Presumably this sucrose is transported via the phloem to the growing internodes at the top of the stem. Sucrose–proton cotransport, diffusion, and possibly sucrose synthase might play a role in sucrose movement, perhaps via the symplasm between the phloem and the endodermis and then via the free space between the endodermis and the parenchyma during sucrose storage or sucrose efflux. It must be emphasized that much of the preceding sentence is supposition and is included to stimulate thought and highlight the remaining questions about sucrose storage in sugar cane stalks. An endodermis forms a diffusion barrier between the stele and cortex in the roots of many plants and in the stems of some plants (Eames and MacDaniels, 1947). Examination by light microscope of sections of tissue did not reveal the presence of a clearly distinguishable classic endodermis around the bundles of sugar cane stems (J. S. Hawker, unpublished). However, further work is needed to determine the location and nature of the diffusion barriers, between the xylem and parenchyme cells, that prevent uncontrolled movement of sucrose into the xylem stream.

One of the fundamental questions about the proposed pathway of sucrose accumulation in sugar cane stalks and other plants is the way in which sucrose-P is involved in transport across membranes. That sucrose-P is the derivative of sucrose that is involved in the movement of sucrose is suggested by the presence and properties of relevant enzymes and by results from uptake experiments in which fructosyl-labelled sucrose and sucrose-P were supplied to stalk tissue. Label from sucrose was randomized, but that from sucrose-P was not. Had sucrose-P been hydrolysed by sucrose phosphatase in the cytoplasm, the resulting sucrose would have been inverted and the label from fructose randomized between the hexose moieties.

More direct evidence of sucrose-P transport across the tonoplast may come from studies of sucrose-P uptake by isolated vacuoles from sugar cane stalks or other tissues. However, there may be difficulties with such an approach, because isolation of vacuoles may damage the transport system, or break a loose association of sucrose phosphatase with the tonoplast, or cause premature hydrolysis of sucrose-P by unattached sucrose phosphatase. Nevertheless, uptake experiments must be attempted.

Of interest when considering sucrose transport in higher plants (although not necessarily relevant) are the recent discoveries of the uptake of sucrose by a phosphoenolpyruvate-dependent phosphotransferase system in the bacteria *Streptococcus mutans* (St. Martin and Wittenberger, 1979), *S. lactis* (Thompson and Chassy, 1981), and *Escherichia coli* containing sucrose plasmids (Schmid *et al.*, 1982; Lengeler *et al.*, 1982). The proposed pathway for *S. lactis* is shown in Fig. 5. The sucrose phosphotransferase system (Suc-PTS in the membrane) is a multicomponent system in which phosphoenolpyruvate acts as the phosphoryl donor instead of the more usual nucleotide triphosphate. The PTS for sugars was first discovered in 1964, and a sucrose-PTS was described in *Bacillus subtilis* in 1968 (see references in St. Martin and Wittenberger, 1979).

Sucrose is phosphorylated at the 6 position of the glucose moiety and hydrolysed by a sucrose-6^G-phosphate hydrolase (β-D-fructofuranosidase, invertase), which is loosely bound to the membrane. The products of the reaction are glucose 6-P and fructose, both of which are further metabolised, as shown in Fig. 5. This is in contrast to green plant sucrose 6^F-P, which is

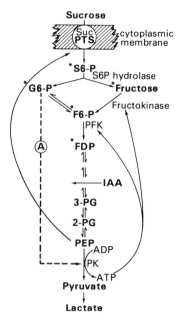

Fig. 5. Proposed pathways for transport and metabolism of sucrose by *S. lactis* K1. The three glycolytic intermediates 3-phosphoglycerate (3-PG), 2-phosphoglycerate (2-PG), and phosphoenolpyruvate (PEP) comprise the endogenous PEP potential of starved cells. Asterisks designate compounds detected by enzymatic analysis and autoradiography after accumulation of sucrose by iodoacetate-inhibited starved cells. A, activator; FDP, fructose 1,6-bisphosphate; F6-P, fructose 6-P; G6-P, glucose 6-P; IAA, iodoacetic acid; PK, pyruvate kinase; PFK, phosphofructokinase; suc-PTS, sucrose-PTS; and S6-P, sucrose-6^G-P. From Thompson and Chassy (1981).

formed from the substrates UDP-glucose and fructose 6-P. Hydrolysis by the specific sucrose phosphatase releases sucrose and P_i. Neither invertase from green plants nor bacterial sucrose-6^G-P hydrolase breaks down sucrose 6^F-P.

Three types of mutants were derived from *E. coli* strains containing the plasmid pUR400; one lacked a specific transport system, another lacked sucrose 6^G-P hydrolase, and another expressed both functions constitutively (Schmid *et al.*, 1982). Transport and sucrose 6^G-P hydrolase were inducible by fructose, sucrose, and raffinose, suggesting that fructose or a derivative acted as an endogenous inducer. Sucrose transport and hydrolase were subject to catabolite repression. The bacterial work has been mentioned here to illustrate a sucrose tranport system involving sucrose-P, despite the difference between the bacterial and green plant sucrose-P. Group translocaters might be involved in sugar cane and/or in grape berries, as described next.

C. Grape Berries

Grapes are produced in greater quantities than any other fruit crop in the world and are used fresh, dried, or for wine making. It is for this reason that they have been studied biochemically. As experimental material they have several disadvantages, among which are their fragile cell walls, high content of organic acids and hence low pH, high content of tannins and phenols, and high activity of vacuolar acid invertase. A few seedless varieties occur that allow experimentation without the complication of seed metabolism.

Berries of the seedless sultana grape (*Vitis vinifera* cv. Sultana syn. Sultanina) were fed ^{14}C-labelled sugars through their pedicels, and the results showed that sucrose was synthesized from either glucose or fructose and that sucrose was hydrolysed and resynthesized within the berries (Hardy, 1968). Glucose was more readily metabolized than fructose, and in this regard it is interesting that it has recently been shown that pea stem hexokinase, having glucose as the preferred substrate, is located on the outer membrane of the mitochondria where its proximity to a supply of ATP would allow more rapid phosphorylation of glucose than of fructose, which is phosphorylated by a cytosolic fructokinase (Tanner *et al.*, 1983). Sucrose is the main sugar translocated in the grapevine, but ripe grapes contain about equal concentrations of glucose and fructose (10 g/100 g fresh weight) and sucrose at less than 1 g/100 g fresh weight (Hawker *et al.*, 1976).

The activities of several enzymes were determined in developing grape berries along with sugars and organic acids. The patterns of activities that were found suggested that the activities could be correlated at least with amounts of enzymes present in the tissues and were not artifacts caused by different chemical and physical properties of the tissues. Figures 6 and 7 show that as the concentration of sugar increased in the berries, the activities of sucrose synthase, sucrose-P synthase, and sucrose phosphatase also increased (Hawker, 1969a).

The activity of sucrose synthase was relatively high during the early stages of growth of the berry before sugar accumulation began, which suggests that it could have a role in sucrose breakdown to provide substrates for respiration and synthesis of building blocks. Acid invertase was also found to increase during the development of the berries, but at an earlier stage than the other enzymes (Fig. 6). The activity of invertase is 100-fold greater than that of sucrose synthase and, despite earlier reports, invertase is likely to be a soluble enzyme located in the vacuole (Hawker, 1969b). Confirmation of its exact location requires examination of grape berry protoplasts and vacuoles in a manner similar to that described for red beet, in which it was shown that acid invertase is located in the vacuoles (Leigh *et al.*, 1979).

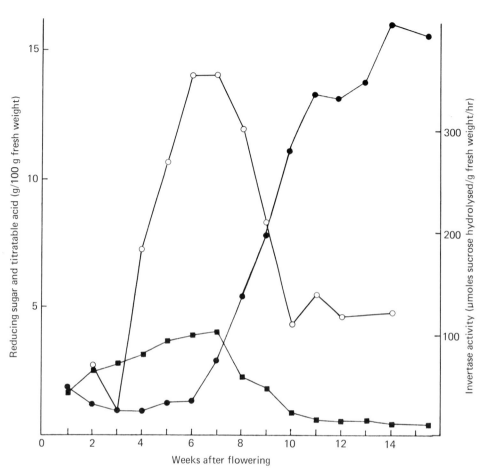

Fig. 6. Concentration of reducing sugar (●), titratable acid (■) and activity of acid invertase (○) in developing sultana grape berries. From Hawker (1969a) © 1969 Pergamon Press.

The pathway of sugar accumulation in grape berries proposed by Hawker and Barras (1972) is shown in Fig. 8. The nature and location of the enzyme involved in the hydrolysis of sucrose prior to uptake is still in doubt, but for want of exact knowledge sucrose synthase has been viewed as the responsible enzyme. Alkaline invertase has not been reported in grape berries.

The involvement of sucrose-P also awaits verification, and its inclusion in the proposed pathway was inspired by study of the sugar cane system and the increase of both sugar accumulation and the relevant enzymes at the same time during the development of the berries. The two pathways (i.e., for sugar cane and grapes) are similar in many respects. Immature sugar cane stalk

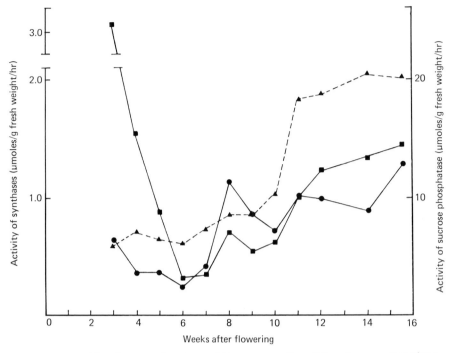

Fig. 7.   Activities of sucrose phosphatase (▲), sucrose synthase (■), and sucrose-P synthase (●) extracted from developing sultana grape berries. Note the different scales on the ordinate axes. From Hawker (1969a) © 1969 Pergamon Press.

tissue contains an acid invertase that hydrolyses some of the accumulated sucrose to glucose and fructose. Mature sugar cane tissue of the best commercial varieties has lost this invertase and stores only sucrose. Commercial grape varieties contain only low levels of sucrose, most of which is probably in transit in the phloem (Hawker, *et al.*, 1976). The high activities of acid invertase in the vacuoles of grape pericarp cells results in the storage of glucose and fructose. The human selective advantage of low reducing sugar in sugar cane is obvious, but the advantage of low sucrose in grapes is not known. Inclusion of grape berries in a chapter on sucrose storage is justified by the proposed pathway of storage rather than by the final stored product.

Uptake studies with slices of berries were thwarted by the bursting of many cells with the release of acid invertase, organic acids, and sugars (J.S. Hawker, unpublished). Pieces of grape skin proved more amenable to study, and sugar uptake by both this tissue and vacuoles prepared from it has been investigated (Brown and Coombe, 1982; 1984, personal communication). The work has not been without its problems, but with the results so far obtained the authors

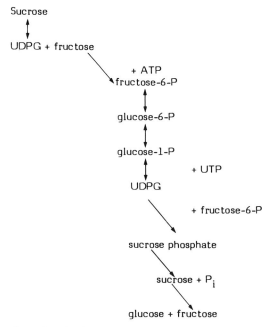

Sucrose

↑↓

UDPG + fructose

+ ATP
fructose-6-P

↑↓

glucose-6-P

↑↓

glucose-1-P

↑↓   + UTP

UDPG

+ fructose-6-P

sucrose phosphate

sucrose + P$_i$

glucose + fructose

Fig. 8. Proposed reactions involved in sugar accumulation in grape berries. From Hawker and Barras (1972).

have speculated that a group translocator is operating at the tonoplast membrane. Although the scheme is indeed speculative, it may prove a useful model for further study of sucrose transport in plants. The group translocator is envisaged as an enzyme complex of glucose-P isomerase and sucrose-P synthase, with vectorial formation of sucrose-P (Fig. 9). The authors have included sucrose phosphatase in their model but cannot comment on the possible sucrosyl transferase function of the enzyme discussed earlier in this chapter. It is interesting that Walker and Leigh (1981) have described a tonoplast-bound pyrophosphatase in red beet, since one of the products of UDP-glucose pyrophosphorylase (PP in Fig. 9) is PP$_i$.

Most fruits store reducing sugars and/or sucrose, but the pathways of accumulation are not known (Willenbrink, 1982). In many fruits, sugar and probably sucrose must move from the phloem to the vacuoles of the storage tissues during ripening. Whether transport is across one membrane via the symplasm or across three membranes via the apoplast (free space) is not known. In other fruits, e.g., banana, starch is hydrolysed in the amyloplasts in the cytoplasm during the climacteric to provide precursors for sucrose which is presumably stored in vacuoles. The form in which the sugar moves is not known.

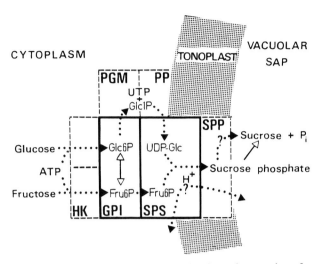

Fig. 9. Elements of a group translocator proposed to be at the tonoplast of grape pericarp cells: an enzyme complex of glucose phosphate isomerase (GPI) and sucrose phosphate synthase (SPS) with vectorial formation of sucrose phosphate. Other enzymes as hexokinase (HK), phosphoglucomutase (PGM), UDPG pyrophosphorylase (PP), or sucrose phosphate phosphatase (SPP) may also be integral components. ——, Enzymes associated with the tonoplast; ————, enzymes possibly integral to the complex; ··· movement of metabolites. From Brown and Coombe (1982).

However, in any pathway, the tonoplast must be crossed. Whether it be by a group translocator system, or by transport of sucrose, or by sugar–proton cotransport, or by another mechanism will only be known after a great deal more research.

D.   Sugar Cane Suspension Cells, Red Beet, and Sugar Beet

These tissues are discussed together because there is recent work described that focuses attention on sugar ion-transport systems across membranes in these plants. In some cases, uptake of sugars into whole cells has been compared with uptake by protoplasts and vacuoles, which has allowed estimates of fluxes across various membranes. Energetic parameters have been measured that have given leads as to mechanisms involved. Research in this field is exacting and time-consuming, with preparation of protoplasts and vacuoles requiring a great deal of skill. The results obtained so far are encouraging, but it is obvious that either many mechanisms exist in various tissues or else experimental conditions must be improved before understanding and agreement on transport systems can be achieved. At the present time there is no unified mechanism, although proton cotransport systems seem the most favoured by workers in the field (Komor *et al.*, 1982b). Whether both symport and antiport proton cotransport systems operate remains to be clarified.

1. Sugar Cane Suspension Cells

Earlier work in Hawaii concerned sugar uptake by whole cells, and this research has been extended recently to include protoplasts and vacuoles (Thom, *et al.*, 1982a,b; Komor *et al.*, 1982a,b). Rapidly growing cells of sugar cane suspension cultures of defined age were used to provide uniform experimental material. The cells were more like apical cells of stem parenchyma than like mature cells in that the cytoplasm occupied about 30% of the cell volume, although they were unlike both immature and mature parenchyma cells of sugar cane stalk in that they contained both soluble acid invertase and soluble neutral invertase in the cytoplasm and no acid invertase in the vacuole. How the suspension cells are related to parenchyma cells *in vivo* is not known, but mechanisms of sugar transport found in such cells are likely to be possible systems operating in whole plants. It is hoped that the results from all sources will eventually be integrated to provide a clear description of the pathway of sucrose accumulation in sugar cane in the field.

Protoplasts were prepared by the usual technique using cell-wall-degrading enzymes and vacuoles were prepared from these by centrifugation of protoplasts onto a cushion of Ficoll at high gravitational force. Yields of 30% were obtained for vacuoles and the vacuoles had about a 10% contamination by cytoplasm. Sugar concentrations and fluxes were determined by measuring the amounts of solutes in protoplasts and vacuoles during the growth cycle of the cells and by feeding labelled sugars to protoplasts and vacuoles prepared from starved cells. In whole cells the vacuolar hexose and sucrose concentrations became 3 to 4 times higher than the cytoplasmic concentrations, being evidence for active hexose and/or sucrose transport at the tonoplast. Label from glucose fed to protoplasts quickly appeared in both glucose and fructose in the cytoplasm and shortly after in the vacuole. Sucrose became labelled first in the cytoplasm and, after a lag, in the vacuolar compartment, eventually at a higher concentration than in the cytoplasm. The conclusion was made that sucrose is synthesized in the cytoplasm and actively transported into the vacuole, but so too are hexoses, further evidence for which was obtained by experiments with the nonmetabolized glucose analogue, 3-*O*-methylglucose.

Experiments with vacuoles showed that hexose uptake at the tonoplast was greater than at the plasmalemma. Sucrose is not transported at the plasmalemma and is transported at a lower rate than hexoses at the tonoplast.

In vivo uptake rates into the vacuole for hexoses agreed well with those for vacuoles. On the other hand, uptake rates of sucrose into vacuoles were only one-third of those found *in vivo*. The authors point out that the discrepancies could be due to damage or deterioration of vacuoles, or to changes in properties of the active transport systems. Matile (1982) draws attention to the usefulness of isolated vacuoles but warns that the removal of the cytoplasm may result in the loss of compounds that are necessary for the usual operation of the tonoplast.

Komor *et al.* (1982b) went on to measure ΔpH and membrane potentials of the plasmalemma and tonoplast of protoplasts and of the tonoplast of isolated vacuoles. With protoplasts pH gradients occurred at both membranes, with the cytoplasm more alkaline than the vacuole. It was concluded that the results might indicate that there are electrogenic pumps at the plasmalemma and tonoplast.

With isolated vacuoles there was a high membrane potential of opposite sign to that at the tonoplast in protoplasts. Neither ΔpH nor membrane potential in vacuoles was affected by uncouplers, whereas both were changed in protoplasts. It was concluded that the gradient between intravacuolar potassium (150 mM) and external medium potassium (5 mM) was responsible for the pH gradient in vacuoles and that in future more care must be taken to mimic cytoplasmic conditions when incubating isolated vacuoles. Furthermore, this could explain the low sucrose uptake by vacuoles. The authors do not suggest how hexoses are transported by isolated vacuoles at the same rate as in protoplasts, while sucrose is not. Might it not be that enzymes or substrates are lost from vacuoles during their isolation from protoplasts? Are there normally enzymes present that transport sucrose-P or does a group translocator operate? At the present time we do not know. Certainly some sucrose is synthesized in the cytoplasm of the sugar cane suspension cells, but again we do not know whether sucrose-P is an intermediate.

2.   Red Beet

Mature beetroot (*Beta vulgaris* L.) contains about 5% sucrose, most of which is in the vacuole (Leigh *et al.*, 1979). During the growth of the beets, acid invertase is present, but in mature beets the activity drops to a low level (Willenbrink, 1982). The activity is 0.05 μmol of sucrose hydrolysed/hr/g fresh weight, whereas mature sugar cane contains about 2 units of invertase, some alkaline and some cell-wall acid invertase. Unlike in sugar cane, sucrose seems not to be hydrolysed prior to uptake. Both the above groups have worked with vacuoles prepared by cutting 0.1-mm-thick beetroot slices in a specially designed apparatus. The presence of the natural dye, betanin, allowed the use of a correlative method to show that sucrose and acid invertase were located in the vacuole. It was also shown that in washed slices (1–3 days) the activity of acid invertase was inversely correlated with the level of sucrose, further proof that vacuolar acid invertase plays a major role in the control of sucrose utilization in plants.

Doll *et al.* (1982) have recently described work on the sugar uptake system of red beet vacuoles carried out by Willenbrink's group over the last few years. One component of sucrose transport across the tonoplast is active, and they speculated that sucrose transport might be linked to the flow of a cation, having previously concluded that the beet root cell contains an electrogenic, ion-translocating ATPase. In the presence of Mg^{2+} and ATP, the addition of sucrose resulted in the flow of H^+ from the vacuole to the medium. Na^+ and K^+ transfer across the tonoplast is also dependent on Mg-ATP. Sucrose stimulates Na^+ but not K^+ uptake. The use of several inhibitors gave results that indicated a relationship between ATPase and sucrose uptake. Bennett *et al.* (1984) have found H^+ translocating ATPase activity in red beet vacuoles but obtained no evidence for active sucrose transport.

Doll *et al.* (1982) are willing to suggest that the "tonoplast-bound ATPase might serve as an electrogenic pump which drives cations into the vacuole, establishing a membrane potential which enables sugar accumulation." Even so, the membrane potential of isolated red beet vacuoles is -50 to -80mV, which agrees with values found by Komor *et al.* (1982b) for isolated vacuoles of sugar cane suspension cells. However, the latter authors have doubts about the validity of their values because *in vivo* (i.e., in protoplasts) the membrane potential is of opposite sign ($+70$ mV). Research with vacuoles still has its pitfalls.

Leigh (1983) describes the evidence for the presence of ATPase on the tonoplast. Work in the latter half of the 1970s described ATPase activity, but this was later claimed to be due to nonspecific acid phosphatase. In 1981 it was shown in both Leigh's and Jacoby's laboratories that red beet vacuoles also contain a partially membrane-bound, Mg^{2+}-dependent, anion-stimulated ATPase with a pH optimum of 8.0. Another enzyme, a membrane-bound, Mg^{2+}-dependent, K^+-stimulated inorganic pyrophosphatase, might function as a proton pump in beet vacuoles. A similar enzyme functions as such in *Rhodospirillum rubrum*, but a role has not been shown for the enzyme in the tonoplast of green plants.

3. Sugar Beet

Considering that sugar beet (*Beta vulgaris* L.) is second only to sugar cane in importance in the worlds' sucrose supply, it is surprising that it is only relatively recently that work has expanded with this plant. Progress in understanding the processes of sucrose uptake by the sugar beet roots is occurring, but many opposing views are current in the interpretation of results using various techniques (Giaquinta, 1979; Saftner and Wyse, 1980; Kholodova et al., 1981; Barbier and Guern, 1982; Silvius et al., 1982). Perhaps sucrose moves from the phloem to the cytoplasm via the apoplast or perhaps via the plasmodesmata and, if the former, crosses the plasmalemma via a nonsaturating site. It is generally agreed that sucrose is not hydrolysed by invertase prior to uptake from the apoplast. Young sugar beets contain acid invertase, which is replaced during sucrose storage by sucrose synthase (Giaquinta, 1979, and references therein). Two forms of sucrose synthase occur in sugar beet roots: one, in the sucrose-storing taproot, has a higher V_{max} and affinity for UDP, whereas with the other, in the fibrous roots, ADP is the better substrate (Silvius et al., 1982). The former enzyme is believed to be important in the development of sucrose storage capacity of the tap roots. Evidence for activity of alkaline invertase in sugar beet has been obtained (Silvius and Snyder, 1979).

Sucrose-P synthase has been found in only low amounts in sugar beet, which has been used to support the conclusion that sucrose-P is not involved in sucrose transport into the vacuole (Giaquinta, 1979). However, from the data in Giaquinta's paper it can be calculated that taken over several days, sucrose accumulation is of the order of 0.2 μmol/hr/g fresh weight and the activity of sucrose-P synthase is between 0.15 and 0.3 μmol/hr/g fresh weight, assuming a protein content of 1.5 mg/g fresh wt. Determining low activities of such a labile enzyme is difficult, and it is possible that in vivo the activity would be higher. Sucrose phosphatase activity of 12 μmol/hr/g fresh weight has also been found in 70-day-old tubers (Hawker and Smith, 1984b). While the presence of enzymes does not prove their operation in vivo, there is nevertheless sufficient activity for sucrose accumulation to occur via sucrose-P. Recent work in which labelled hexose, sucrose, hexose-P, or sucrose-P has been supplied to vacuole preparations from sugar beet, red beet, and immature stem tissue of sugar cane has not resulted in sucrose accumulation (Hawker and Smith, 1984c and unpublished results). M. Thom (personal communication) has incubated vacuoles of sugar cane suspension cells with labelled sucrose-P and found no uptake of label. The lack of measurable sucrose-P transport could be due to damage of the vacuoles or to the requirement for the operation of a group translocator at the tonoplast (Brown and Coombe, 1982, 1984). As shown in Fig. 9, sugars other than sucrose-P would be required for the mechanism to operate. Whether sucrose moves as sucrose from cytoplasm to vacuole driven by a proton motive force or whether sucrose phosphatase or sucrose synthase act as carriers to transport sucrose-P or sucrose awaits further experimentation.

Most workers assume that sucrose moves across the tonoplast unchanged. They point out that there is general agreement that sugar transport by plant cells in a number of cases is dependent on a gradient of electrochemical potential for protons, as there surely is. It has been suggested from experiments with disks of intact tissue that "downhill electrochemical potential gradient across the tonoplast is coupled with sucrose uptake via a proton antiport system and that sucrose transport is coupled to a K^+ gradient" (Saftner and Wyse, 1980). Saftner *et al.* (1983) have concluded from sucrose efflux and uptake studies with disks of sugar beet root that uptake into the vacuole was mainly by active transport whereas transport into the cytoplasm was passive.

While it is agreed that the downhill electrochemical potential gradient for protons is directed from the vacuole to the cytoplasm, there are conflicting results for the transtonoplast potential difference in plant cells. This conflict seems to be due to differences in the technique used for measuring the values. Microelectrodes give small positive values for tonoplasts of whole cells or isolated vacuoles, whereas cationic probes give negative values with isolated vacuoles but positive values with protoplasts (Barbier and Guern, 1982; Komor *et al.*, 1982b). It was concluded that the tonoplast potential difference is much less well understood than the plasmalemma potential difference and that its nature and regulation need further study. Leigh (1983) has discussed the measurement and significance of potential differences across the tonoplast.

Using microelectrodes to measure tonoplast potential differences of vacuoles isolated from different varieties of sugar beet, Barbier and Guern (1982) showed a correlation between the capacity to store sucrose and the potential differences of their tonoplasts. The authors would not speculate about a mechanism but did suggest that the results warrant further study.

E. Other Tissues

The distribution of sugars in cotyledons from germinating apple seeds was determined by comparing the composition of whole tissue, isolated protoplasts, and isolated vacuoles (Yamaki, 1982). Half the total sorbitol and nearly all of the sucrose was found in the extracellular free space, but inositol and fructose were present mainly in the cytoplasm and vacuoles. Glucose was present at much lower concentrations than sorbitol and sucrose, and about half of it was in the free space. The author suggests that the difference in distribution of the different sugars supports the claim that much of the sugar is in the free space. An alternative explanation would be that specific transport of sucrose, sorbitol, and glucose occur during isolation of the protoplasts. Both explanations support apoplastic transport, while the former provides support for the presence of sugars in the free space, as already discussed for sugar cane stalks. Yamaki (1982) did not express the sugar content of free space as a concentration, a value needed to evaluate another possible explanation of the results, i.e. protoplast release from nonrepresentative cells of the cotyledons. We hope that further work will extend this interesting approach.

In developing soybean seeds, sucrose is unloaded from the seed coat phloem and subsequently released into the free space separating the seed coat and the embryo. Accumulation into the embryo is without extracellular hydrolysis and is mainly by active transport up to 50 mM sucrose. Above 50 mM, a nonsaturable component of uptake is superimposed on the active uptake. Transport has a limited dependence on external proton concentration, is not dependent on external alkaline cations, is not affected by glucose, and seems to be a sucrose–proton cotransport system (Lichtner and Spanswick, 1981; Thorne, 1982).

Excised tomato roots take up sucrose more rapidly than glucose or fructose. The uptake proceeds against a sucrose concentration gradient and shows saturation kinetics suggesting an active transport process (Chin *et al.*, 1981).

A procedure has been developed to determine the *in vivo* rate of sucrose uptake by cotyledons of *Phaseolus vulgaris* (Patrick, 1981), and analyses of developing almond fruits suggested that sucrose moves from the testa to the kernel during the accumulation of protein and lipid (Hawker and Buttrose, 1980). Sucrose uptake of *Lilium longiflorum* pollen has been proposed to occur by way of a system of proton–sucrose cotransport (Dehusses *et al.*, 1981). However, the possibility of hydrolysis of sucrose prior to its uptake by pollen is still very real, and pollen of *Lilium auratum* has recently been shown to contain a cell-wall acid invertase and a soluble acid invertase (Singh and Knox, 1984). Gentianose occurs in the vacuoles, while sucrose is located mainly in the cytoplasm of storage roots of *Gentiana lutea* L. (Keller and Wiemken, 1982).

Studies on mesophyll cells of *Pisum sativum* L. have included the first report of sugar uptake by isolated plant vacuoles (Guy *et al.*, 1979, 1981). The effect of pH, light, and proton conductors has also been determined using protoplasts isolated from the same species. The evidence is consistent with the hypothesis that the driving force for plasmalemma transport of sucrose is the electrochemical potential gradient for protons, but only in the light. Protonation alone does not seem to be sufficient to fully activate the sucrose transport system because a light-dependent factor is also required. There is a component of uptake that is active in the dark, and vacuoles also show sucrose transport. It was suggested that energization of the dark transport at the plasmalemma and the transport at the tonoplast might be directly fuelled by ATP. Mg-ATP has been shown to stimulate sucrose uptake by cowpea leaf vacuoles (Knuth *et al.*, 1983). Sucrose transport at the plasmalemma in the pea leaf system was threefold greater at pH 5.5 than at pH 8.0, and light caused a doubling of the dark rate. It is interesting that light has recently been shown to cause a fivefold increase in the transport of glycerate into isolated spinach chloroplasts (Robinson, 1982). Although uptake of glycerate is not increased by a decrease in pH, other evidence suggests that a proton cotransport system may be involved.

Recent general papers describing vacuoles and emphasizing their storage and lysosomal roles as well as their enzymic composition have appeared along with the other papers on vacuoles in the same volume of the journal (Boller, 1982; Matile, 1982). A brief review has emphasised the use of higher plant isolated vacuoles in solute transport studies (Leigh, 1983).

V. REGULATION OF SUCROSE SYNTHESIS AND STORAGE

A. Regulation in Storage Organs

The amount of sucrose stored in a sugar cane stalk depends on the size of the stalk, the average net flux of sucrose to the stalk, and the time over which storage occurs. Influx into the stalk relies on sucrose produced by photosynthesis being translocated via the phloem to the free space (apoplast) of the stalk, with translocation unlikely to be limiting. Sucrose influx into the storage parenchyma cells is made up of two types of transport. One is diffusion, which depends on the concentration of sucrose in the free space, and the second is active uptake, which depends on the activity of invertase in the free space and perhaps metabolic compartment, the rate of carrier-mediated transport of sugars across the plasmalemma, and the rate of transport of sucrose-P across the tonoplast. It seems likely that the activity of invertase controls the rate of active uptake of sucrose (Glasziou and Gayler, 1972). Once sucrose has reached high concentrations in the parenchyma vacuoles, it is possible that the partially competitive inhibition of sucrose phosphatase by sucrose could act as another control mechanism, especially if sucrose phosphatase is functioning as a sucrosyl transferase. Changes in sucrose-P synthase activities might be another coarser control of sucrose transport. In mature commercial sugar cane stalks storing high amounts of sugar, sucrose is virtually the only sugar present. As far as is known, diffusion is the only mechanism for movement of sugar from this type of tissue. Hence the concentration of sucrose in the storage compartment is the difference between the sum of inward diffusion plus active uptake, and outward diffusion. Net diffusion depends on the relative concentrations of sucrose in the free space and storage compartment of the stalks. As described earlier, active uptake was sufficient to maintain only a 5% (w/v) gradient in ripe cane containing up to 23% sucrose, but the concentration of free space sucrose approached the concentration of sucrose in the storage compartment. Cane that had resumed growth and needed a large supply of carbohydrate at the apex had depleted levels of sucrose in the free space of mature cane, and consequently had a larger gradient between storage compartment and free space resulting in more rapid efflux of sucrose (Hawker, 1965). Thus the supply or demand of sugar by the rest of the plant has an almost direct effect on levels of sucrose in the stalk. Nevertheless, sucrose buildup in cane is a slow process, taking 14 weeks to increase from 5 to 20% in one experiment (Hawker, 1965).

In a study of several varieties of sugar cane, it was shown that there was a correlation between sucrose storage in the field and the amount of free space in the tissues as determined by microscopy and sugar diffusion techniques (Oworu *et al.*, 1977a,b). The best sugar storers had the highest free space, a finding that suggests that the free space could well have contained high concentrations of sucrose. It should be pointed out that the authors did not take into account cut and/or damaged cells in their determinations of free space, possibly explaining their unusually high values of above 55% with some varieties. In 1-mm-thick slices, 40% of the tissue was apparent free space in the variety Pindar (J.S. Hawker, unpublished). Hawker (1965) used microphotography, equilibration with labelled sucrose, equilibration with glucose, and vacuum infiltration of slices 1, 2, and 3 mm thick to obtain a value of 18% for free space in intact tissue. By the use of slices of different thickness, allowance was made for damaged cells, providing the valid value for intact tissue free space (Hawker, 1965). Comparison of several varieties of cane is a useful approach, and a recent study has suggested that leaf weight per unit weight of cane, small storage cells, and the efficiency of translocation through the dewlap may influence the photosynthetic rate of sugar cane leaves, and presumably the amount of sucrose stored by a sugar cane stalk (McDavid and Midmore, 1980). Another approach to determine promising sugar cane varieties would be a comparative study of influx, efflux, and free space sugar levels, with a view to determining maximum concentrations of sucrose theoretically possible in the stalk.

The control of sugar accumulation in grape berries depends partly on the activity of the enzymes involved in its accumulation, but the limiting step is not known. Phytohormones are important in the development of the berries, and the beginning of ripening can be altered by the application of chemicals, but what actually induces the synthesis of the enzymes of sucrose metabolism again is not known. Generally speaking, under reasonable conditions, if the sink (the grape berry) has grown and there is sugar available from photosynthesis, the berry will accumulate it. Slow ripening of grape berries in marginal climatic areas is usually caused by a delay in the onset of ripening rather than a sluggish sugar transport system.

Control of sucrose storage in sugar beets is not a clear phenomenon (Willenbrink, 1982). The development of the ratio of storage root to fibrous roots may be regulated by isozymes of sucrose synthase, which eventually control the amount of sucrose stored (Silvius *et al.*, 1982). It is tempting to accept the notion that plants store any carbohydrate surplus to their needs for growth, but this simple idea seems not to be true (Willenbrink, 1982). Certainly in sugar cane, decreased growth rates due to lower temperatures result in increased sugar concentrations, a fact used commercially in many areas of the world. There is competition between storage volume and storage concentration. However, when considering sugar cane varieties, some of which store only low amounts of sucrose, it is obvious that factors other than a surplus of sucrose are responsible for high sugar storage. Possibly "energy overflow" via the alternative pathway is involved (see Lambers, 1982).

Source–sink considerations are of continuing interest (Herold, 1980). If sucrose in the stalk of sugar cane accumulates to a concentration that eventually causes the occurrence of inhibiting levels in the leaves, then a further control on storage is imposed.

Recent source–sink studies on eggplant (*Solanum melongena* L.) have included analysis of enzymes of sucrose metabolism in plants with and without fruit (Claussen, 1983a,b; Claussen and Lenz, 1983). The fruit contain relatively low levels of sucrose, glucose, and fructose (about 1% by fresh weight of each), but neither the distribution of sugars within the fruit nor the pathway of accumulation are known. The growth rates of leaves, shoots and roots decreased as fruit growth occurred. At the same time the activity of sucrose synthase showed a similar pattern, such that positive correlations existed between the growth rates and enzyme activity. In plants without fruit, sucrose and starch content of the leaves was higher and the activity of sucrose-P synthase in the leaves and the net photosynthetic rate was lower than in plants with fruit. In sugar cane leaves, sucrose-P synthase and photosynthesis increased 3 and 2.5-fold, respectively, when cane was moved from a sugar storage environment to a growth environment (Hawker, 1967b). Another example of enzyme induction and repression is found in immature stalk tissue where vacuolar acid invertase synthesis is repressed by glucose in the cytoplasm (Glasziou and Gayler, 1972). As stalks mature, vacuolar acid invertase synthesis virtually ceases and acid invertase in sugar beets drops to a low level during development. However, in grape berries, vacuolar acid invertase was induced 2 to 3 weeks earlier than other enzymes of sucrose metabolism, and the activity remained high.

Within cells, synthesis of sucrose is probably controlled by the activities of phosphofructokinase and fructose 1,6-bisphosphatase, both of which respond to several biochemical effectors such as PEP, ATP, and AMP (Avigad, 1982). Recently, Woodrow *et al.* (1982) have partially purified from roots a novel fructose bisphosphatase that has a possible role in cytosolic gluconeogenesis during the conversion of starch to sucrose. Also, a phosphofructokinase from tomato fruits has been partially purified and studied (Issac and Rhodes, 1982). Further discussion of metabolic control in leaves and the role of fructose 2,6-bisphosphate is presented next.

B. Comparisons with Metabolism of Sucrose in Leaves

Even though leaves are not commercial sources of sucrose, they provide the sucrose for translocation to commercial storage organs. More importantly in the present context, the control of sucrose synthesis in leaves has received some attention in recent years. It seems likely that fine control may be achieved by inhibitors of fructose 1,6-bisphosphatase (FBP) and sucrose-P synthase and coarser control by the regulation of the amount of sucrose-P synthase present. Sucrose might be involved in feedback inhibition of sucrose phosphatase and/or sucrose-P synthase. The following discussion shows that this statement is an oversimplification and illustrates the complexity of the data and current thoughts on the topic.

Two FBPases are found in leaves, one in the stroma, which is involved in the Calvin cycle and another in the cytoplasm, which forms part of the pathway of sucrose synthesis. The cytosolic enzyme from wheat protoplasts is inhibited by micromolar concentrations of fructose, 2,6-bisphosphate, which, in addition, increases the sensitivity of the enzyme to inhibition by AMP and P_i (Stitt *et al.*, 1982). A cytosolic PP_i-dependent phosphofructokinase was also shown to be active only in the presence of fructose 2,6-bisphosphate. The enzyme is thought to play a reciprocal role in sucrose metabolism to FBPase because of its location in the cytosol, its reciprocal modulation by fructose 2,6-bisphosphate, and its likely dependence on sucrose-P synthase (via UDP-glucose pyrophosphorylase) as a source of PP_i. The enzyme has been found in many photosynthetic tissues and is activated by fructose 2,6-bisphosphate (Carnal and Black, 1983). Whether the enzyme has a role in storage organs is not known.

Further work with wheat protoplasts (Stitt *et al.*, 1983) showed that when photosynthesis is limited by low light or CO_2, gluconeogenesis and sucrose synthesis are inhibited. The restriction of sucrose synthesis was due to enzyme regulation brought about by a shortfall in the supply of organic carbon from the chloroplasts rather than to a lack of substrates for gluconeogenesis. It was suggested that FBPase was regulated synergistically by fructose 2,6-bisphosphate and P_i or AMP, resulting in a decrease in the rate of conversion of triose-P to hexose-P. Continued sucrose synthesis would result in a decrease in fructose 6-P and glucose 6-P, an increase in P_i, and a possible decrease in the activity of sucrose-P synthase by the decreased fructose 6-P/P_i and glucose 6-P/P_i ratios (Amir and Preiss, 1982; Doehlert and Huber, 1983). The result of the regulation was that stromal metabolites were maintained at concentrations high enough to allow operation of the Calvin cycle. In addition sucrose synthesis can be inhibited by feedback inhibition so that triose-P is used for starch synthesis. Sucrose and P_i are known to inhibit sucrose phosphatase (Hawker, 1967a), and this could lead to a buildup of sucrose-P, which is known to inhibit sucrose-P synthase (Amir and Preiss, 1982). On the other hand, Foyer *et al.* (1983) do not consider inhibition of sucrose phosphatase a likely control mechanism. Stitt *et al.* (1984) consider that fructose 2,6-bisphosphate increases when sucrose accumulates in leaves, and this could control sucrose synthesis by inhibiting cytosolic FBPase. These authors have also shown that the regulation of sucrose synthesis is not primarily dependent on the presence or absence of light. Sucrose synthesis can be inhibited in the light and proceed in the dark from starch stored in the light (Fondy and Geiger, 1982). However, in leaf protoplasts under some conditions it seems that sucrose synthesis is largely prohibited in darkness, because sucrose synthesis ceased almost immediately when the light was switched off (Foyer *et al.*, 1983). Light is known to activate several leaf enzymes (Cséke *et al.*, 1982; Anderson *et al.*, 1979; Heuer *et al.*, 1982; Leegood and Walker, 1982), but no one has found any effects of light on enzymes of sucrose synthesis.

If sucrose causes feedback inhibition in leaves, it seemed reasonable that sucrose pretreatment would inhibit photosynthesis and sucrose synthesis of leaf protoplaasts (Foyer *et al.*, 1983). The rate of CO_2 fixation and sucrose synthesis was not affected by sucrose pretreatment in mature wheat and barley protoplasts, but both were inhibited by sucrose pretreatment in protoplasts from mature spinach leaves. Since neither sucrose-P synthase, FBPase, nor UDP-glucose pyrophosphorylase showed inhibition by sucrose, it is unlikely that sucrose caused feedback inhibition of the sucrose pathway unless via sucrose inhibition of sucrose phosphatase as discussed above. The lack of effect of sucrose on wheat and barley cannot be explained.

Huber's group has developed the idea that the activity and properties of sucrose-P synthase in leaves of different plants determine the partitioning of carbon between sucrose and starch. From a study of several species and cultivars, it was found that there was a negative correlation between leaf starch and sucrose-P synthase activity (Huber, 1983). In addition, a positive correlation between sucrose content and enzyme activity was found in several cultivars of peanut. Sucrose was also shown to be inhibitory in about half the peanut cultivars by up to 42%. Such differences in response to sucrose could explain the conflicting results that have appeared in the literature over many years.

In some species of plants, sorbitol fully or partially replaces sucrose as the form of carbon translocated from leaves to the rest of the plant. Even though sorbitol is a 6-carbon sugar alcohol, comparison of research on this compound with research on sucrose may provide leads or answers to problems pertaining to sucrose translocation and storage (Carey *et al.*, 1982; Grant and ap Rees, 1981; Loescher *et al.*, 1982; Negm and Loescher, 1981; Yamaki, 1982).

Interesting as the work on control of sucrose synthesis in leaves is, it is not yet at the stage to be applied to storage organs. Nevertheless, obvious similarities occur, and leaf studies will no doubt add to our understanding of sucrose storage.

C.   Phytohormones, Sucrose as a Regulator, and Plant Regulators

1.   Phytohormones

Gibberellin stimulates invertase activity in *Avena* internodes, sugar cane internodes, tomato roots, and cucumber hypocotyls and also decreases the osmotic potential of the epidermal cells of intact cucumber hypocotyls. The latter effect is enhanced by the presence of sucrose (see Katsumi *et al.*, 1982). Invertase activity in immature sugar cane tissue is also stimulated by auxin and repressed by glucose (Glasziou, 1969). The same review discusses other effects of hormones on invertase in other tissues. There are a few reports of effects of hormones on enzymes of sucrose metabolism, but usually a long time lapsed between treatment and effect (Pontis, 1977). Abscisic acid was shown to inhibit net phloem loading of sucrose by cotyledons, and it was suggested that the result was due to an enhancement of sucrose efflux (Vreugdenhil, 1983). Saftner and Wyse (1984) have found that ABA can stimulate and IAA can inhibit active sucrose uptake by disks of sugar beet root. The authors suggest that there is a close relationship between active uptake and the physiological activities of IAA and ABA.

2.   Sucrose as a Regulator

Although regulation of sucrose synthesis by sucrose in tissues is a contentious subject, there are many examples in the literature on tissue culture of regulation of other processes by sucrose (Pontis, 1977; Avigad, 1982). Sucrose concentration, sometimes in association with hormones, affects the degree of differentiation of callus tissues from various plants. Starch synthesis is modified in other callus tissue, and chlorophyll synthesis is suppressed in carrot callus tissue containing no invertase. Some pollen grains will only germinate when supplied with sucrose as a substrate. Recently Barthe and Bulard (1982) have shown that dormancy of apple seeds was deeper in the presence of sucrose than in its absence and that sucrose in association with ammonium ions inhibits flowering of several species of *Lemna* (Ives and Posner, 1982). Sucrose increased the rate of root initiation on cultured rose shoots (Hyndman *et al.*, 1982), and probably affected growth in the region of cell elongation of detached pea epicotyls (Singh and MacLachlan, 1983). While these effects do not provide answers on the regulation of sucrose synthesis and storage, they do show that the sucrose molecule can influence cell metabolism and not just function as a convenient relatively inert form of transportable carbon.

3. Plant Regulators

Control of sugar storage commercially has become economically feasible in sugar cane (Nickell, 1978, 1982). Investigators in Australia began studies on chemical ripeners in the field, and in the early 1960s there was a cooperative programme between Hawaii and Australia. However, it soon emerged that Australia's cool and dry weather prior to harvest was conducive to optimum natural ripening. It was considered that under the cool dry conditions, growth of the stalks almost ceased, whereas photosynthesis continued and supplied sucrose to fill the existing stalks. Research continued in the United States and several compounds were found to increase the sugar content of the cane in the field. In 1980 two compounds were registered in the United States as commercial products for this purpose. The chemicals are glyphosine [N,N-bis(phosphonylmethyl)glycine], common name Polaris, and phosphomonomethyl glycine, common name Polado, generic name glyphosate. The former compound has been used extensively, but the latter compound is likely to replace it because of its greater activity and lower costs (Nickell, 1982).

Three other chemicals have been registered in the United States for field evaluation. They are Ripenthol, Chlormequat, and Embark. Some problems, such as phytotoxicity and low activity, have been encountered. Nevertheless, it seems certain that chemical sugar cane ripeners will be used in many parts of the world. Whether they influence sugar accumulation directly or indirectly is still not known.

One of the early chemical ripeners tested on sugar cane in Australia was methyl 2-(ureidooxy)propionate (MUP). It appeared promising but was not developed further with sugar cane. Trials with two varieties of grapevines in glasshouses showed that at a concentration of 0.1% (w/v) MUP hastened the ripening of grape berries by about 1 week when applied 4 to 5 weeks after flowering (Hawker et al., 1981). Subsequent work in the field confirmed the hastening effect with two other varieties of grapes. *Vitis vinifera* L. cultivars Mataro and Sultana ripened 2 weeks and 1 week earlier, respectively, than controls when sprayed with 0.1% solution of MUP about halfway through the first rapid growth phase and again 2 weeks later. Terminal and lateral shoot growth was inhibited and the start of the second rapid growth phase of the grape berries was advanced, but the subsequent *rate* of sugar accumulation was not affected. It just occurred 2 weeks or 1 week later in the controls, and hence the date on which the sugar:acid ratio reached 30 (the stage to harvest for wine making) occurred 2 weeks to 1 week earlier in the treated grapes. In view of the facts that growth regulators can decrease grapevine shoot growth without hastening ripening, and other hormones and regulators can affect the rate of development of grape berries, it seemed more likely that MUP had a direct effect on the berries rather than acting by decreasing the size of the competitive sink, i.e. the growing shoots.

There is no evidence as to the steps involved in the control brought about by MUP. Abscisic acid and ethylene applied precisely during the slow growth phase of berry development hastened ripening by a few days, but that may be only coincidence. However, MUP advances the accumulation of glucose and fructose via sucrose in the grapes. The control may not be at a step close to the uptake of sucrose, but again we do not know. Nevertheless, the overall result is the partial control of sugar storage and once potential toxicological and environmental hazards are overcome, a chemical might be of value to commercial vignerons. The findings with MUP show that hastening the ripening of grapes by the application of a regulator is feasible.

D. Effect of Salinity

The mechanisms by which glycerol, sucrose, and starch are linked in the photosynthetic halophilic alga *Dunaliella* are being investigated and might be of relevance to the control of partitioning of carbon in higher plants.

Most higher plants are nonhalophytes, and mechanisms of salt tolerance in these is of high current interest as the world's agricultural areas become more saline (Greenway and Munns, 1980). Reduction in leaf growth is a common plant response to salinity. Concentrations of NaCl up to 180 mM in the external medium caused decreased rates of leaf elongation in salt-tolerant barley seedlings (*Hordeum vulgare* cv. Beecher). NaCl treatment had no effect on glucose, fructose, or starch concentration but caused an increase of glucose and fructose released by invertase (presumably sucrose) and unidentified soluble carbohydrate (Delane *et al.*, 1982; Munns *et al.*, 1982). The authors concluded that photosynthesis was not limiting growth, that the reason for reduction of growth was located in the elongating tissue of growing leaves, and that these tissues suffered from water deficit rather than from inhibitory effects of ions on metabolism.

Hawker and Walker (1978a) found that the rate of expansion and the invertase activity of leaves of two salt-sensitive species, *Phaseolus vulgaris* and *Zea mays*, decreased with increasing concentrations of NaCl from 0 to 50 mM. Sucrose concentrations were higher and reducing sugar concentrations were lower in leaves of the plants grown at the higher concentrations of NaCl.

Leaves of *Hordeum vulgare* were hardly affected by NaCl up to 50 mM. The *in vitro* activity of invertase from barley, beans or maize leaves were not affected by NaCl at concentrations up to 200 mM. Subsequently it was shown that in bean leaves it was acid invertase that decreased with NaCl treatment; alkaline invertase activity was not affected (Hawker, 1980). It was not possible to determine whether it was a specific effect on invertase of Na$^+$ or Cl$^-$ or both, or a general osmotic effect. However, acid invertase is often associated with growing tissues, and the concurrent effect of NaCl on both growth and invertase activity point to a causal relationship. The fact that alkaline invertase (and also cellulase) was not affected shows that NaCl was not causing a general reduction in enzyme activity.

In another series of experiments, partial confirmation of the above effects of salt was obtained. An inverse relationship was found between sucrose content and acid invertase activity in leaves of *Phaseolus vulgaris* on treatment of plants with salts (Rathert, 1982). KCl was the most effective *in vivo*. In the salt-tolerant plant sugar beet, salts caused little change in invertase activity or sucrose content. It was suggested that the effect of ions on carbohydrate levels and on acid invertase activity might be used as marker for salt tolerance in crops. However, in a moderately salt-tolerant crop, grapevine, no invertase response to salt treatment could be demonstrated (Hawker and Walker, 1978b).

Potassium ions activate starch-bound ADP-glucose starch synthase from plants, e.g. from cassava tubers (Hawker and Smith, 1982), but only relatively low concentrations are required (10–25 mM), so that this effect is not usually considered a salinity effect. Enzymes of sucrose metabolism besides invertase either have not been tested or have not shown salinity effects.

E. Low-Temperature Sweetening

Many plants show an increase in sugar content when subjected to temperatures below 10°C. A role in frost hardiness has been suggested for this phenomenon in nature. An important, unfortunate, and unwanted consequence of the effect is that it renders one of the world's most important foods, the potato, unpalatable and useless for processing. Brown discoloration of chips and crisps is caused by reaction between reducing sugars and amino acids in sweetened potatoes.

ap Rees *et al.* (1982) have presented evidence that fine control of phosphofructokinase and possibly pyruvate kinase is important in the control of the conversion of starch to sucrose when harvested potato tubers are transferred from 10 to 2°C. Within 5 days the sucrose content began to increase and by 20 days had increased to 5 times the original level. Sucrose was synthesized by sucrose-P synthase. Reducing sugar content rose to about one-third of the sucrose during the same period. An examination of apparent equilibrium constants of enzymes of the glycolytic cycle showed that phosphofructokinase and pyruvate kinase catalysed nonequilibrium steps. Since phosphofructokinase is the first nonequilibrium step unique to glycolysis, it was suggested that hexose phosphates derived from a series of reactions from starch entered glycolysis at a rate controlled by this enzyme. The enzyme is affected by cold, with a decrease in activity at $2°C$ that is much greater than the effect of low temperature on sucrose-P synthase. The cold lability of the enzyme was due to weakened hydrophobic interactions, resulting in the spontaneous dissociation of the active oligomeric form. The reduced entry of hexose phosphates into glycolysis allowed diversion of extra hexose phosphates into sucrose catalysed by sucrose-P synthase. The authors pointed out that many unknowns still remain about cold-induced sweetening and suggested that other mechanisms could be operating in addition to the fine control described. The nature of the products traversing the amyloplast membranes is one question, although triose phosphates are likely candidates (see section on sucrose synthase). It could be argued that the triose phosphates coming from the amyloplasts would reduce the necessity of the action of phosphofructokinase. Then the role of fructose bisphosphatase in the production of hexose phosphates from the triose phosphates would become important. Is that enzyme a control point as in leaves? Much remains to be learnt. However, it is interesting in a general appreciation of the control of sucrose synthesis in plants that it appeared that a step other than sucrose-P synthase was one of the control points.

VI. FUTURE DEVELOPMENTS

Even though there has been a reasonable amount of research carried out on the enzymes of sucrose metabolism, there are still some properties of the enzymes that differ from report to report. In some cases there is evidence that the differences are due to species differences, but in others it may be that extraction and purification techniques result in enzymes with variable properties. Limited research into the transport of sucrose and the control of transport has proposed basically two mechanisms, one involving sucrose-P and the other an active transport of sucrose in conjunction with transport of protons. Regulation of sucrose synthesis and transport has been studied and control mechanisms have been proposed, but insufficient evidence has been obtained for general acceptance. It is generally agreed that sucrose is an early product of photosynthesis, that it is synthesized in the cytoplasm external to the chloroplasts, that it is transported in the phloem, and that in some storage organs it is accumulated at high concentrations in the vacuoles of the parenchyma cells. All the other processes and control mechanisms require further study. Further purification and study of the properties of both sucrose-P synthase and sucrose phosphatase is required. The use of isolated protoplasts and vacuoles has proven and will prove beneficial in transport studies, although preparation of vacuoles and uptake experiments with vacuoles are not without their problems. Recently it has been demonstrated that barley mesophyll vacuoles accumulate sucrose *in vivo* (Kaiser *et al.* 1982). Sucrose uptake into isolated vacuoles from these cells may occur by mediated diffusion (G. Kaiser, personal communication). Eventually we may learn what determines the partitioning of triose phosphates, which are produced by the chloroplasts during photosynthesis in whole plants, betwen sucrose and other compounds, and whether significant amounts of sucrose move into leaf vacuoles for temporary storage before being transported to the phloem. The role of sinks and mechanism of action of the control exerted by sinks on these processes needs to be understood.

Studies on sucrose synthase mRNA of developing maize kernels (Wöstemeyer *et al.* 1981) and on isoenzymes of rice sucrose synthase have led to the suggestion that eventually rice grain yield might be enhanced by genetic engineering (Su, 1982). In a fruit, avocado, three mRNAs increased markedly during ripening, indicating that the process may be linked to the expression of specific genes (Christoffersen *et al.* 1982). The functional significance of all three mRNAs is not known, but they may be involved in specific enzyme synthesis because one codes for cellulase (G. G. Laties, personal communication). If translation of specific mRNAs occurs in other fruit during ripening, it might be that one or more controls the synthesis of enzymes of sugar synthesis or transport, e.g. sucrose-P synthase in the grape berry. Changes in the genetic composition might result in earlier or faster accumulation of sugar. Once the pathways and controls of sugar accumulation are known in sugar cane and sugar beet, perhaps it will be feasible to enhance the sugar transport rates in one species by gene transfer to the other.

The field of genetic engineering and other newer techniques of breeding are advancing rapidly (Meredith, 1982, and following papers). The potential DNA recombinant vector, cauliflower mosaic virus, now appears not to be useful as such. *Agrobacterium tumefaciens* harbors a Ti plasmid that might be used to carry specific genes to dicotyledonous plants. However, recent work suggests that the Ri plasmid of *A. rhizogenes* might prove to be equally useful as a genetic engineering vector (Chilton *et al.*, 1982). In monocotyledonous plants, transposable elements (jumping genes) may play a role in directed gene transfer in the laboratory. Genetic engineering of higher plants in order to control sucrose accumulation may not be achieved for a considerable time yet.

Irrigated sugar cane on the Ord River in North Western Australia yields up to 20 tons sucrose/hectare (J. S. Gallagher, personal communication).

Irrigated grapes in Australia on permanent trellises with minimal pruning and mechanical harvesting can yield 10 tons glucose and fructose/hectare on land less fertile than required by sugar cane (A. J. Antcliff, personal communication). Average yields of sucrose from commercial sugar beet production in some countries are about 10 tons/hectare (Anonymous, 1979). There exists a challenge to plant biochemists to provide the information needed by genetic engineers for the production of plants yielding higher amounts of sugar for food and/or power alcohol production.

REFERENCES

Akazawa, T. and Okamoto, K. (1980). *In* "The Biochemistry of Plants" (J. Preiss, ed.), Vol. 3, pp. 199–220. Academic Press, New York.

Amir, J. and Preiss, J. (1982). *Plant Physiol.* **69,** 1027–1030.

Anderson, J. M. (1983). *Plant Physiol.* **71,** 333–340.

Anderson, L. E., Chin H. and Gupta, V. K. (1979). *Plant Physiol.* **64,** 491–494.

Anonymous (1979). *Foreign Agric. Circ.* **FS I-79.** USDA.

Anonymous (1980). *Prod. Yearb, FAO* **34,** 167–170.

ap Rees, T. (1974). *MTP Int. Rev. Sci.: Plant Biochem.* **11,** 89–127.

ap Rees, T. (1977). *Symp. Soc. Exp. Biol.* **31,** 7–32.

ap Rees, T., Dixon, W. L., Pollock, C. J. and Franks, F. (1982). *In* "Recent Advances in the Biochemistry of Fruits and Vegetables" (J. Friend and M. J. C. Rhodes eds.), pp. 43–61. Academic Press, New York.

Avigad, G. (1982). *In* "Encyclopedia of Plant Physiology, New Series" (F. A. Loewus and W. Tanner, eds.), Vol. 13A, pp. 217–347. Springer-Verlag, Berlin and New York.

Barbier, H. and Guern, J. (1982). *In* "Plasmalemma and Tonoplast: Their function in the Plant Cell" (D. Marmé, E. Marrè and R. Hertel, eds.), pp. 233–240. Elsevier, Amsterdam.

Barthe, P. H. and Bulard, C. (1982). *New Phytol.* **91,** 517–529.

Bennett, A. B., O'Neill, S. D., and Spanswick, R. M. (1984) *Plant Physiol.* **74,** 538–544.

Boller, T. (1982). *Physiol. Veg.* **20,** 247–257.

Brown, S. C. and Coombe, B. G. (1982). *Naturwissenschaften* **69,** 43–44.

Brown, S. C. and Coombe, B. G. (1984). *Physiol. Veg.* **22,** 231–240.

Cardini, C. E., Leloir, L. F. and Chiriboga, J. (1955). *J. Biol. Chem.* **214,** 149–155.

Carey, E. E., Dickinson, D. B., Wei, L. Y. and Rhodes, A. M. (1982). *Phytochemistry* **21,** 1909–1911.

Carnal, N. W. and Black, C. C. (1983). *Plant Physiol.* **71,** 150–155.

Chilton, M., Tepfer, D. A., Petit, A., David, C., Casse-Delbart, F. and Tempé, J. (1982). *Nature (London)* **295,** 432–434.

Chin, C., Lee, M. and Weinstein, M. (1981). *Can. J. Bot.* **59,** 1159–1163.

Christoffersen, R. E., Warm, E. and Laties, G. G. (1982). *Planta* **155,** 52–57.

Claussen, W. (1983a). *Z. Pflanzenphysiol.* **110,** 165–173.

Claussen, W. (1983b). *Z. Pflanzenphysiol.* **110,** 175–182.

Claussen, W. and Lenz, F. (1983). *Z. Pflanzenphysiol.* **109,** 459–468.

Cséke, C., Nishizawa, A. N. and Buchanan, R. B. (1982). *Plant Physiol.* **70,** 658–661.

Delane, R., Greenway, H., Munns, R. and Gibbs, J. (1982). *J. Exp. Bot.* **33,** 557–573.

Dehusses, J., Gumber, S. C. and Loewus, F. A. (1981). *Plant Physiol.* **67,** 793–796.

Doehlert, D. C. and Huber, S. C. (1983). *FEBS Lett.* **153,** 293–297.

Doll, S., Effelsberg, U. and Willenbrink, J. (1982). *In* "Plasmalemma and Tonoplast: Their Functions in the Plant Cell" (D. Marmé, E. Marrè and R. Hertel, eds.), pp. 217–224. Elsevier, Amsterdam.

Duffus, C. M. (1979). *In* "Recent Advances in the Biochemistry of Cereals" (D. L. Laidman and R. G. Wyn Jones, eds.), pp. 209–238. Academic Press, New York.

Eames, A. J. and MacDaniels, L. H. (1947). *In* "An Introduction to Plant Anatomy," pp. 158–163. McGraw-Hill, New York.

Eschrich, W. (1980). *Ber. Dtsch. Bot. Ges.* **93**, 363–378.

Evert, R. F. (1982). *BioScience* **32**, 789–795.

Felkner, F. C. and Shannon, J. C. (1980). *Plant Physiol.* **65**, 864–870.

Fondy, B. R. and Geiger, D. R. (1982). *Plant Physiol.* **70**, 671–676.

Foyer, C., Rowell, J. and Walker, D. A. (1983). *Arch. Biochem. Biophys.* **220**, 232–238.

Geiser, M., Weck, E., Döring, H. P., Werr, W., Courage-Tebbe, U., Tillman, E. and Starlinger, P. (1982). *EMBO J.* **1**, 1455–1460.

Giaquinta, R. T. (1979). *Plant Physiol.* **63**, 828–832.

Giaquinta, R. T. (1983). *Annu. Rev. Plant Physiol.* **34**, 347–387.

Gibson, D. M. and Shine, W. E. (1983). *Proc. Natl. Acad. Sci. U. S. A.* **80**, 2491–2494.

Glasziou, K. T. (1969). *Annu. Rev. Plant Physiol.* **20**, 63–88.

Glasziou, K. T. and Gayler, K. R. (1972). *Bot. Rev.* **38**, 471–490.

Grant, C. R. and ap Rees, T. (1981). *Phytochemistry* **20**, 1505–1511.

Greenway, H. and Munns, R. (1980). *Annu. Rev. Plant Physiol.* **31**, 149–190.

Gross, K. C. and Pharr, D. M. (1982a). *Plant Physiol.* **69**, 117–121.

Gross, K. C. and Pharr, D. M. (1982b). *Phytochemistry* **21**, 1241–1244.

Guy, M., Reinhold, L. and Michaeli, D. (1979). *Plant Physiol.* **64**, 61–64.

Guy, M., Reinhold, L., Rahat, M. and Seiden, A. (1981). *Plant Physiol.* **67**, 1146–1150.

Harbron, S., Foyer, C. and Walker, D. A. (1981). *Arch. Biochem. Biophys.* **212**, 237–246.

Hardy, P. J. (1968). *Plant Physiol.* **43**, 224–228.

Hatch, M. D. (1964). *Biochem. J.* **93**, 521–526.

Hawker, J. S. (1965). *Aust. J. Biol. Sci.* **18**, 959–969.

Hawker, J. S. (1967a). *Biochem. J.* **102**, 401–406.

Hawker, J. S. (1967b). *Biochem. J.* **105**, 943–946.

Hawker, J. S. (1969a). *Phytochemistry* **8**, 9–17.

Hawker, J. S. (1969b). *Phytochemistry* **8**, 337–344.

Hawker, J. S. (1980). *Aust. J. Plant Physiol.* **7**, 67–72.

Hawker, J. S. and Barras, D. R. (1972). *Bull. O.I.V.* **45**, 849–851.

Hawker, J. S. and Buttrose, M. S. (1980). *Ann. Bot. (London)* [N. S.] **46**, 313–321.

Hawker, J. S. and Hatch, M. D. (1966). *Biochem. J.* **99**, 102–107.

Hawker, J. S. and Hatch, M. D. (1975). *In* "Methods in Enzymology" (W. A. Wood, ed.), Vol. 42, pp. 341–347. Academic Press, New York.

Hawker, J. S. and Smith, G. M. (1982). *Aust. J. Plant Physiol.* **9**, 509–518.

Hawker, J. S. and Smith, G. M. (1984a). *Phytochemistry* **23**, 245–249.

Hawker, J. S. and Smith, G. M. (1984b). *In* "Advances in Agricultural Biotechnology. Advances in Photosynthesis Research." Proceedings of the VIth International Congress on Photosynthesis, Brussels, Belgium. Aug. 1–6, 1983. (C. Sybesma, ed.) Vol. III, pp. 501–504.

Hawker, J. S. and Smith, G. M. (1984c). Proceedings of the Australian Biochemical Society **16**, 73.

Hawker, J. S. and Walker, R. R. (1978a). *Aust. J. Plant Physiol.* **5**, 73–80.

Hawker, J. S. and Walker, R. R. (1978b). *Am. J. Enol. Vitic.* **29**, 172–176.

Hawker, J. S., Ruffner, H. P. and Walker, R. R (1976). *Am. J. Enol. Vitic.* **27**, 125–129.

Hawker, J. S., Hale, C. R. and Kerridge, G. H. (1981). *Vitis* **20**, 302–310.

Herold, A. (1980). *New Phytol.* **86**, 131–144.

Heuer, B., Hansen, M. J. and Anderson, L. (1982). *Plant Physiol.* **69**, 1404–1406.

Ho, L. C. and Baker, D. A. (1982). *Physiol. Plant.* **56**, 225–230.

Huber, S. C. (1981). *Z. Pflanzenphysiol.* **102**, 443–450.

Huber, S. C. (1983). *Plant Physiol.* **71**, 818–821.

Huber, S. C. and Moreland, D. E. (1981). *Plant Physiol.* **67**, 163–169.

Huber, S. C. and Pharr, D. M. (1981). *Plant Physiol.* **68**, 1294–1298.

Humphreys, T. and Echeverria, E. (1980). *Phytochemistry* **19**, 189–193.

Hyndman, S. E., Hasegawa, P. M. and Bressan, R. A. (1982). *Plant Cell, Tissue Organ Cult.* **1**, 229–238.

Issac, J. E. and Rhodes, M. J. C. (1982). *Phytochemistry* **21**, 1553–1556.

Ives, J. M. and Posner, H. B. (1982). *Plant Physiol.* **70**, 311–312.

Jacob, J., Prevot, J. and D'Auzac, J. (1982). *Phytochemistry* **21**, 851–853.

Jenner, C. F. (1982). *In* "Encyclopedia of Plant Physiology, New Series" (F. A. Loewus and W. Tanner, eds.), Vol. 13A, pp. 700–747. Springer-Verlag, Berlin and New York.

Kaiser, G., Martinoia, E. and Wiemken, A. (1982). *Z. Pflanzenphysiol.* **107**, 103–113.

Katsumi, M., Kazama, H., Yamada, J. and Matsumura, M. (1982). *Plant Cell Physiol.* **23**, 953–958.

Keller, F. and Wiemken, A. (1982). *Plant Cell Rep.* **1**, 274–277.

Kholodova, V. P., Bolyakina, Y. P., Meshcheriakov, A. B. and Orlova, M. S. (1981). *In* "Structure and Function of Plant Roots (R. Brouwer *et al.*, eds.), pp. 209–213. Junk, The Hague.

Knuth, M. E., Keith, B., Clark, C., Garcia-Martinez, J. L. and Rappaport, L. (1983). *Plant Cell Physiol.* **24**, 423–432.

Komor, E. (1982). *In* "Encyclopedia of Plant Physiology, New Series" (F. A. Loewus and W. Tanner, eds.), Vol. 13A, pp. 635–676. Springer-Verlag, Berlin and New York.

Komor, E., Thom, M. and Maretzki, A. (1982a). *Plant Physiol.* **69**, 1326–1330.

Komor, E., Thom, M. and Maretzki, A. (1982b). *Physiol. Veg.* **20**, 277–287.

Lambers, H. (1982). *Physiol. Plant.* **55**, 478–485.

Leegood, R. C. and Walker, D. A. (1982). *Planta* **156**, 449–456.

Leigh, R. A. (1983). *Physiol. Plant.* **57**, 390–396.

Leigh, R. A., ap Rees, T., Fuller, W. A. and Banfield, J. (1979). *Biochem. J.* **178**, 539–547.

Leloir, L. F. and Cardini, C. E. (1955). *J. Biol. Chem.* **214**, 157–165.

Lengeler, J. W., Mayer, R. J. and Schmid, K. (1982). *J. Bacteriol.* **151**, 468–471.

Lichtner, F. T. and Spanswick, R. M. (1981). *Plant Physiol.* **68**, 693–698.

Loescher, W. H., Marlow, G. C. and Kennedy, R. A. (1982). *Plant Physiol.* **70**, 335–339.

Lott, R. V. and Barrett, H. (1967). *Vitis* **6**, 257–268.

McDavid, C. R. and Midmore, D. J. (1980). *Ann Bot.* (*London*) [N. S.] **46**, 479–483.

Macdonald, F. D. and ap Rees, T. (1983a). *Biochim Biophys. Acta* **755**, 81–89.

Macdonald, F. D. and ap Rees, T. (1983b). *Phytochemistry* **22**, 1141–1143.

Mangat, B. S. and Kerson, G. (1982). *Phytochemistry* **21**, 1259–1261.

Matile, P. (1982). *Physiol. Veg.* **20**, 303–310.

Maynard, J. W. and Lucas, W. J. (1982a). *Plant Physiol.* **69**, 734–739.

Maynard, J. W. and Lucas, W. J. (1982b). *Plant Physiol.* **70**, 1436–1443.

Meredith, C. P. (1982). *Calif. Agric.* **36**(8), 5.

Morell, M. and Copeland, L. (1984). *Plant Physiol.* **74**, 1030–1034.

Munns, R., Greenway, H., Delane, R. and Gibbs, J. (1982). *J. Exp. Bot.* **33**, 574–583.

Negm, F. B. and Loescher, W. H. (1981). *Plant Physiol.* **67**, 139–142.

Nickell, L. G. (1978). *Chem. Eng. News* **56**, 18–34.

Nickell, L. G. (1982). *In* "Chemical Manipulation of Crop Growth and Development" (J. S. McLaren, ed.) pp. 167–189. Butterworth, London.

Oworu, O. O., McDavid, C. R. and MacColl, D. (1977a). *Ann. Bot.* (*London*) [N. S.] **41**, 393–399.

Oworu, O. O., McDavid, C. R. and MacColl, D. (1977b). *Ann. Bot.* (*London*) [N. S.] **41**, 401–404.

Patrick, J. W. (1981). *Aust. J. Plant Physiol.* **8**, 221–235.
Patrick, J. W. and Turvey, P. M. (1981). *Ann. Bot. (London)* [N. S.] **47**, 611–621.
Pontis, H. G. (1977). *Int. Rev. Biochem.* **13**, 79–117.
Pontis, H. G., Babio, J. R. and Salerno, G. (1981). *Proc. Natl. Acad. Sci. U. S. A.* **78**, 6667–6669.
Prado, F. E., Fleischmacher, O. L., Vattuone, M. A. and Sampietro, A. R. (1982). *Phytochemistry* **21**, 2825–2828.
Preiss, J. (1982). *Annu. Rev. Plant Physiol.* **33**, 431–454.
Pressey, R. (1969). *Plant Physiol.* **44**, 757–764.
Rathert, G. (1982). *J. Plant Nutr.* **5**, 97–109.
Robinson, S. P. (1982). *Plant Physiol.* **70**, 1032–1038.
Saftner, R. A. and Wyse, R. E. (1980). *Plant Physiol.* **66**, 884–889.
Saftner, R. A. and Wyse, R. E. (1984). *Plant Physiol.* **74**, 951–955.
Saftner, R. A., Daie, J. and Wyse, R. E. (1983). *Plant Physiol.* **72**, 1–6.
St. Martin, E. J. and Wittenberger, C. L. (1979). *Infect. Immun.* **24**, 865–868.
Salerno, G. L. and Pontis, H. G. (1978). *Planta* **142**, 41–48.
Salerno, G. L. and Pontis, H. G. (1980). *In* "Mechanisms of Polysaccharide Polymerization and Depolymerization" (J. J. Marshall, ed.), pp. 31–42. Academic Press, New York.
Sauter, J. J. (1981). *Z. Pflanzenphysiol.* **103**, 183–187.
Schmid, K., Schupfner, M. and Schmitt, R. (1982). *J. Bacteriol.* **151**, 68–76.
Silvius, J. E. and Snyder, F. W. (1979). *Plant Physiol.* **64**, 1070–1073.
Silvius, J. E., Snyder, F. W. and Kremer, D. F. (1982). *Plant Physiol.* **70**, 316–317.
Singh, M. B. and Knox, R. B. (1984). *Plant Physiol.* **74**, 510–515.
Singh, R. and MacLachlan, G. (1983). *Plant Physiol.* **71**, 531–535.
Stitt, M., Mieskes, G., Soling, H. and Heldt, H. W. (1982). *FEBS Lett.* **145**, 217–222.
Stitt, M., Kürzel, B. and Heldt, H. W. (1984). *Plant Physiol.* **75**, 554–560.
Streeter, J. G. (1982). *Planta* **155**, 112–115.
Su, J. (1982). *Proc Natl. Sci. Counc., Repub. China, Part B* **6**, 172–180.
Su, J. and Preiss, J. (1978). *Plant Physiol.* **61**, 389–393.
Tanner, G. J., Copeland, L. and Turner, J. F. (1983). *Plant Physiol.* **72**, 659–663.
Thom, M., Maretzki, A. and Komor, E. (1982a). *Plant Physiol.* **69**, 1315–1319.
Thom, M., Komor, E. and Maretzki, A. (1982b). *Plant Physiol.* **69**, 1320–1325.
Thompson, J. and Chassy, B. M. (1981). *J. Bacteriol.* **147**, 543–551.
Thorne, J. H. (1982). *Plant Physiol.* **70**, 953–958.
Tupý, J. and Primot, L. (1982). *J. Exp. Bot.* **33**, 988–995.
Vreugdenhil, D. (1983). *Physiol. Plant.* **57**, 463–467.
Walker, R. R. and Leigh, R. A. (1981). *Planta* **153**, 150–155.
Whiting, G. C. (1970). *In* "The Biochemistry of Fruits and Their Products" (A. C. Hulme, ed.), Vol. 1, pp. 1–31. Academic Press, New York.
Whittingham, C. P., Keys, A. J. and Bird, I. F. (1979). *In* "Encyclopedia of Plant Physiology, New Series" (M. Gibbs and E. Latzko, eds.), Vol. 6, pp. 313–326. Springer-Verlag, Berlin and New York.
Willenbrink, J. (1982). *In* "Encyclopedia of Plant Physiology, New Series" (F. A. Loewus and W. Tanner, eds.), Vol. 13A, pp. 684–699. Springer-Verlag, Berlin and New York.
Woodrow, I. E., Kelly, G. J. and Latzko, E. (1982). *Z. Pflanzenphysiol.* **106**, 119–127.
Wöstmeyer, J., Behrens, U., Merckelbach, A., Muller, M. and Starlinger, P. (1981). *Eur. J. Biochem.* **114**, 39–44.
Yamaki, S. (1982). *Plant Cell Physiol.* **23**, 881–889.

Chapter 2

D-Galactose-Containing Oligosaccharides

P. M. DEY

Department of Biochemistry
Royal Holloway College, (University of London)
Egham Hill, Egham
Surrey, England

I. INTRODUCTION

The D-galactose-containing oligosaccharides occur in plants generally as α-galactosyl compounds. The β-anomeric derivatives are rare; for example, lactose (4-*O*-β-D-galactopyranosyl-D-glucopyranose) has been found in only a limited number of plant sources (Kuhn and Löw, 1949; Venkataraman and Reithel, 1958; Bouchardat, 1871). The distribution of α-galactosyl-containing oligosaccharides in the plant kingdom is wide and probably ranks next to

BIOCHEMISTRY OF STORAGE
CARBOHYDRATES IN GREEN PLANTS

sucrose. However, some galactosyl compounds, such as clusianose, floridoside, and umbelliferose, are limited to certain plant families and have taxonomic importance (Hiller and Kothe, 1969; Heywood, 1972; Kremer and Vogl, 1975; Hopf and Kandler, 1976; Dey, 1980c; Kandler and Hopf, 1982). Much exploratory work is required before other oligosaccharides, such as the raffinose family of oligosaccharides, can be evaluated as chemotaxonomic markers. The distribution and concentration of the oligosaccharides may vary in different plant organs (e.g., leaves, roots, rhizomes, and seeds) (Kandler and Hopf, 1982).

Methods applied for isolating the oligosaccharides are important in assessing their *in vivo* distribution and levels. Many of the oligosaccharides are readily hydrolysed and degraded by endogenous hydrolytic enzymes or in an environment of low pH and high temperatures, yielding secondary oligosaccharides and other products. It is therefore important to introduce appropriate measures during sugar extraction in order to minimise the degradative processes. Such processes include (a) incorporation of enzymic inhibitors, (b) homogenizing the plant material while frozen in liquid nitrogen, and (c) plunging the plant material in boiling alcohol. It is clearly important to distinguish between the primary and secondary oligosaccharide. The primary oligosaccharides are generally considered to be those that are synthesised *in vivo* and are metabolically important, whereas secondary oligosaccharides originate as degradation products from carbohydrate molecules, such as higher homologous oligosaccharides, heterosides, and polysaccharides.

The raffinose family of oligosaccharides and other galactosyl oligosaccharides generally serve as storage carbohydrates in plants and can act as both short-term and long-term storage materials (Dey, 1980c; Kandler and Hopf, 1982). It is considered that these oligosaccharides are utilized during the initial stages of seed germination, prior to polysaccharide mobilization. On the other hand, some galactosyl compounds provide frost hardiness to winter-hardy plants (Dey, 1980c; Kandler and Hopf, 1982), and these are also discussed in this chapter. However, isofloridoside, a galactosyl glycerol, acts as a reserve material responsible for maintaining osmotic balance in some algal species (Kauss, 1979). In this chapter the discussion is limited only to α-galactosyl polyols and other derivatives, some of which may act as precursors of α-galactosyl reserve oligosaccharides.

II. METABOLISM OF D-GALACTOSE

A. Source of D-Galactose

In plants, photosynthesis results in the formation of a number of phosphorylated monosaccharides. These may be either enzymatically hydrolysed, causing accumulation of free sugars, or transformed in other ways, one of which is the conversion into sugar nucleotides. The glycosyl esters are important as glycosyl donors in the biosynthesis of complex carbohydrates. UDP-D-Galactose [uridine-5′-(α-D-galactopyranosyl diphosphate)] is an important intermediate for the synthesis of D-galactose-containing saccharides and lipid derivatives. An outline of the photosynthetic generation of UDP-D-galactose is shown in Fig. 1. D-Galactose in the free form has also been detected in some plant tissues (Hattori and Shiroya, 1951; Cerbulis, 1955b; Jeremias, 1958). For example, Courtois (1968) showed D-galactose on the surface of prunes that seemed to have been formed as a result of degradative processes that may have taken place in the overripe fruit. In addition, free D-galactose was shown to accumulate when some seed embryos were incubated under the conditions of seed germination; germination of whole seeds did not show such accumulation (Dey, 1978). In the latter case, D-galactose was probably immediately metabolised and transformed into various intermediates and products. In callus cultures of cucumber hypocotyls grown on carbon sources that contained α-galactosidic linkages, D-galactose was detected in free space washes of calli (Gross, 1981). This was mainly due to the presence of extracellular α-galactosidase.

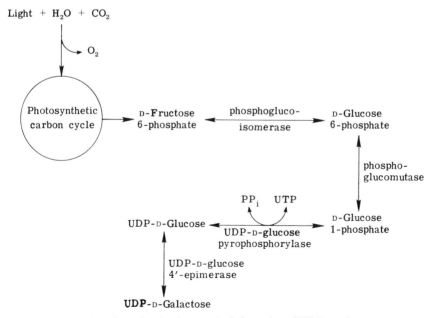

Fig. 1.   A pathway for the photosynthetic formation of UDP-D-galactose.

B.   Enzymes of D-Galactose Metabolism

Exogenously supplied D-galactose is readily metabolised by various plant tissues. For example, barley embryo tissue, which contains raffinose (O-α-D-galactosylsucrose) and is devoid of stachyose (di-O-α-D-galactosylsucrose), accumulated stachyose following administration of D-galactose and exposure to air (MacLeod and McCorquodale, 1958a). Alternatively, when D-[1-^{14}C]galactose was taken up by barley stems, labelled D-glucose appeared in a very short time, more than 80% being D-[1-^{14}C]glucose. Channelling of carbon from D-[^{14}C]galactose into various metabolic pools has been demonstrated; thus [^{14}C]sucrose was found in manna leaves (Hassid et al., 1956), and labelled L-ascorbic acid and D-galacturonic acid were found in strawberries (Loewus and Jang, 1958; Loewus et al., 1958) and various constituents in other plants (Roberts and Butt, 1969; Hoffman et al., 1971).

Various ways by which D-galactose may be initially transformed are shown in Fig. 2. In plants, the important initial step involves phosphorylation at C-1, catalysed by D-galactokinase (EC 2.7.1.6). Although this first enzyme of the Leloir pathway of D-galactose metabolism has been detected in a number of plant tissues and can be predicted to occur in all plant storage organs which contain α-galactosyl-substituted storage carbohydrates, only the enzymes from mung bean (Chan and Hassid, 1975), fenugreek (Foglietti and Percheson, 1976), and broad beans (Dey, 1983) have been purified with their properties studied. The phosphorylation of D-galactose occurs in the presence of ATP and Mg^{2+}. The relative activity of the broad bean enzyme with respect to various monosaccharides followed the order D-galactose > 2-deoxy-D-galactose > D-galactosamine. However, D-fucose, L-arabinose, L-galactose, and D-glucose were not phosphorylated. The metal ion requirement for the enzymic activity followed the order Mg^{2+} > CO^{2+} > Mn^{2+} > Ni^{2+} > Ca^{2+}. Whereas ATP acted as an efficient phosphate donor, ADP, GTP, and UTP were ineffective. Product inhibition of D-galactokinase was observed; D-galactose 1-phosphate and ADP were competitive and noncompetitive inhibitors, respecctively. The high K_i values imply that these metabolites are unlikely to play direct roles in the *in vivo* regulation of the enzymic activity.

The level of D-galactokinase in broad beans increased during the early stages of germination (Dey, 1983). The raffinose and stachyose levels decreased (Table I), due to α-galactosidase-catalysed removal of the terminal D-galactose residues (Pridham and Dey, 1974). There was, however, no detectable accumulation of free D-galactose; this was due to the sufficiently high level of D-galactokinase at all stages of germination.

The phosphorylation of D-galactose at C-6 in plants is questionable; D-galactose 6-phosphate is more likely to arise from D-galactose 1-phosphate via

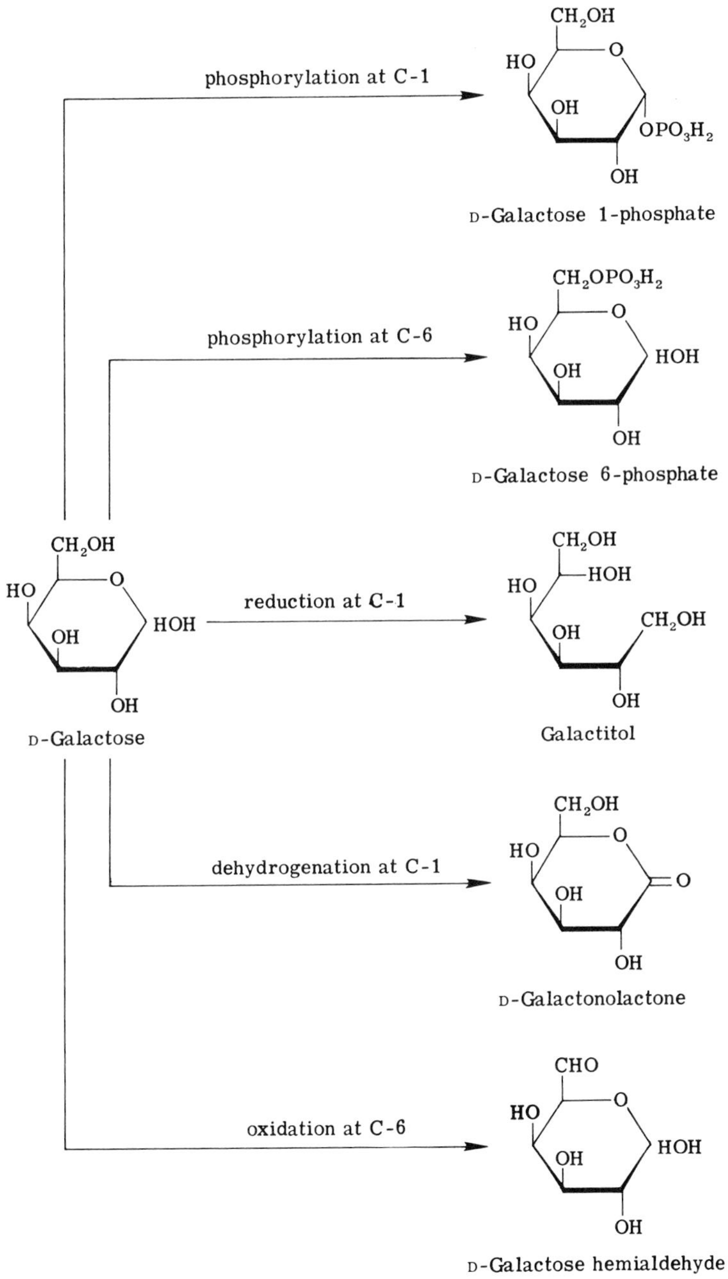

CH₂OH

HO

O

OH

OPO₃H₂

OH

D-Galactose 1-phosphate

phosphorylation at C-1

CH₂OPO₃H₂

HO

O

HOH

OH

OH

D-Galactose 6-phosphate

phosphorylation at C-6

CH₂OH

HO

O

HOH

OH

OH

D-Galactose

reduction at C-1

CH₂OH

HO

HOH

OH

CH₂OH

OH

Galactitol

dehydrogenation at C-1

CH₂OH

HO

O

OH

O

OH

D-Galactonolactone

oxidation at C-6

CHO

HO

O

HOH

OH

OH

D-Galactose hemialdehyde

Fig. 2. Various ways of initial transformation of D-galactose.

Table I

The Levels of D-Galactose-Containing Oligosaccharides and Galactokinase during Germination of Broad Bean (*Vicia faba*) Seeds[a]

Germination time (hr)	Amount of oligosaccharide (μg/g dry weight of seed)		Excepted amount of D-galactose liberated (nmol/g dry weight of seed)	D-Galactokinase activity (units[b]/g dry weight of seeds)
	Raffinose	Stachyose		
0	75	130	—	230
6	70	85	108	240
12	52	63	100	255
18	43	26	104	280
24	31	15	60	310
36	23	7	41	190
48	15	Traces	39	98

[a] From Dey (1983).
[b] Expressed as nmol D-galactose 1-phosphate formed/min at 25°C.

the action of a mutase. In microorganisms, D-galactose 6-phosphate is utilized through the tagatose pathway. Other alternative initial steps of D-galactose transformation, for example, reduction, dehydrogenation and oxidation, are generally encountered in animal and microbial systems.

The secondary steps of D-galactose metabolism in plants are outlined in Fig. 3. The second step of the Leloir pathway, in which D-galactose 1-phosphate is converted to UDP-D-galactose, is catalysed by D-galactose-1-phosphate uridyltransferase (EC 2.7.7.12; UDP-D-glucose:D-galactose-1-phosphate uridyltransferase):

$$\text{D-Galactose 1-phosphate} + \text{UDP-D-glucose} \rightleftharpoons \text{UDP-D-galactose} + \text{D-glucose 1-phosphate}$$

This enzyme, commonly found in microbial and animal systems, was absent from several D-galactose-metabolising plant tissues (Maretzki and Thorn, 1978; Gross and Pharr, 1982). However, it was detected in soybean cotyledons (Pazur and Shadaksharaswamy, 1961; Pazur *et al.*, 1962) and in mutants of isolated cultured tobacco cells (Kapitsa and Evdonina, 1981). Although D-galactose metabolism and its regulation in plants are not well understood, at least in some instances the above mentioned enzyme is substituted by UDP-D-galactose pyrophosphorylase (EC 2.7.7.10; UTP:D-galactose-1-phosphate uridyl tranferase), for example in canna leaves and wheat seedlings (Hassid *et al.*, 1956), *Avena* coleoptiles (Ordin and Bonner, 1957), bean seedlings (Cooper and Greenshields, 1961), cotton seedlings (Shiroya, 1963), corn roots and barley coleoptiles (Göering, *et al.*, 1968; Roberts and Butt, 1969; Roberts

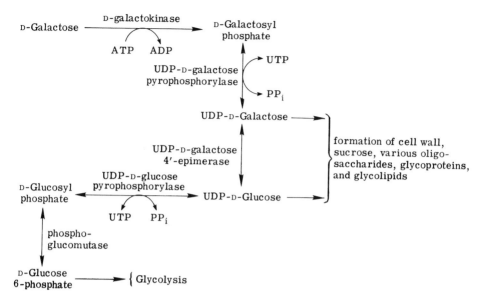

Fig. 3. Metabolism of D-galactose in plants.

et al., 1971), bambara ground nut (Amuti and Pollard, 1977), sugar cane cell cultures (Maretzki and Thorn, 1978), and cucumber peduncles (Smart and Pharr, 1981; Gross and Pharr, 1982).

UDP-D-Galactose pyrophosphorylase of the fruit peduncles of cucumber was shown to be associated with UDP-D-glucose pyrophosphorylase activity (Smart and Pharr, 1981). The purified enzyme, however, showed preference for D-galactose 1-phosphate as compared to the D-glucose derivative. In a manner similar to the human liver enzyme (Turnquist *et al.*, 1974) and some microbial enzymes (Lee *et al.*, 1978; Lobelle-Rich and Reeves, 1982), the cucumber enzyme exhibited both activities in a single protein, which catalysed the following reactions:

$$\text{D-Galactose 1-phosphate} \underset{\text{UTP} \quad \text{PP}_i}{\longleftrightarrow} \text{UDP-D-galactose}$$

$$\text{D-Glucose 1-phosphate} \underset{\text{UTP} \quad \text{PP}_i}{\longleftrightarrow} \text{UDP-D-glucose}$$

The activity of the enzyme for the formation of UDP-D-galactose was inhibited by pyrophosphate with a K_i of 0.58 ± 0.10 mM (Smart and Pharr, 1981). On the other hand, the K_m for pyrophosphate in the conversion of UDP-D-glucose to D-glucose 1-phosphate was 0.18 mM. Thus PP$_i$ will play a controlling role by inhibiting the formation of UDP-D-galactose and facilitating the formation of D-glucose 1-phosphate. However, the presence of other PP_i-forming and -utilizing systems will profoundly influence the effective concentration of this metabolite.

The next important enzyme of D-galactose metabolism in UDP-D-galactose 4′-epimerase (EC 5.1.3.2), which catalyses the interconversion of UDP-D-galactose and UDP-D-glucose. This reaction is the primary site for the formation of D-glucose when this sugar is metabolised for energy requirements. On the other hand, when D-galactose is required for the biosynthesis of structural components of the cell, it is derived from D-glucose at the nucleotide sugar level.

Since the discovery of this enzyme, it has been isolated from a number of microbial, plant, and animal sources (Glaser, 1972; Feingold and Avigad, 1980; Feingold, 1982). The microbial enzymes have been most extensively studied, and detailed descriptions of the mechanism of action are presented in recent reviews (Feingold and Avigad, 1980; Feingold, 1982). This enzyme is expected to be widely distributed in plants, playing a key role in D-galactose metabolism; however, the enzyme has been thus far studied from only a few plant sources (Neufeld et al., 1957; Fan and Feingold, 1969, 1970; Dalessandro and Northcote, 1977a,b,c; Clermont and Percheson, 1979). The only examples of isolation and purification of the enzyme from seeds are from fenugreek (Clermont and Percheson 1979) and broad bean (Dey, 1984b).

The reaction catalysed by UDP-D-galactose 4′-epimerase is NAD^+-dependent; the cofactor may be loosely bound, as with the enzyme from animal sources, or tightly bound, as in microbial and plant enzymes. Thus, plant enzymes are generally active in the absence of exogenously added NAD^+. The removal of bound NAD^+ by charcoal treatment causes loss of activity of the enzymes from wheat germ (Fan and Feingold, 1969) and broad beans (Dey, 1984b); reactivation occurs on addition of NAD^+. The charcoal-treated broad bean enzyme was therefore affinity purified by using an NAD^+–Sepharose column. The broad bean enzyme was resolved by gel filtration into a stable major form (MW 78,000) and two unstable minor forms (MW 40,000 and 159,000). The activity of the major form was strongly inhibited by NADH with a competitive K_i of 5 μM. Thus, the *in vivo* concentration ratio of $NAD^+/NADH$ is likely to play a controlling role in the interconversion of UDP-D-galactose and UDP-D-glucose.

These nucleotide sugars serve as the source of glycosyl moieties in various biosynthetic processes, such as in the formation of polysaccharides, cell-wall components, sucrose, and various oligosaccharides, glycoproteins, and glycolipids (Fig. 3). They are also known to be involved in the formation of lipid-linked (dolichol phosphate) glycosyl intermediates, which are well-known donors of sugar residues in irreversible transfer reactions whereby many glycoproteins and structural polysaccharides are synthesised. These lipid-linked intermediates may also be important in the transfer of glycosyl molecules across membranes. On the other hand, the nucleotide sugars can be converted to their glycosyl phosphates via either irreversible hydrolysis by phosphodiesterase (nucleotide pyrophosphatase)—for example, UDP-D-glucose → UMP + D-glucose 1-phosphate—or reversible pyrophosphor-ylase-catalysed reaction, UDP-D-galactose + $PP_i \rightleftharpoons$ UTP + D-galactose 1-phosphate.

The enzyme for the latter reaction has already been discussed. UDP-D-glucose pyrophosphorylase (EC 2.7.7.9; UTP:D-glucose-1-phosphate uri-dyltransferase), however, is a widely distributed and well-investigated plant enzyme (Turnquist and Hanson, 1976; Feingold, 1982) catalysing the reaction:UDP-D-glucose + $PP_i \rightleftharpoons$ UTP + D-glucose 1-phosphate (Fig. 3). At the pH optimum of pH 8.0–9.0, the concentration ratio

$$\frac{[\text{D-glucose 1-phosphate}][\text{UTP}]}{[PP_i][\text{UDP-D-glucose}]},$$

was found to be 0.15, but UDP-D-glucose as a substrate was an inhibitor showing allosteric kinetics under physiological conditions (Gustafson and Gander, 1972; Hopper and Dickinson, 1972; Dickinson *et al.*, 1973). Thus, the *in vivo* generation of D-glucose 1-phosphate will depend strongly on the cellular concentrations of the various metabolites discussed above. From Fig. 3 it may be inferred that UDP-D-glucose constitutes an important branch-point metabolite that provides the basis for various biosynthetic pathways or may enter the glycolytic pathway via the formation of D-glucosyl phosphate.

C. Toxicity of D-Galactose

D-Galactose at high concentrations may be toxic to plants and has been shown to inhibit cellular growth (Quednow, 1930; Wynd, 1933; O'Kelley, 1955; Ferguson and Street, 1957; Ordin and Bonner, 1957; Baker and Ray, 1965; Ernst, 1967; Malca *et al.*, 1967; Göering *et al.*, 1968; Ernst *et al.*, 1971; Roberts *et al.*, 1971; Maretzki and Thom, 1978; Kapista and Evdonina, 1981). As D-galactose is a major breakdown product of D-galoctosyloligosaccharides, the toxicity of this monosaccharide may be physidogically important. At a concentration of 10 mM, D-galactose inhibited the growth of *Avena* coleoptile sections, and this was not due to osmotic effects as mannitol at 20 mM showed no inhibitory effect (Ordin and Bonner, 1957). D-Galactose toxicity is also recognised in animal systems (Hansen and Gitzelmann, 1975), for example, in galactosemic human subjects or in experimental animals kept on high D-galactose diet; in these instances, accumulation of D-galactose 1-phosphate and galactitol was noted (Schwarz *et al.*, 1956; Mayes *et al.*, 1970). D-Galactose 1-phosphate was suggested to affect D-glucose metabolism in the steps involving phosphoglucomutase, UDP-D-glucose pyrophosphorylase, and D-glucose-6-phosphate dehydrogenase (Ginsberg and Neufeld, 1957; Oliver, 1961; Sidburg, 1969). Thus, in plants, D-galactose 1-phosphate probably interferes with D-hexose metabolism (Fig. 4); galactitol does not accumulate and may not be regarded as playing an inhibitory role (Roberts *et al.*, 1971). Metabolite toxicity in plants is not uncommon: for example, in trehalose toxicity in *Cuscuta reflexa* (Veluthambi *et al.*, 1981, 1982a,b) the biosynthesis of cell-wall polysaccharides is affected. In D-galactose feeding experiments with corn roots, two metabolities, namely, D-galactose 1-phosphate and UDP-D-galactose, were shown to accumulate (Table II) (Roberts *et al.*, 1971). The inhibition of growth commenced at 0.1 mM D-galactose, and under this condition the level of UDP-D-galactose was 123 nmol/g wet weight of tissue. This level is much higher than that found in untreated mung bean seedlings (14 nmol/g tissue) (Ginsberg *et al.*, 1956). The specific radioactivity of UDP-D-galactose compared to the radioactivity of supplied D-galactose was not much changed. This is indicative of a high pool size of the metabolite, and there was only little dilution due to its synthesis from endogenous sources. The highest level of UDP-D-galactose was at 1 mM D-galactose, but declined considerably upon increasing the concentration to 10 mM. On the other hand, there was a rapid and continuous build-

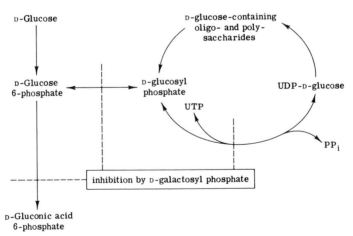

Fig. 4.   Influence of D-galactosyl phosphate on the metabolism of D-hexoses. Possible points of inhibition are shown with dotted lines.

Table II

The Effect of Different Concentrations of D-[1-^{14}C]Galactose on the Accumulation of Some Metabolities in Corn Roots[a]

D-Galactose (mM)	Uptake into ethanol-soluble fraction	D-Galactose	D-Galactose 1-phosphate	UDP-D-galactose	Others
0.1	362	127	102	123	10
1.0	2345	820	890	540	95
5.0	2659	1142	1065	345	107
10.0	3375	1585	1453	270	67

[a] The amounts of the metabolities are expressed as nmol/g fresh weight of tissue. Data calculated after Roberts *et al.* (1971).

up of D-galactose 1-phosphate. Experiments with corn coleoptiles showed similar formation of products of D-galactose feeding. Göering *et al.* (1968) have also demonstrated accumulation of D-galactose 1-phosphate in plant tissues which were fed with toxic levels of D-galactose; the formation of D-glucose 1-phosphate was affected in these tissues. It was further shown that the synthesis of cell-wall polysaccharides, especially of α-cellulose, was impaired, confirming the interference of D-galactose 1-phosphate with various steps in carbohydrate metabolism (Ordin and Bonner, 1957; Göering *et al.*, 1968; Roberts *et al.*, 1971; Hughes and Street, 1974).

The ability of isolated cells of tobacco culture to sustain growth in lactose medium was examined by Kapista and Evdonina (1981). The extracellular β-galactosidase was responsible for the hydrolysis of this disaccharide. The wild-type cells were unable to grow, whereas the lac$^+$ gal$^-$ and lac$^+$ gal$^+$ mutants showed normal growth in lactose medium. The membrane transport properties of the mutants were different than the original line; the mutants showed high affinity for D-glucose, whereas the wild type did not display much selectivity between D-glucose and D-galactose. Purely D-galactose medium was toxic to the lac$^+$ gal$^-$ mutant, but this hexose was completely assimilated without accumulation of the toxic D-galactose 1-phosphate if D-glucose was also present in the medium. On the other hand, lac$^+$ gal$^+$ mutant was able to grow in D-galactose medium. It was found that the UDP-D-galactose pyrophosphorylase level in all three types of tobacco cells was comparable. However, the cells of lac$^+$ gal$^+$ mutant contained an additional enzyme, UDP-D-glucose:D-galactose-1-phosphate uridyltransferase, the level of which was nearly fourfold that of the pyrophosphorylase. Thus, it seems that in the lac$^+$ gal$^+$ mutant the combined action of the two enzymes not only decreases the level of toxic D-galactose 1-phosphate but also converts it to D-glucose 1-phosphate, which is essential for normal cell metabolism.

Maretzki and Thom (1978) arrived at a different conclusion from their study of a sugar cane cell mutant that was capable of growing in a D-galactose-containing medium. The levels of UDP-D-galactose pyrophosphorylase in the mutant and the original cells were similar, and the enzyme UDP-D-glucose:D-galactose 1-phosphate uridyltransferase was absent in both. However, the level of UDP-D-galactose 4'-epimerase was 10-fold higher in the mutant. Thus the authors were inclined to regard UDP-D-galactose as the toxic metabolite.

From electron microscope study, Ernst *et al.* (1971) demonstrated that inhibition of growth of orchid seedlings (*Phalaenopsis* cv.) by D-galactose is accompanied by invagination of nuclear envelope and rupture of tonoplast. It was suggested that D-galactose (or its metabolites) affect the factors that are responsible for the maintenance of membrane permeability. In *Avena* coleoptiles (Anker *et al.*, 1973; Anker, 1974), D-galactose inhibits the synthesis of auxin, thus probably contributing to the retardation of growth. On the other hand, D-galactose-induced evolution of ethylene is known to retard the growth of mung-bean seedlings (Colclasure and Yapp, 1976). In conclusion, it may be said that the factors involved in D-galactose toxicity are diverse and the underlying mechanism is not yet clear.

III.   α-D-GALACTOSIDES OF POLYOLS

Polyols are defined as polyhydric alcohols with three or more hydroxyl groups, for example, glycerol, alditols, and inositols. Such acyclic polyols as D-glucitol (sorbitol), D-mannitol, and glycerol are known to occur in both free and combined forms. D-Glucitol is translocated in the phloem of some higher plants and is also physiologically important in the leaves (Vincent and Delachanal, 1889; Lohmar, 1957; Webb and Burley, 1962; Hansen, 1967). D-Mannitol occurs in several algal species (Bidwell, 1958; Feige, 1973, 1974) and in some plant exudates (Lohmar, 1957). Its physiological importance in some plants is in providing winter-hardiness (Seybold, 1969). Glycerol occurs most widely in its combined form, for example, in lipids, glycosyl derivatives, and other metabolites. Among the cyclic polyols, inositols are the most commonly occurring examples in plants (Anderson and Wolter, 1966). *Myo*-inositol is physiologically important: it plays a role in the biosynthesis of hemicelluloses and oligosaccharides, occurs as a storage product (phytic acid), a coenzyme for sugar transport, a precursor of phosphoinositides, and esterified to indoleacetic acid (Angyal and Anderson, 1959; Hoffmann-Ostenhof, 1969; Miller, 1973; Gander, 1976). For further details on plant polyols review articles by Stacey (1974), Grisebach (1980), Loewus and Loewus (1980), Bieleski (1982) and Loweus and Dickinson (1982) are recommended.

A number of glycosides of polyhydric alcohols have been detected and identified in plant sources (Culberson, 1969; Craigie, 1974; Culberson *et al.*, 1977; Dey, 1980c). They are generally synthesised via pathways involving nucleotide sugar derivatives as the glycosyl donor and the polyhydric alcohol as the acceptor. The primary step of their mobilization involves the action of glycoside hydrolase, for example, α-galactosidase.

A. Floridoside

Algae are the main source of floridoside (2-*O*-α-D-galactopyranosylglycerol), in particular the red algae (members of Rhodophyta) (Kylin, 1918; Stanek *et al.*, 1965; Craigie *et al.*, 1968; Bisson and Kirst, 1979; Dey, 1980c; Kirst, 1980; Reed *et al.*, 1980; Avigad, 1982). The structure of the compound (**1**) was deduced by standard chemical and physical methods (Colin and Guéguen, 1930; Putman and Hassid, 1954; Aplin *et al.*, 1967).

Floridoside

(1)

The chemotaxonomic importance of floridoside in red algae was suggested by Augier (1947). From the studies of photosynthetic assimilation of $^{14}CO_2$ in red algae, Majak *et al.* (1966) demonstrated that floridoside was the major alcohol-soluble carbohydrate synthesised in various species from six orders of Rhodophyta (Bangiales, Bonnemaisoniales, Ceramiales, Crytonemiales, Gigartinales and Nemalionales). The floridoside content generally ranges from 1.5 to 8%. However, no [^{14}C]floridoside was detected in members of the Nemalionales and only low concentrations of the compound were found in those of the Ceramiales. In similar experiments, Kremer and Vogl (1975) found that in all 63 species of Rhodophyta excepting those belonging to the order Ceramiales, floridoside was intensely labelled. Similar conclusions were drawn by Nagashima (1976) and Kremer and Feige (1979). The isomeric isofloridoside was also synthesised in some quantity in members of the Bangiales but not in those of the Ceramiales and Rhodymeniales (Nagashima, 1976). From *in vivo* labelling experiments, it was demonstrated that transfer of ^{14}C from labelled floridoside to isofloridoside was slow (Craigie *et al.*, 1968).

Thus, despite the structural similarities of the two compounds, they are unrelated metabolically in the Rhodophyta.

The kinetics of $^{14}CO_2$ assimilation by some red algae gave evidence for the occurrence of the normal Calvin pathway of photosynthetic fixation of carbon (Craigie *et al.*, 1968; Kremer, 1978a,b,c; Bisson and Kirst, 1979). Both constitutents of floridoside, namely, D-galactose and glycerol, were labelled. Bean and Hassid (1955) identified 1(3)-glycerol phosphate, UDP-D-galactose, and floridoside phosphate as some of the intermediary products in $^{14}CO_2$ assimilation experiments. These observations point to the existence of a pathway as shown in Fig. 5 that would result in the formation of floridoside phosphate (Dey, 1980c); hydrolysis of this compound by a phosphatase would then yield floridoside. Kremer and Kirst (1981) have now demonstrated the time-dependent formation of floridoside phosphate in an incubation mixture containing UDP-D-galactose, [^{14}C]glycerol 1(3)-phosphate, and a cell-free extract of *Cystoclonium purpureum*. The enzyme responsible for this reaction was termed floridoside phosphate synthase (UDP-D-galactose:*sn*-glycerol-3-phosphate 2-D-galactosyltransferase). From *in vitro* experiments, the authors also demonstrated the conversion of dihydroxyacetone phosphate to glycerol 1(3)-phosphate; the cell-free extract thus possessed glycerol 1(3)-phosphate dehydrogenase activity:

$$\text{Dihydroxyacetone phosphate} \xrightleftharpoons[\text{NADH} \quad \text{NAD}^+]{} \text{glycerol 1(3)-phosphate}$$

Further work on the purification and characterization of floridoside phosphate synthetase is awaited. Analogous to the formation of isofloridoside from

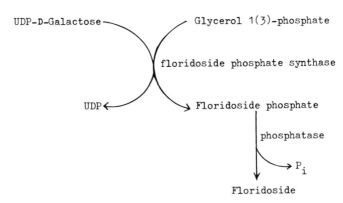

Fig. 5. Pathway for the biosynthesis of floridoside.

its phosphate derivative (Spang *et al.*, 1981), the participation of a specific phosphatase in the biosynthetic pathway of floridoside may also be envisaged.

The *in vivo* degradation of floridoside is, most probably, catalysed by α-galactosidase. This enzyme is present in *Porphyra umbilicalis* (Peat and Rees, 1961) and is specific for floridoside. Although isofloridoside is also present in this alga, the enzyme hydrolyses it very slowly; the enzyme has only a weak action on such α-D-galactosides as melibiose and raffinose.

D-Galactosylglycerols (floridoside and isofloridoside) are considered to be involved in osmoregulation in red algae (Kauss, 1968, 1969). Kauss (1968, 1969) showed that high salt concentration in the growth medium induced the synthesis of floridoside in *Iridophycus flaccidum* and both D-galactosylglycerols in *Porphyra perforata*. However, there are some inconsistencies in the data showing higher floridoside levels with increasing salinity, and there was doubt on whether the compounds in red algae were solely responsible for compensation to osmotic stress (Reed *et al.*, 1980). Later studies with some red algal species, however, positively showed variations in the concentration of D-galactosylglycerols in response to external salinity, but the concentrations of the compounds can only account for up to 10% of the internal osmotic potential (Bisson and Kirst, 1979; Kirst and Bisson, 1979, 1980). In *Porphyra purpurea*, the level of floridoside increased with hypersalinity and decreased with hyposalinity of the media (Reed *et al.*, 1980). The responses were identical in light or dark, and the changes were completed within 24 hr. These responses were much slower than those reported by Kauss (1968) in relation to isofloridoside. The amount of the compound involved in the synthetic or the degradative process was theoretically insufficient to fully counter the changes in the cell volume (Reed *et al.*, 1980); however, the compound probably plays some role as a reserve material. Ions such as Na^+, K^+, and Cl^-, in addition to D-galactosyl glycerols, were also involved in osmotic adaptation in some marine algae (Kirst and Bisson, 1979, 1980; Wiencke and Laeuchli, 1981). In *Porphyra umbilicalis*, the two D-galactosylglycerols and K^+ were probably located in the cytoplasm, whereas Cl^- accumulated in vacuoles (Wiencke and Laeuchli, 1981). Kremer (1979), on the other hand, showed that the percentage of ^{14}C-labelling of isofloridoside as a photoassimilatory product was unaffected by osmotic stress. However, the duration of his experiments was only up to 1 hr, as compared to the 24 hr required for full adaptation.

Further work on the changes in activities of enzymes responsible for the degradation and biosynthesis of floridoside in response to the changes in external salinity of rhodophytes is necessary for a better understanding of the mechanism of osmotic adaptation. The extent to which floridoside acts as a reserve material is not known.

B. Isofloridoside

Isofloridoside (1-O-α-D-galactopyranosylglycerol) was first isolated by Lindberg (1955a,b) from *Porphyra umbilicalis*, a member of the Rhodophyta. Later, it was shown to be widely distributed amongst other algal species (Su and Hassid, 1962; Craigie *et al.*, 1968; Kauss, 1968, 1969, 1979; Nagashima *et al.*, 1969; Impellizzeri *et al.*, 1975; Court, 1980; Reed *et al.*, 1980; Nagashima and Fukuda, 1981). Difficulties were encountered in detecting and characterizing the compound, mainly due to (a) its presence in low concentration, (b) difficulties in crystallizing the compound and resolving it from floridoside, and (c) low radioactive incorporation into the compound during photosynthetic assimilation of $^{14}CO_2$. However, the structure (2) was established through the chemical synthesis of the compound (Wickberg, 1958).

$$
\begin{array}{c}
CH_2OH \\
HO \diagup \quad O \\
\diagdown \ OH \\
\\
OH \\
CH_2O \\
HOCH \\
CH_2OH
\end{array}
$$

Isofloridoside

(2)

The biosynthesis of isofloridoside in *Porphyra umbilicalis* during photosynthesis was demonstrated by Shibuya (1961). Photosynthesis in the presence of $^{14}CO_2$ for 2 hr in the light followed by 18 hr in the dark demonstrated continued synthesis of [^{14}C]isofloridoside. Thus this compound is not a direct product of photosynthesis but is formed from its precursor, which is synthesised in the presence of light (Craigie *et al.*, 1968). The bulk of work on the metabolism of isofloridoside has been carried out by using the golden-brown alga *Poterioochromonas malhamensis* Peterfi (syn: *Ochromonas malhamensis* Pringsheim), in which it is involved in osmotic regulation (Kauss, 1979).

A cell-free extract of *P. malhamensis*, when incubated with UDP-D-galactose and glycerol 1(3)-phosphate, was able to synthesise isofloridoside phosphate (Kauss and Schobert, 1971). The specific enzyme responsible for this synthesis was termed isofloridoside phosphate synthase (UDP-D-galactose:sn-glycerol-3-phosphoric acid 1-α-D-galactosyltransferase) (Fig. 6).

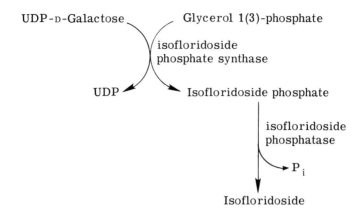

Fig. 6.   Pathway for the biosynthesis of isofloridoside.

Dephosphorylation of the compound was affected by a specific phosphatase; this enzyme was unable to act on D-glucose 1-phosphate, D-glucose 6-phosphate, D-fructose 6-phosphate, and glycerol 1(3)-phosphate (Spang *et al.*, 1981). Isofloridoside is degraded to D-galactose and glycerol by α-D-galactosidase. This enzyme was purified from *P. malhamensis* and its properties were characterised (Dey and Kauss, 1981). It has an alkaline pH optimum (~ 7.0) as compared to acidic optima (~ 3–6) displayed by most plant α-D-galactosidases (Dey and Pridham, 1972; Dey and Del Campillo, 1984). The enzyme showed no apparent hydrolysis of isofloridoside phosphate.

In *P. malhamensis*, the physiological role of isofloridoside is to serve as a carbohydrate reserve and as a factor responsible for controlling the osmotic balance of the cell (Kauss, 1967, 1968, 1969, 1974, 1979). The unicellular alga lacks a cell wall and therefore the regulation of the osmotic pressure inside the cell is important for its survival in surroundings of fluctuating salt concentrations. Under the conditions of low ionic concentration outside the cell, a $(1 \rightarrow 3)$-β-D-glucan is the main product of photosynthesis, while isofloridoside is synthesised at higher ionic concentrations (Kauss, 1967). Osmotic solutes cause the cells to shrink within 1–3 min; however, the cell volume recovers during the following 1–2-hr period (Schobert *et al.*, 1972). The recovery is accompanied by an increase in the internal concentration of isofloridoside. The removal of the osmotic stress (by dilution of the external medium with water) causes a fall in the level of this compound. During this process the carbon from both glycerol and D-galactose moieties of isofloridoside is incorporated into the storage $(1 \rightarrow 3)$-β-D-glucan in the cell (Kauss, 1967; Schobert *et al.*, 1972). Pulse-chase experiments using D-[^{14}C]glucose as the carbon source showed radioactive incorporation into isofloridoside. This took place irrespective of the external osmotic conditions, thus demonstrating a rapid turnover of the compound in the cell (Kauss, 1973). The turnover was faster and the pool size bigger under conditions of rapidly changing external osmotic pressure. Thus, the interconversion of the high molecular weight glucan and isofloridoside is important for the osmotic regulation of the cell. Their roles as reserve materials is also evident from the above discussions. A pathway (Fig. 7) for this interconversion in *P. malhamensis* was postulated by Kauss (1977, 1979).

Much research has been carried out on the modulation of the activity of isofloridoside phosphate synthase in *P. malhamensis*. Application of an increased osmotic pressure to the cell resulted in an elevated level of extractable enzyme activity, whereas such application to the extract itself had no effect on the enzyme activity. This reflects the presence of a sensing mechanism in the cell (probably membrane based), which translates the osmotic shock into an elevated enzymatic activity. As the time involved for the response was found to be very short, the process does not seem to be due to *de novo* synthesis of the enyzme; instead, it is more likely due to either changes in the protein conformation or some chemical modification. *In vitro* experiments demonstrated that the regulation of the enzymatic activity involves its proteolytic modification (Kauss and Quader, 1976; Kauss *et al.*, 1978, 1979). The enzyme system involved in the activation of the isofloridoside phosphate synthase is itself activated as a result of cell shrinkage at a higher external

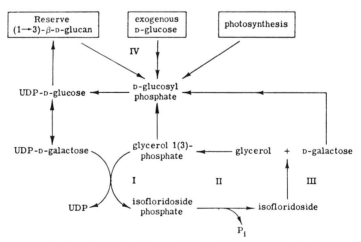

Fig. 7. Postulated pathway for the metabolism of isofloridoside in *Poterioochromonas malhamensis*. The enzymes identified are I, isofloridoside phosphate synthase; II, isofloridoside phosphatase; III, α-galactosidase; IV, $(1 \rightarrow 3)$-β-D-glucan phosphorylase.

osmotic pressure. The results obtained by Kauss (1981, 1983) suggest that this response is Ca^{2+}-controlled, and calmodulin is probably involved in a sequence of reactions:

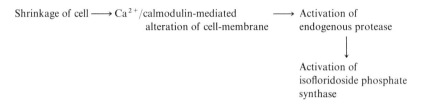

Shrinkage of cell ⟶ Ca^{2+}/calmodulin-mediated ⟶ Activation of
alteration of cell-membrane   endogenous protease

Activation of
isofloridoside phosphate
synthase

Isofloridoside phosphate synthase is considered to take part in the short-term regulation of isofloridoside level in the cell. On the other hand, the two degradative enzymes α-D-galactosidase and D-galactokinase are probably involved with long-term regulation. The increased levels of these enzymes upon application of higher osmotic pressure to the cells was due to *de novo* protein synthesis (Dey, 1980a; Kreuzer and Kauss, 1980). The specific phosphorylase that takes part in the initial step of the conversion of the reserve (1→3)-β-D-glucan (see Fig. 7) was also purified and shown to be partially activated by AMP (Albrecht and Kauss, 1971).

C. Galactinol

Galactinol [*O*-α-D-galactopyranosyl-(1→1)-*myo*-inositol] was first isolated by Brown and Serro (1953) from sugar beet juice. It was later detected in various products of the sugar refining process (Walker *et al.*, 1965; Dutton, 1966; Schiweck and Buesching, 1969), cambial sap of trees (Huller and Smith, 1966; Oesch, 1969), leaves of plants (Senser and Kandler, 1967a,b; Imhoff, 1970, 1973; Beveridge *et al.*, 1972), roots and rhizomes (Senser and Kandler, 1967a; Pressey and Shaw, 1969), and seeds (Petek *et al.*, 1966; Tanner, 1969a; Lehle *et al.*, 1970; Sioufi *et al.*, 1970; Theander and Aman, 1976; Sosulski *et al.*, 1982).

Galactinol was found to occur commonly in plant organs that also possessed the raffinose family of oligosaccharides (Tables III and VI, and see Section IV). The level of galactinol in the leaves was comparable to, or even higher than in some instances, that of raffinose. However, in the storage organs such as roots, rhizomes, and seeds, the level was much lower than that of stachyose, which is the major storage component of the raffinose-family. Serro and Brown (1954) adopted a simple paper-chromatographic method for the isolation of galactinol; however, an alternative method (Dutton, 1966) is also available that involves lead acetate precipitation, ion-exchange chromatography, and thin-layer chromatography (TLC). Senser and Kandler (1967a) successfully utilized a two-dimensional paper-chromatographic technique for separating and identifying [^{14}C]galactinol from the leaf extracts of numerous plants. The high-pressure liquid chromatography (HPLC) technique has been recently employed for both detecting and quantitating galactinol in molasses (Schiweck, 1982).

The structure of galactinol (**3**) proposed by Brown and Serro (1953) was confirmed by Kabat *et al.* (1953). Senser and Kandler (1967a,b) obtained crystalline galactinol from the leaves of *Lamium maculatum* and confirmed the structure by infrared (IR) spectroscopy.

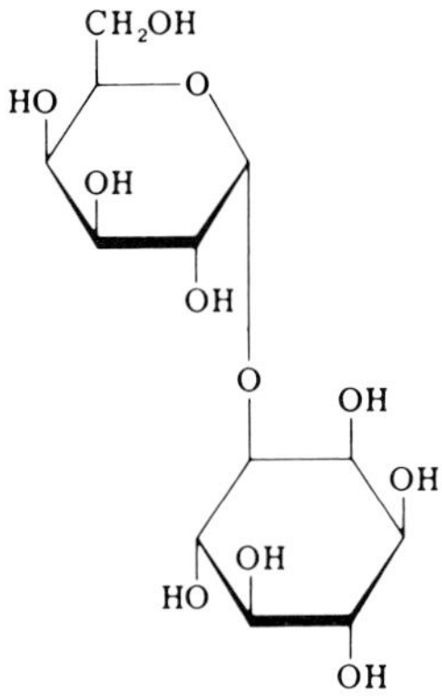

Galactinol

(3)

Studies on photosynthesis in the presence of $^{14}CO_2$ have demonstrated the formation of labelled galactinol in the leaves of a large number of plant species (Table IV) (Senser and Kandler, 1967a). A high level of radioactive incorporation occurs into this compound as compared with the incorporation into raffinose and stachyose. From kinetic studies of such incorporation in photosynthesizing leaves of *Catalpa bignonioides* and *Lamium maculatum* (Kandler, 1967; Senser and Kandler, 1976b), it became apparent that the labelling of galactinol was faster, and its pool becomes rapidly enriched with ^{14}C as compared to those of the raffinose family of oligosaccharides. However, if photosynthesis in the presence of $^{14}CO_2$ was followed by a chase of unlabelled CO_2, labelling of galactinol decreased and that of the oligosacchardies increased. During the initial stages of photosynthesis, there are more enrichment of radioactivity in the D-galactose moieties of the oligosaccharides than in the sucrose part of the molecules. These observations indicate that (a) a ripid turnover of the galactinol pool takes place and (b) the

Table III

The Distribution of Galactinol, Sucrose, and the Raffinose Family of Oligosaccharides in Various Organs of Some Plant Species[a]

Plant species	Galactinol	Sucrose	Raffinose	Stachyose	Verbascose
Andromeda japonica					
Leaf	0.34	2.26	0.55	1.16	—
Stem	0.28	2.69	0.64	2.76	—
Buddleia davidii					
Leaf	0.18	3.68	0.22	0.16	—
Catalpa bignonioides					
Leaf, young	1.24	2.58	0.29	0.55	—
Leaf, old	2.32	5.67	0.86	3.61	—
Lamium maculatum					
Leaf	0.48	1.15	0.66	0.68	—
Root and rhizome	0.99	2.07	0.89	9.62	1.36
Lycopus europaeus					
Leaf	2.81	3.43	2.49	2.12	0.82
Root	2.12	3.60	3.27	7.70	5.50
Marrubium vulgare					
Leaf	0.57	2.00	0.65	1.34	—
Root and hypocotyl	0.22	2.65	0.33	1.84	0.62
Oenothera pumila					
Leaf	0.88	4.75	0.42	0.87	—
Stem	0.10	7.32	0.23	0.15	—
Origanum vulgare					
Leaf	0.97	8.28	0.78	1.40	—
Root and rhizome	0.88	4.86	0.87	1.67	0.92
Prunella grandiflora					
Leaf	0.72	2.47	0.52	1.88	—
Root and rhizome	0.68	3.20	0.93	2.78	1.42

[a] The quantities are expressed in mg/g fresh weight of tissue. After Senser and Kandler (1967a).

synthesised galactinol is utilized for the synthesis of the raffinose family of oligosaccharides. There was little evidence to show that galactinol is translocated to other parts of the plant; however, the raffinose family of oligosaccharides synthesized in the leaves are translocated to the stem. Thus, galactinol serves as an intermediate in the biosynthesis of the storage oligosaccharides.

During the initial stages of photosynthesis in $^{14}CO_2$ the galactinol molecule is labelled only in the D-galactosyl group and even after 4 hr the label in the *myo*-inositol is only half that in the D-galactose moiety (Kandler, 1967; Senser and Kandler, 1967b). Thus, the D-galactosyl group which is derived from the photosynthetic carbon cycle is rapidly transformed into this D-galactoside. On

Table IV

The Distribution (%) of ^{14}C Among Oligosaccharides and Inositol during Foliar Photosynthesis in the Presence of $^{14}CO_2$ in Some Plant Species

Plant species	Duration of photosynthesis (min)	Galactinol	Sucrose	Raffinose	Stachyose	Inositol
Aristolochiales						
Aristolochiaceae						
Aristolochia clematis	60	10.3	86.7	2.3	0.1	0.6
Asarum europaeum	30	1.3	98.5	0.1	—	0.1
Celastrales						
Celastraceae						
Evonymus alatus	30	18.5	64.6	4.8	11.8	0.3
E. hamiltonianus	30	5.4	81.0	6.3	7.2	0.1
E. phellomanus	30	31.2	50.7	6.9	11.0	0.2
E. verrugineus	30	13.2	77.6	2.5	6.6	0.1
Cistales						
Violaceae						
Viola odorata	60	2.3	95.4	2.0	0.2	0.1
V. canadensis	60	0.6	98.4	0.9	0.1	—
Ericales						
Ericaceae						
Calluna vulgaris	30	2.5	96.3	1.0	0.2	—
Erica carnea	60	21.0	70.9	6.7	1.2	0.2
Rhododendron russatum	30	6.7	87.9	4.2	0.6	0.6
Vaccinium myrtillus	30	0.8	97.5	1.4	0.2	0.1
Gentianales						
Buddleiaceae						
Buddleia davidii	60	25.4	58.6	5.6	10.1	0.3

Oleaceae					
Forstiera neomexicana	30	6.9	88.9	0.7	3.5	—
Syringa villosa	30	26.3	54.4	3.1	16.2	—
Lamiales[a]						
Lamiaceae						
Ajuga reptans	60	13.3	52.3	8.6	24.1	0.1
Calamintha illyrica	30	6.2	75.7	8.9	8.5	0.1
C. vulgaris	30	8.5	65.5	4.0	19.5	0.2
Dracocephalum iberica	60	6.3	78.8	6.9	7.4	0.2
D. moldavica	60	12.1	57.0	8.1	21.4	0.1
D. sibiricum	30	32.5	35.6	5.8	23.9	—
Elscholtzia stauntonii	30	13.5	77.2	6.8	2.2	0.1
Galeopsis dubia	60	10.9	47.5	19.5	20.4	0.2
Hyssopus officinalis	60	20.1	53.7	7.6	16.8	0.3
Lamium album	60	8.5	46.2	14.3	29.4	0.1
L. maculatum	60	6.9	55.6	11.0	22.6	0.6
Lavendula latifolia	60	13.8	70.4	13.2	2.0	0.6
Lycopus europaeus	30	4.9	73.2	9.9	12.0	—
Marrubium vulgare	30	8.5	65.5	4.0	19.5	0.2
Mentha longifolia	30	6.2	76.5	7.5	9.2	—
Origanum vulgare	30	8.4	77.8	7.7	5.4	0.3
Phlomis tuberosa	60	8.4	65.3	6.7	19.6	—
Prunella grandiflora	60	12.2	55.3	14.1	17.7	—
Salvia cruinea	30	18.9	64.8	6.9	9.0	—
Salvia farinnacea	30	11.5	48.9	5.7	32.4	0.2
Scutellaria alpina	30	18.5	54.0	9.5	17.3	0.7
S. baicalensis	60	12.5	69.3	5.2	12.7	0.3
Stachys officinalis	60	16.6	58.2	6.4	17.1	0.3
Thymus villosa	60	20.6	54.3	9.4	14.1	0.3
Teucrium montanum	60	6.6	81.5	4.3	7.5	0.1
Verbenaceae						
Verbena hybridum (var. *compactum*)	30	7.0	60.1	9.4	21.3	1.0

(Table continues)

Table IV (continued)

Plant species	Duration of photosynthesis (min)	Galactinol	Sucrose	Raffinose	Stachyose	Inositol
Myrtalis						
Onagracea						
Epilobium palustre	30	7.1	70.4	14.9	7.4	0.2
Jussieua elegans	60	3.6	67.5	9.8	19.1	—
Oenothera tetraptera	30	1.4	93.7	4.7	0.2	—
Punicaceae						
Punica granata	30	.11.0	65.4	4.5	19.1	—
Trapaceae						
Trapa natans						
Surface leaves	30	0.8	98.7	0.3	0.1	0.1
Submerged leaves	30	0.1	99.6	0.2	0.1	—
Papaverales						
Papaveraceae						
Dicentra spactabilis	30	0.7	99.0	0.2	—	0.1
Glaucium flavum	30	1.1	98.2	0.4	—	0.3
Papaver nudicaulis	30	0.6	99.2	0.1	—	0.1
Passifloralis						
Cucurbitaceae						
Cucurbita pepo	30	3.4	81.8	2.8	12.0	—
C. sativus	60	9.6	84.8	1.8	3.8	—
Ecballium elaterum	60	8.0	47.5	13.0	31.2	0.3
Thladiantha dubia	60	18.0	35.9	13.2	32.8	0.1
Trichosanthes japonica	60	13.5	58.8	9.2	18.5	—

Rosales					
Rosaceae					
Agrimonia leucantha	30	3.0	96.2	0.8	—	—
Alchemilla mollis	30	8.6	84.9	5.3	0.1	1.1
Filipendula hexapetala	60	2.2	94.2	3.0	0.1	0.5
Fragaria vesca	30	4.9	94.3	0.6	0.1	0.1
Rosa fendleri	30	6.6	90.5	1.3	0.1	1.5
Rutales					
Rutaceae					
Citrus trifoliatus	30	0.7	99.1	0.1	0.1	—
Dictamnus alba	30	2.6	92.7	2.7	2.0	—
Ruta graveolens	30	1.7	95.4	2.6	0.1	0.2
Scrophulales					
Scrophulariaceae					
Verbascum longifolium (var. *pannosum*)	30	13.7	79.8	1.5	4.1	0.9
Wulfenia carinthiaca	60	0.8	97.0	1.5	—	0.7

[a] Radioactivity in verbascose accounted for up to 3% in a number of species. After Senser and Kandler (1967a).

the other hand, *myo*-inositol is newly synthesised at a much lower rate and serves as a D-galactosyl carrier.

The enzyme responsible for the biosynthesis of galactinol, galactinol synthase (UDP-D-galactose:*myo*-inositol D-galactosyltransferase), was first identified by Frydman and Neufeld (1963) in maturing peas. The enzyme catalyses the reaction

$$\text{UDP-D-Galactose} + \textit{myo}\text{-inositol} \xrightarrow{\text{Mn}^{2+}} \text{galactinol} + \text{UDP}$$

and requires Mn^{2+} for its activity. This ion could not be replaced by Mg^{2+}. The occurrence of this reaction was demonstrated in several plant organs (Tanner and Kandler, 1966, 1968; Tanner, 1969a,b; Lehle *et al.*, 1970; Webb, 1973; Pharr *et al.*, 1981; Webb, 1982). UDP-D-Galactose could not be replaced by ADP-D-galactose. The pH optimum of the pea enzyme was 5.6, versus the neutral optima of the enzymes involved in the biosynthesis of the raffinose family of oligosaccharides (Tanner *et al.*, 1967; Tanner and Kandler, 1968; Lehle *et al.*, 1970; Lehle and Tanner, 1973b; Whittingham *et al.*, 1979). The pH *in situ* is also near neutral. However, galactinol synthase from the leaves of *Cucumis sativus* (Cucumber) has since been purified 40-fold and it was shown that the Mn^{2+} concentration in the reaction mixture profoundly influences the pH optimum. At low Mn^{2+} concentration (0.2 mM), the optimum is 7.0, while at a high concentration (7.0 mM), the value is 5.5 (Pharr *et al.*, 1981).

The synthase from the leaves of *Cucurbita pepo* (straight-necked squash) was purified approximately 400-fold, and its also exhibited alkaline pH optima (pH 7.5 and 8.0) (Webb, 1982). The requirement for Mn^{2+} and a sulphydryl protectant for full enzymic activity was essential. Chelating agents as well as uridine nucleotides and UDP-D-glucose were all inhibitory, the latter possibly being involved in *in vivo* regulation of the enzymic activity. In contrast to the results obtained with pea leaf chloroplasts (Imhoff, 1973), the cucumber chloroplast homogenate showed the absence of galactinol synthase activity. The pea leaf enzyme was inhibited by raffinose, with inhibition more marked at higher *myo*-inositol concentrations. However, all experiments with the enzyme from this source were carried out at pH 5.6 in the presence of 10 mM MnCl$_2$. Thus, any regulatory role of raffinose at physiological pH on the biosynthesis of its precursor D-galactosyl donor cannot be judged well.

Higher homologues of galactinol have been detected in some plant sources. For example, di-*O*-α-D-galactosyl-*myo*-inositol [*O*-α-D-galactopyranosyl-(1→6)-*O*-α-D-galactopyranosyl-(1→1)-*myo*-inositol] was detected in the seeds of *Brassica campestris*, *B. nepus* (Carruthers *et al.*, 1963; Siddiqui *et al.*, 1973; Theander and Aman, 1976), *Trigonella foenumgraecum* (Sioufi *et al.*, 1970), and *Vicia sativa* (Petek *et al.*, 1966). The presence of tri- and tetra-*O*-α-D-galactosyl-*myo*-inositols in plants was also predicted, and these were suspected to play roles in the metabolism of D-galactose-containing oligosaccharides (Courtois and Percheson, 1971).

D. Clusianose

Clusianose (1-*O*-α-D-galactopyranosylhamamelitol) was first detected in the leaves of *Primula clusiana* (Beck, 1969) and later demonstrated in various species of four subsections—*Cyanopsis, Arthritica, Chamaecallis* and *Rhopsidium*—of the main section *Auricula* belonging to the genus *Primula* (Sellmair *et al.*, 1969). The compound accumulates in the leaves of several species in levels as high as those of sucrose; thus its chemotaxonomic relevance was suggested (Sellmair *et al.*, 1969, 1977). Structure **4** for clusianose was proposed (Beck, 1969).

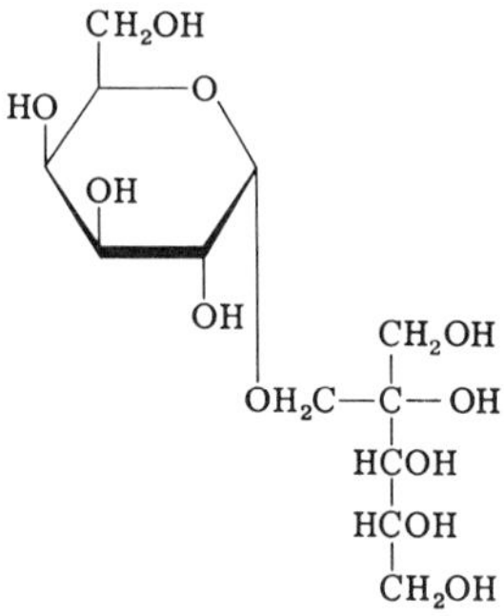

Clusianose

(4)

Clusianose is formed during foliar photosynthesis in $^{14}CO_2$ at a much lower rate than sucrose. However, while the sucrose level decreased in the subsequent dark period, the clusianose level remained unaltered (Beck, 1969). The compound is metabolically very stable and is not translocated in the plant (Sellmair and Kandler, 1970). The kinetics of ^{14}C-labelling (Sellmair *et al.*, 1969) showed much slower incorporation of radioactivity into the hamamelitol residue of the molecule as compared to the D-galactosyl group. This demonstrates the origin of the D-galactosyl donor directly from the pool of photosynthetic intermediates. The probable route of biosynthesis is by transfer of the D-galactosyl group to free hamamelitol:

Clusianose is probably an end metabolic product and accumulates in the leaves, especially in the older ones, and provides frost resistance (Sellmair and Kandler, 1970). Its function as a reserve carbohydrate is not clear. The exact nature of the D-galactosyl donor and the specficity of the enzyme involved in clusianose formation is not known. α-Galactosidase is probably involved in the degradation of this compound.

E. Pinitol D-Galactoside

Pinitol D-galactoside [O-α-galactopyranosyl-(1→2)-D-4-O-methyl-*chiro*-inositol] was isolated from the seeds of subterranean clover (*Trifolium subterraneum*). It was also detected in the seeds of 36 pasture legumes (Beveridge *et al.*, 1977). Schweizer *et al.* (1978) showed its presence in a few other legume seeds, such as soybean, chickpea, and lentils. Structure 5 is assigned to the compound.

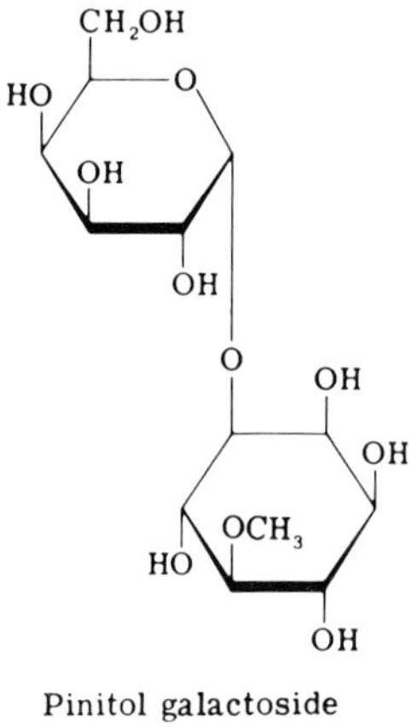

Pinitol galactoside

(5)

The compound was absent from leaves and stems, although D-pinitol was detected in significant amounts. Schweizer *et al.* (1978) found pinitol galactoside along with galactinol in chickpeas but was unable to detect galactinol in soybean. It was suggested that like galactinol, pinitol galactoside may also act as a D-galactosyl donor in the biosynthesis of the raffinose family of oligosaccharides (Beveridge *et al.*, 1977).

F. Ononitol D-Galactoside

Ononitol D-galactoside [O-α-D-galactopyranosyl-(1→1)-D-4-O-methyl-*myo*-inositol] was isolated from adzuki beans (*Vigna angularis*) and its chemical structure was assigned (Yasui, 1980). The compound was also present in the seeds of cowpea (*Vigna sinensis*) but absent from pea (*Pisum sativum*) and soybean (*Glycine max*). Its significance as a chemotaxonomic maker was suggested. In a recent publication, Yasui and Ohno (1982) reported its presence in 60 plant species from 21 different families. The physiological role of this compound, either as a D-galactosyl donor or a reserve material, is not yet clear.

IV. α-D-GALACTOSIDES OF SUCROSE

Sucrose is the most abundantly occurring disaccharide in the plant kingdom. This sugar constitutes the major transportable form of carbohydrate being synthesized in the leaves during photosynthesis. Sucrose is also a good source of D-glucose and D-fructose, which can be transformed into various glycosyl donors, thus providing an important base for the biosynthesis of various complex carbohydrates. A detailed discussion of sucrose metabolism has been presented in Chapter 1.

Most oligosaccharides based on sucrose are those that contain α-D-galactosyl groups. D-Glucosyl and D-fructosyl groups may also be linked to sucrose, for example, joining of (a) D-glucose to D-glucosyl group by β-(1→6)-linkage produces gentianose, (b) D-fructose to D-glucosyl group by β-(2→6)-linkage produces neokestose, (c) D-fructose to the D-fructosyl group by β-(2→6)-linkage produces kestose, and (d) D-fructose to the D-fructosyl group by β-(2→1)-linkage produces isokestose. These products are potential precursors of D-glucose/D-fructose-containing carbohydrates of higher DP. Various ways of linking D-galactose to sucrose are shown in Table V. The linking may occur with D-fructosyl groups yielding trisaccharides, or subsequent higher homologous oligosaccharides may also be formed. In the latter case, further D-galactosyl groups are joined by α-(1→6)-linkages to existing D-galactosyl groups of the oligosaccharides. However, D-galactose may also be found

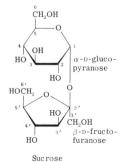

Sucrose

Table V

Mono-*O*-α-D-Galactosylsucrose Derivatives

α-D-Galactosyl linkage with	Product
C-2 of D-glucose	Umbelliferose
C-3 of D-glucose	Unnamed
C-6 of D-glucose	Raffinose
C-1 of D-fructose	Unnamed
C-3 of D-fructose	Unnamed
C-6 of D-fructose	Planteose

joined to both groups of sucrose, as in the tetrasaccharides lychnose and isolychnose.

A. Trisaccharides

Of the eight possible isomers of mono-O-α-D-galactosylsucrose, only six have been isolated (Table V). These are discussed in this section.

1. Raffinose

This trisaccharide [O-β-D-galactopyranosyl-(1→6)-α-D-glucopyranosyl-(1→2)-β-D-fructofuranoside] is the first member of a homologous series of oligosaccharides, generally termed the raffinose family, in which successive D-galactosyl groups are joined to each other by O-α-D-galactosyl-(1→6)-linkages. The raffinose family of oligosaccharides is very widely distributed in the plant kingdom, perhaps occuping a place next to sucrose with respect to abundance in the plant kingdom. At least among higher plants, these sugars may even be ubiquitous (Jermias, 1962; Dey, 1980c). Raffinose was first isolated and crystallized from *Eucalyptus manna* (Johnston, 1843). It has since been detected in leaves, rhizomes, roots, seeds, and stems of numerous leguminous and other plants. Raffinose often co-occurs with the members of its higher homologous series and galactinol. However, the relative ratios of the members may vary among plant species, as in legume seeds (see Table VI). The higher homologues are generally found in storage organs in high concentrations (see Tables III, IV, and VI). The distribution of these oligosaccharides in plants has been discussed in several articles (Dey, 1980c; Eskin *et al.*, 1980; Kandler and Hopf, 1980, 1982; Rathbone, 1980). The various techniques utilized for the estimation of raffinose and its separation from related sugars are well documented (Gross, 1974; Schiweck, 1978, 1982; Rathbone, 1980).

The structure of raffinose (**6**) was established by using both chemical and enzymatic methods. A detailed description of this can be found in a recent review (Rathbone, 1980). Very little is known of the cellular localization of raffinose as compared to that of sucrose (Leigh *et al.*, 1979; Willenbrink and Doll, 1979; Doll and Willenbrink, 1980; Hawker, this volume), although the latter serves as precursor of the trisaccharide. In barley grains, raffinose is present in increasing concentrations toward the center of the kernel (Gohl *et al.*, 1977). Raffinose occurs in leaves only at low concentrations; however, it accumulates in the storage organs during the process of plant development. The level increases as the tissue looses water, which is a characteristic feature of maturation of seeds (Fig. 8) and the hardening process of winter-hardy plants.

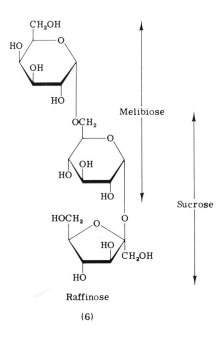

(6)

The formation of raffinose and its family of oligosaccharides in higher plants during photosynthesis has been described earlier (Dey, 1980c). The results of radioactivity incorporation into sucrose, raffinose, and stachyose in the leaves of various plants undergoing photosynthesis in the presence of $^{14}CO_2$ are shown in Table IV. The maximum incorporation was undoubtedly into sucrose, followed by either raffinose or stachyose; there was no phylogenic relation in the order of labelling of the latter two sugars. In some plant species the maximum label was not in stachyose (Senser and Kandler, 1966) suggesting that this sugar is probably the end product of photosynthesis and is not metabolised until the leaves are placed in the dark. The kinetics of synthesis of the oligosaccharides are typical of reserve substances (Kandler, 1967). The distribution of label in the sugar molecules suggests that they are synthesised by the transfer of D-galactosyl groups from D-galactose-containing intermediates derived from the photosynthetic pool.

The basic pathway for the biosynthesis of raffinose is via a trans-D-galactosylase-catalysed reaction. There are three possible ways of achieving trans-D-galactosylation: (a) using UDP-D-galactose, (b) using galactinol, and (c) using α-D-galactosidic substrates in α-D-galactosidase-catalysed reactions. The last reaction is depicted as

$$\text{D-Galactosyl—O-R} + H_2O \rightleftharpoons \text{D-galactose} + \text{ROH}$$

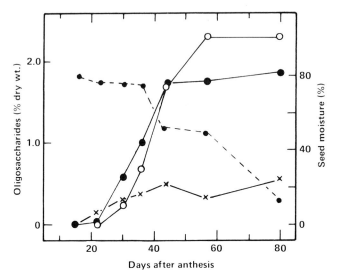

Fig. 8. The formation of the raffinose family of oligosaccharides during development of field pea (*Pisum sativum*) seeds: ○, verbascose; ●, stachyose; ● – – – ●, moisture; x, raffinose. Adapted from Holl and Vose (1980.)

where R is the aglycone residue. The enzyme can transfer a D-galactosyl group to sucrose, yielding raffinose (Dey, 1969, 1979); however, hydrolysis of the starting D-galactosyl compound is favoured. Unless the localised concentration of substrates in the vicinity of the enzyme is very high, the synthetic reaction seems unfeasible.

The participation of UDP-D-galactose in the biosynthesis of raffinose was first demonstrated in maturing seeds of *Phaseolus lunatus* (Korytnyk and Metzler, 1962). The reaction

$$\text{UDP-D-Galactose} + \text{sucrose} \rightleftharpoons \text{raffinose} + \text{UDP}$$

was reported to be catalysed by enzyme preparations from *Vicia faba* seeds (Bourne *et al.*, 1962, 1965; Pridham and Hassid, 1965), *Glycine max* seeds (Gomyo and Nakamura, 1966), and the isolated chloroplasts of *Pisum sativum* leaves (Imhoff, 1973). However, this pathway was disputed, as only crude enzyme preparations were utilized (Lehle and Tanner, 1973b) and the results could not be reproduced.

The wide occurrence of galactionol in plant organs that possess raffinose and related oligosaccharides provides a good argument for its role as a D-galactosyl donor in the biosynthesis of the trisaccharide. This nonnucleotide D-galactosyl donor has a negative free energy of hydrolysis, approximately 4.2 kJ/mol higher than that for the hydrolysis of the terminal

Table VI

Amounts of α-D-Galactosides and Sucrose in Some Legume Seeds (% of Dry Matter)[a]

Legume seeds	Galactinol	Pinitol galactoside	Sucrose	Raffinose	Stachyose	Verbascose
Chick pea (*Cicer Arietinum*)	0.39	0.34	2.69	0.45 (1.10)	1.72 (2.50)	0.10
Cow pea (*Vigna unguiculata*)	0.12	—	2.64	0.41 (0.40)	4.44 (4.80)	0.48 (0.50)
Faba bean (*Vicia faba*)	0.22	0.17	2.00	0.22	0.67	1.45
Field bean (*Phaseolus vulgaris*)	0.15	—	3.01	0.26 (0.20)	2.16 (1.20)	0.03 (4.00)
Lentil (*Lens culinaris, L. esculenta*	Traces	0.34	3.36	0.31 (0.90)	1.47 (2.70)	0.47 (1.40)
Lima bean (*Phaseolus lunatus*)	0.10	Traces	18.5	0.46	2.76	0.31
Mung bean (*Vigna radiata*)	0.19	Traces	0.96	0.23	0.95	1.83
Lupin (*Lupinus albus*)	0.23	—	2.63	0.82	4.11	0.48
Peas (*Pisum sativum*)	0.17	—	1.85	0.60 (0.60; 0.30)	1.71 (1.90; 1.70)	2.30 (2.20)
Soybean (*Glycine max*)	Traces	0.59	6.35	1.15 (0.80)	2.85 (5.40)	—

[a] Data are taken from Sosulski *et al.* (1982), except that those presented in parentheses are from Cristofaro *et al.* (1974).

D-galactosyl group of the oligosaccharide stachyose (Tanner, 1969a). Thus, like sucrose. ($\Delta G^\circ = 27.6\,\mathrm{kJ/mol}$), galactinol may be considered to act as a good glycosyl donor. The galactinol route for the biosynthesis of raffinose and its family of oligosaccharides is now generally accepted. Earlier demonstration of UDP-D-galactose as the D-galactosyl donor may have been due to the following sequence of reactions:

$$\text{UDP-D-Galactose} + myo\text{-inositol} \longrightarrow \text{galactinol} + \text{UDP}$$

$$\text{Galactinol} + \text{sucrose} \rightleftharpoons \text{raffinose} + myo\text{-inositol}$$

In *in vitro* experiments, in which enzyme extracts from wheat germ or mature seeds of *Vicia faba* were incubated with galactinol and sucrose, the formation of raffinose was demonstrated (Lehle *et al.*, 1970). The enzyme was termed galactinol: sucrose 6-α-D-galactosyltransferase, and it was purified from only one plant source, *V. faba* seeds (Lehle and Tanner, 1973b). Several related enzymic activities were measured during the purification of this enzyme (Table VII). Two forms of the enzyme were resolved in the final step, both being apparently free of stachose synthesizing activity and α-galactosidase. The sensitivity of assay of the latter enzyme (activity expressed as μmol/hr) is, however, low when compared to that of the radiometric assay of the transferase (nmol/hr). Both forms of the finally purified transferase also displayed an exchange activity that catalysed the reaction

$$\text{Raffinose} + [^{14}\text{C}]\text{sucrose} \rightleftharpoons [^{14}\text{C}]\text{raffinose} + \text{sucrose}$$

Juding from the separation results, it is quite likely that two different enzymes are involved in catalysing the transfer and the exchange reactions, respectively. The results could also be explained on the basis of a single protein possessing both activities but displaying relatively different stability characteristics. The enzyme preparations, however, showed apparent lack of homogeneity on disc gel electrophoresis. Moreno and Cardini (1966) demonstrated the exchange reaction using an enzyme preparation from wheat germ, and the preparation was devoid of raffinose-synthesizing activity.

The purified, but unresolved, enzyme preparation (Table VII, step 5) showed a good degree of acceptor specificity for the raffinose-synthesizing reaction in which $[^{14}\text{C}]$galactinol was used as the D-galactosyl donor (Lehle and Tanner, 1973b). Sucrose was the best acceptor, giving rise to raffinose and some free D-galactose. The radioactivity in the trisaccharide was five-fold higher than that in the free monosaccharide; thus the transfer reaction exceeds hydrolysis. Fructose, glucose, galactose, glycerol, cellobiose, lactose, melibiose, raffinose, and stachyose failed to act as acceptors. The efficiency of the compounds tested as D-galactosyl donors, using $[^{14}\text{C}]$sucrose as the acceptor and yielding labelled raffinose, followed the order (with cpm incorporated into the products are shown in parentheses): raffinose

Table VII

Purification of Galactinol:sucrose 6-α-D-galactosyltransferase from *Vicia faba* Seeds[a]

Purification step	Total protein (mg)	Raffinose synthesis[b] Total activity (nmol/hr)	Specific activity (nmol/hr/mg)	Exchange reaction[c] Total activity (nmol/hr)	Specific activity (nmol/hr/mg)	Stachyose synthesis[d] Total activity (nmol/hr)	Specific activity (nmol/hr/mg)	α-Galactosidase activity[d] Total activity (μmol/hr)	Specific activity (μmol/hr/mg)
1. Crude extract	22,000	1,576	0.071	1,980	0.091	2,148	0.098	1,540	70
2. Protamine sulfate	9,350	1,750	0.187	2,072	0.222	2,306	0.246	272	29
3. Ammonium sulfate	2,688	1,398	0.520	1,730	0.647	1,488	0.553	161	58
4. DEAE–cellulose	182	557	3.067	614	3.376	0	0	5	29
5. Sephadex G-200	50	262	5.273	359	7.168	0	0	0	0
6. Hydroxyapatite									
Peak I	4.5	3.1	0.703	32.9	7.326	0	0	0	0
Peak II	2.3	68.5	29.800	58.5	25.345	0	0	0	0

[a] After Lehle and Tanner (1973b).

[b] Assayed according to galactinol + [14C]sucrose → [14C]raffinose + myo-inositol.

[c] Assayed according to raffinose + [14C]sucrose → [14C]raffinose + sucrose.

[d] Assayed according to [14C]galactinol + raffinose → [14C]stachyose + myo-inositol.

[e] Assayed according to p-nitrophenyl-α-D-galactoside + H_2O → D-galactose + p-nitrophenol.

(33,405) > galactinol (24,012) > *o*-nitrophenyl α-D-galactoside (ONPG) (13,110) > stachyose (1005) > melibose (995) > UDP-D-galactose (0). It is unique that ONPG, a nonphysiological D-galactosyl donor, is half as efficient as galactinol. In ONPG, the glycosidic bond may be weakened due to hydrogen-bond formation between the nitro group and the 2-hydroxyl group of the glycone (Nath and Rydon, 1954). In this context it is worth noting that the V_{max} values (shown as μmol/min/mg in parentheses) for several of the above compounds using α-galactosidase II of *V. faba* (Dey and Pridham, 1969) follow the order: raffinose (4.18) > ONPG (2.80) > stachyose (1.36) > galactinol (0.72) > melibiose (0.41). It is of interest to note from the results of Lehle and Tanner (1973b) that while the enzyme causes considerable hydolysis of galactinol in the absence of sucrose (the D-galactosyl acceptor), ONPG is not hydrolysed. However, as the incubation of the enzyme with ONPG and [^{14}C] sucrose yields [^{14}C] raffinose, the equivalent amount of *o*-nitrophenol must also be liberated.

The mechanisms of both synthesis and exchange reactions are likely to involve two steps as shown below:

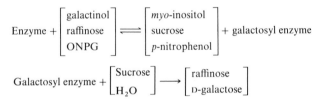

Both reactions display similar profiles of heat inactivation (Lehle and Tanner, 1973b), and the D-galactosyl donor (galactinol and raffinose) protects the activities more efficiently than the acceptor (sucrose). This observation favours the above mechanism.

Raffinose may be utilized in plants via (a) its use as a precursor for the synthesis of stachyose in a reaction involving galactinol and a specific transferase, galactinol + raffinose ⇌ stachyose + *myo*-inositol and (b) degradation by the action of invertase. Figure 8 shows the appearance of raffinose in maturing peas slightly earlier than that of stachyose. The profile of synthesis may be taken as in support of option (a). The accumulation of raffinose and its family oligosaccharides in maturing seeds is of general occurrence (Dey, 1980c, Kanlder and Hopf, 1980). On the other hand, raffinose level decreases during seed germination (Pridham and Dey, 1974; Aaman, 1979; Gendraud and Cloux, 1979; Silva and Luh, 1979; Dey, 1980c; El-Shimi *et al.*, 1980; Labaneiah and Luh, 1981). An appreciable loss of this and related sugars occurs during the preliminary soaking period of seeds. In cotton seeds, a supply of oxygen is necessary for the depletion of the oligosaccharides and, therefore, for germination (Shiroya, 1963). A high level of α-galactosidase is generally found in resting seeds which also contain the raffinose family of oligosaccharides (Dey and Pridham, 1972; Pridham and Dey, 1974; Dey, 1978, 1980c). Thus, the co-occurrence of these oligosaccharides and the enzyme must imply the existence of some form of compartmentation of either the enzyme or the substrate. In this respect it should be noted that several seed α-galactosidases are glycoproteins (Dey and Del Campillo, 1984), and some display D-glucose/D-mannose specific lectin activity (Dey *et al.*, 1982a,b, 1983). Various theoretical models for the localization of such enzymes are shown in Fig. 9. In addition, several α-galactosidases have been found bound to cell-wall components (Dey and Del Campillo, 1984) and in protein bodies (Plant and Moore, 1982; Dey, 1984a).

α-Galactosidases often hydrolyse more readily the lower members of homologous series of α-D-galactosyl-containing oligosaccharides (Dey and Pridham, 1972; Dey, 1978, 1980c). For example, the order for the raffinose family is raffinose > stachyose > verbascose > ajugose. The occurrence of multiple forms of the enzyme in seeds is very common, and in several instances these forms display relative specificity for a specific member of the raffinose family (Dey and Pridham, 1969; Dey and Dixon, 1974; Dey, 1981; Gaudreault and Webb, 1983). The liberated D-galactose is rapidly metabolised (see Section II).

During seed germination, some resynthesis of the oligosaccharides also occurs simultaneously as they are degraded (Dupéron, 1964; Wahab and Burris, 1975). Thus, seeds of *P. vulgaris* and *G. max* when imbibed in the presence of radioactively labeled D-fructose, D-galactose, D-glucose, and sucrose in aerated media incorporated label into raffinose.

The D-fructosyl group of raffinose and its higher homologues can be cleaved by both acid and alkaline invertase; however, the rate of hydrolysis is much slower than that of sucrose. The relative hydrolytic rates of the raffinose family of oligosaccharides follow the order raffinose > stachyose > verbascose (Courtois *et al.*, 1956b). Alkaline invertase gives slower rates of hydrolysis than the acid form. The acid invertase is preponderant largely in rapidly grow-ing tissues such as germinating seeds, whereas the alkaline form is present in the dormant seeds (Walter, 1963; Dey and Del Campillo, 1984). Thus the occurrence of melibiose (6-*O*-α-D-galactosyl-D-glucose), the hydrolytic pro-duct of raffinose, will depend strongly on the type of plant organ and its physiological state. The relative concentration of α-galactosidase and inver-tase in the tissue will also influence the level of melibiose (Avigad, 1982). In most cases melibiose represents a secondary sugar.

The primary role of raffinose and its family of oligosaccharides in the leaves, vegetative organs, and seeds is to serve as storage carbohydrates which could be for short- or long-term purposes. It is, however, not certain whether the mechanisms of synthesis and deposition of the sugars in these organs are

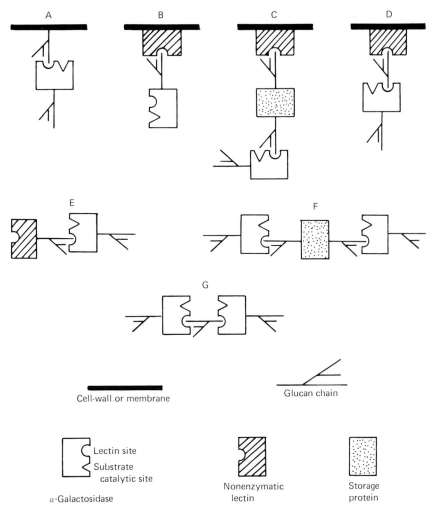

Fig. 9. Some theoretical models for the existence of α-galactosidase in a localized form in *Vicia faba* seed. The enzyme is a D-mannose-rich glycoprotein and also displays D-glucose/D-mannose-specific lectin activity (Dey *et al.*, 1982a,b). Various ways of binding the enzyme to cell wall/membrane components are depicted in **A–D**. The nonenzymatic lectin could be either a simple protein (as shown in B, C, D) or a glycoprotein; in the latter case, other modes of attachment are also possible (see also Bowles, 1979; Metcalfe *et al.*, 1983). Binding of the enzyme to the matrix via other interactions, such as ionic forces, hydrogen bondings, hydrophobic interactions, etc., is not shown here. The existence of the enzyme in the form of complex conjugates in the protein body, vacuole or cytosol is depicted in E–G.

similar. The raffinose family of oligosaccharides may be translocated to different tissues as such but can also be transformed and deposited as different sugars (Kandler and Hopf, 1980, 1982). Translocation of raffinose along with other oligosaccharides and sugar alcohols has been discussed in detail elsewhere (Peel, 1975; Ziegler, 1975; Zimmerman and Ziegler, 1975). Labelled sugars from photosynthesising organs (photosynthesis in the presence of $^{14}CO_2$) are translocated to distant parts of plants (Webb and Gorham, 1964; Trip *et al.*, 1965; Senser and Kandler, 1967b; King and Zeevaart, 1974; Turgeon and Webb, 1975; Dickson and Nelson, 1982; Hendrix, 1982).

Raffinose and its higher homologues serve among the factors responsible for frost resistance in winter-hardy plants. This topic has been dealt with in several reviews (Alden and Hermann, 1971; Levitt, 1972; Larcher *et al.*, 1973; Dey, 1980c; Kandler and Hopf, 1982). The accumulation of raffinose is linked to frost-hardiness of plants and is regulated by both temperature and photoperiod (Kandler and Hopf, 1980). Cryoprotection of chloroplast thylakoids was demonstrated by raffinose, sucrose, and glucose (Lineberger and Steponkus, 1980). Santarius and Milde (1977) showed that, in the frost-hardy cells of cabbage leaves, the dehardening process causes a decrease in the concentrations of sucrose and raffinose, especially in the chloroplasts.

Raffinose has been related to viability of seeds (Ovcharov and Koshelev, 1974). Corn seeds with high moisture content showed a low level of the oligosaccharide after prolonged storage. The decrease in the level was large in seeds that had lost viability. Raffinose and sucrose were entirely absent from the nonviable seeds. It is considered that soluble oligosaccharides are the carbohydrates that are mobilized first during seed germination, prior to the metabolism of polysaccharides such as starch and galactomannans (Dey, 1978, 1980c).

The importance of raffinose in the industrial field is in the manufacture of sucrose from sugar beets. The small proportion of raffinose ($\sim 0.05\%$) compared to sucrose ($\sim 16\%$) in this source is enough to lessen the crystallization and yield of sucrose. α-Galactosidases are used for minimizing the raffinose content in molasses during the manufacturing process. A recent review deals with the application of immobilized α-galactosidase in the sugar beet industry (Linden, 1982).

Raffinose is also an important factor in causing flatulence, thus limiting the use of legumes in human diet. Members of the raffinose family of oligosaccharides are not digested by humans, mainly due to the lack of α-galactosidase in the intestinal mucosa (Gitzelmann and Auricchio, 1965; Ruttloff *et al.*, 1967). The microflora in the lower intestinal tract then metabolize the oligosaccharides and cause a lowering of pH and production of gases such as methane, hydrogen and carbon dioxide. Further details on this subject and various approaches to lessen the flatulence factors in food legumes have been described elsewhere (Cristofaro *et al.*, 1974; Olson *et al.*, 1975; Rackis 1975; Dey, 1980c; Rathbone, 1980; Arora, 1983).

2. Umbelliferose

This trisaccharide [*O*-α-D-galactopyranosyl(1 → 2)-α-D-glucopyranosyl-β-D-fructofuranoside] is an isomer of raffinose and is found in the roots of plant species belonging to the family Umbellifereae (Boerheim-Svendsen, 1954, 1958; Wickström and Boerheim-Svendsen, 1956). The presence of this oligosaccharide was further demonstrated in various plant organs such as leaf, stem, root, flower-head and ripe and unripe fruits of several species of this family (Boerheim-Svendsen, 1956). Its ubiquity amongst umbellifers is now well acknowledged (Crowden *et al.*, 1969; Hiller, 1972; Hopf and Kandler, 1976). The distribution of umbelliferose and sucrose in various parts of umbellifers is shown in Table VIII. The level of the oligosaccharide in the leaf is generally low except in a few species where the level is substantially high. However, the concentration in the storage organs and fruits is higher; in the fruits of some species the level is higher than that of sucrose.

Two other plant families, Araliaceae and Pittosporaceae, also contain umbelliferose (Hopf, 1973; Kandler and Hopf, 1982). These families are suggested as being closely related (Hegnauer, 1969; Dahlgren, 1975), and they have many similarities with the Umbelliferae (Takhtajan, 1959; Hegnauer, 1964, 1973; Cronquist, 1968).

Umbelliferose was purified on a preparative scale from the roots of *Angelica archangelica*, and its structure (**7**) was elucidated (Wickström and Boerheim-Svendsen, 1956) by using coffee bean α-galactosidase and a chemical procedure involving periodate oxidation, methylation, and acid hydrolysis.

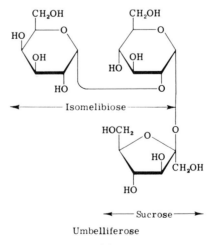

Umbelliferose

(7)

Table VIII

Distribution of Umbelliferose and Sucrose in the Leaves, Storage Organs (Rhizomes and Roots), and Fruits of Various Umbellifers[a]

| Species | Umbelliferose | | | Sucrose | | |
	Leaf	Storage organ	Fruit	Leaf	Storage organ	Fruit
Aethusa cynapium	1.0	—	2.1	38.1	—	5.5
Aegopodium podagraria	15.9	6.6	1.1	17.9	19.1	6.3
Anethum graveolens	—	—	10.5	—	—	2.1
Anthriscus cerefolium	6.0	1.6	11.6	50.6	25.2	4.2
Apium graveolens	2.6	0.9	7.4	29.6	12.5	5.2
Archangelica officinalis	7.4	9.5	27.4	17.0	9.2	26.1
Astrantia mairor	0.8	1.3	10.2	51.0	34.2	39.4
Azorella trifurcata	1.5	3.1	—	14.6	19.0	—
Bifora radians	—	—	8.3	—	—	3.5
Bowlesia incana	27.2	2.1	—	7.4	12.8	—
Bupleurum fruticosum	0.6	0.6	21.9	34.1	6.4	2.2
Carum carvi	1.1	3.0	5.6	32.1	29.0	10.3
Chaerophyllum bulbosum	19.0	5.3	6.0	45.0	11.8	18.2
Circuta virosa	—	—	10.2	—	—	8.2
Conium maculatum	0.8	0.8	2.8	51.5	24.3	13.2
Coriandrum sativum L .	—	—	11.1	—	—	6.5
Cryptotaenia japonica	—	—	10.7	—	—	20.2
Daucus carota	—	0.8	3.5	—	34.3	8.3
Dorema ammoniacum	20.9	8.0	15.5	43.6	17.2	22.6
Eryngium giganteum	1.3	1.4	4.4	44.1	6.8	19.3
Falcaria vulgaris	1.2	2.6	—	31.5	24.1	—
Foeniculum vulgare	1.0	0.7	6.5	50.3	28.6	6.7
Hacquetia epictis	1.0	0.7	—	52.9	21.9	—
Heracleum sphondylium	1.2	1.4	20.5	50.8	9.6	11.0
Hydrocotyle dissecta	—	—	0.6	—	—	4.4
Laserpitium siler	3.1	3.3	9.9	18.2	17.0	6.3
Ligusticum mutellina	1.8	3.4	12.1	42.6	10.9	20.1
Levisticum officinale	6.8	5.3	9.4	6.0	7.9	10.5
Myrrhis odorata	18.2	—	9.7	50.8	—	1.4
Oenanthe crocata	1.3	0.7	9.8	42.8	36.2	14.9
Orlaya grandiflora	—	—	8.8	—	—	5.7
Pastinaca sativa	1.2	—	8.3	52.7	—	10.7
Petroselinum crispum	1.5	—	8.3	55.4	—	7.2
Peucedanum ostruthium	25.6	11.4	9.8	34.2	6.8	8.2
Pimpinella siifolia	4.6	4.7	10.7	47.5	8.0	10.4
Pleurospermum austriacum	23.4	8.8	24.8	48.9	5.6	30.8
Scandix pecten-veneris	—	—	6.7	—	—	1.1
Seseli glaucum	0.7	1.5	8.3	26.8	24.5	11.9
Silaus teniufolius	1.1	1.5	—	47.6	42.4	—
Siler trilobus	2.5	—	—	52.4	—	—
Sium sisarum	2.9	1.3	12.1	41.6	37.5	24.3
Smyrnium olusatrum	1.7	0.7	4.8	52.7	26.4	15.5
Trinia kitaibellii	5.1	1.6	—	28.6	16.4	—

[a] Expressed as mg/g fresh weight of tissue; —, not examined. Data taken from Hopf and Kandler (1976).

The photosynthetic fixation of $^{14}CO_2$ into carbohydrates and assimilation of labelled sugars was extensively studied by Hopf and Kandler (1976) using the plant species *Aegopodium podagraria, Carum carvi,* and *Foeniculum vulgarie.* In the leaves of *A. podagaria* the label from $^{14}CO_2$ was distributed equally in umbelliferose and sucrose. However, whereas the percentage label in sucrose increased in distant parts and organs of the plant, the label in umbelliferose remained low. This indicates that the trisaccharide is poorly translocated. Its turnover is also slower than that of sucrose. Thus, accumulation of umbelliferose in storage organs and ripening fruits (Table VIII) must be the result of its *de novo* synthesis. The labelled sugar assimilation studies with fruits of *C. carvi* demonstrated the biosynthesis of umbelliferose in the endosperm. The basic component for umbelliferose formation is sucrose, which is translocated from the leaves and other sites of its synthesis. From the kinetic studies of distribution of label in the constituent monsaccharides of umbelliferose during foliar $^{14}CO_2$ fixation in *A. podagaria,* it was proposed that the biosynthesis proceeds via transfer of an α-D-galactosyl group from an activated precursor to a sucrose molecule (Hopf and Kandler, 1976). The possible D-galactosyl donors are galactinol and UDP-D-galactose. In the mature leaves of *A. podagaria,* photosynthetic $^{14}CO_2$ fixation did not result in the formation of $[^{14}C]$galactinol and yet $[^{14}C]$umbelliferose was formed. It is of interest that no raffinose is found in these leaves. However, in the immature young leaves there is ^{14}C incorporation into galactinol, but in this case raffinose is also labelled in addition to umbelliferose. As described earlier, galactinol is a well-recognised D-galactosyl donor for the biosynthesis of raffinose. These results indicate that umbelliferose is synthesised from sucrose and UDP-D-galactose.

The following pathway of umbelliferose synthesis was confirmed using an enzyme preparation from the leaves of *A. podagaria* (Hopf and Kandler, 1974).

$$UDP\text{-}D\text{-}[^{14}C]Galactose + sucrose \rightleftharpoons [^{14}C]umbelliferose + UDP$$

The specificity of the enzyme-catalysed reaction is shown in Table IX; of the 10 acceptors tested, only sucrose yielded a product, and of the 5 donors, UDP-D-galactose was the only effective compound. No higher homologs of umbelliferose were formed in this reaction, although small quantities of the tetrasaccharide were detected in the tissue extract of *A. podagraria* (Kandler and Hopf, 1982). Thus, in contrast to raffinose synthesis, galactinol does not play any role in the biosynthesis of umbelliferose. Likewise, *p*-nitrophenyl α-D-galactoside would not be expected to act as a D-galactosyl donor. The enzyme is also unable to catalyse an exchange reaction between $[^{14}C]$sucrose and umbelliferose.

The enzyme has a pH optimum of ~ 7.5 and is stable when stored at $-20°C$. It was found that phenolic materials extracted in the crude enzyme

Table IX

The Acceptor and Donor Specificities of UDP-D-Galactose:Sucrose 2-α-D-Galactosyltransferase (Umbelliferose Synthase) from the Leaves of *Aegopodium podagraria*[a]

Acceptor	Donor	Newly synthesised labelled oligosaccharide
Cellobiose	UDP-D-[^{14}C]Galactose	None
D-Galactose	UDP-D-[^{14}C]Galactose	None
D-Glucose	UDP-D-[^{14}C]Galactose	None
Lactose	UDP-D-[^{14}C]Galactose	None
Maltose	UDP-D-[^{14}C]Galactose	None
Melibiose	UDP-D-[^{14}C]Galactose	None
Raffinose	UDP-D-[^{14}C]Galactose	None
Trehalose	UDP-D-[^{14}C]Galactose	None
Umbelliferose	UDP-D-[^{14}C]Galactose	None
Sucrose	UDP-D-[^{14}C]Galactose	Umbelliferose
[^{14}C]Sucrose	UDP-D-Galactose	Umbelliferose
[^{14}C]Sucrose	Galactinol	None
[^{14}C]Sucrose	D-Galactosyl phosphate	None
[^{14}C]Sucrose	Raffinose	None
[^{14}C]Sucrose	Umbelliferose	None

[a]After Hopf and Kandler (1974).

preparation cause much loss of activity. Incorporation of polyvinylpyrrolidone in the extracting medium or passage of the enzyme preparation through a Sephadex G-25 column was, therefore, important for obtaining an active and stable enzyme. However, the enzyme preparation was contaminated by such activities as α-galactosidase, UDP-D-galactose 4'-epimerase, invertase and sucrose synthase. This caused labelling of sucrose on longer incubation of the enzyme with UDP-D-[^{14}C]galactose and sucrose. As the pH optimum of the hydrolases is ~ 5.5, these enzymes did not cause rapid breakdown of the newly synthesised umbelliferose (Hopf and Kandler, 1974).

The main pathway for the utilization of umbelliferose *in vivo* is by the action of α-galactosidase (to yield D-galactose and sucrose), followed by that of invertase (to yield D-glucose and D-fructose). Umbelliferose is fairly resistant to invertase action (Sömm and Wickström, 1965). In kinetic studies using yeast invertase, umbelliferose showed no competition with sucrose. This is probably because of steric hindrance in the vicinity of the *β*-D-fructofuranosyl linkage, caused by the presence of the D-galactosyl group at C-2 of the D-glucosyl residue of the molecule (**7**). However, Kandler and Hopf (1980) have reported the presence of isomelibiose in the roots and rhizomes of some umbelifers.

Evidently, the invertase from these sources shows some action on umbelliferose yielding this compound. On the other hand, isomelibiose could not be detected in germinating seeds (Hopf, 1973). However, the level of umbelliferose remained constant in the germinating seeds of *C. carvi*; the sugar is translocated to the seedlings (Kandler and Hopf, 1982). This overall observation may have been due to a balance between the breakdown and *de novo* synthesis of the trisaccharide.

The level of umbelliferose increases in the developing ovary and ripening fruit of *C. carvi*. During this period a $(1\rightarrow4)$-β-D-mannan, earlier termed as reserve cellulose, is deposited in the endosperm (Hopf and Kandler, 1977). Trehalose and sucrose levels also increase, but as the development advances, these decline. In the ripe fruit, only small quantities of trehalose remain (Hopf and Kandler, 1976). The main function of umbelliferose is generally considered to be that of a reserve carbohydrate.

3. Planteose

Planteose [O-α-D-galactopyranosyl-$(1\rightarrow6)$-β-D-fructofuranosyl-α-D-glucopyranoside] was first identified and extracted from the seeds of *Plantago major* and *P. ovata* (Wattiez and Hans, 1943). Following this, the trisaccharide was found in all 35 species of *Plantago* examined (Gorenflot and Bourdu, 1962). Other plant species in which it occurs have been described in earlier reviews (French, 1954; Dey, 1980c; Kandler and Hopf, 1980, 1982). The orders that include the plant species that contain planteose (Kandler and Hopf, 1982) are Gentianales (Apocynaceae, Asclepiadaceae, Buddleyaceae, Loganiaceae), Hippuridales (Hippuridaceae), Lamiales (Lamiaceae, Verbenaceae), Loasales (Loasaceae), Oleales (Oleaceae), Scrophulariales (Bignoniaceae, Pedaliaceae, Plantagenaceae, Scropulariaceae) and Solanales (Boraginaceae, Convolvulaceae, Nolanaceae, Polemoniaceae, Solanaceae). The orders from which this oligosaccharide is absent are Asterales (Asteraceae), Campanulales (Campanulaceae, Lobeliaceae), Dipsacales (Caprifoliaceae, Dipsacaceae, Valerianaceae) and Gentianales (Gentianaceae, Rubiaceae). Thus, as suggested by Jukes and Lewis (1974), planteose seems to have chemotaxonomic significance.

The following structure of planteose (**8**) was established by French *et al.* (1953a,b) via the use of chemical and enzymatic methods. Partial acid hydrolysis yields D-glucose and planteobiose. Hydrolysis by α-galactosidase from *Vicia faba* liberated D-galactose and sucrose; however, yeast invertase was unable to cleave the D-fructofuranosyl group (Dey, 1980b). The latter observation was presumably due to the substituted β-D-fructofuranoside group. Electrophoretic and paper-chromatographic methods were used for separating raffinose from planteose; the latter gave a specific fluorescence under ultraviolet (UV) light after spraying with *p*-aminobenzoic acid reagent (Dey, 1980b).

Planteose
(8)

In *Fraxinus excelsior* seed, planteose is found in the endosperm (Jukes and Lewis, 1974). The deposition of the oligosaccharide occurs during the development and maturation of seeds (Table X) (Amuti and Pollard, 1977; Dey, 1980b). It is interesting to note that in some species sucrose, raffinose and stachyose are found in high amounts in the roots and stems while planteose is present in highest concentration in the seeds (Bourdu *et al.*, 1963). Thus it seems that planteose is synthesised *de novo* in the seeds from the translocated precursor, sucrose. The seeds of *Sesamum indicum* were reported to contain raffinose and stachyose in addition to planteose (Wankhede and Tharanathan, 1976); however, later examination demonstrated the absence of the raffinose family of oligosaccharides (Dey, 1980b). It seems that the physiological state of the seed is a determining factor for detection of a specific oligosaccharide (Dey, 1980b). Seeds of a number of plant species have been shown to contain planteose as the major oligosaccharide (Wattiez and Hans, 1943; French *et al.*, 1953a, 1959; Cerbulis, 1955a; French. 1955; Herissey, 1957; Hatanaka, 1959; Jukes and Lewis, 1974).

As regards the biosynthesis of planteose, it was earlier suggested that planteose was formed during the post-harvest period of seeds, through α-D-galactosyl transfer to sucrose, catalysed by α-galactosidase (Courtois *et al.*, 1961). However, such a reaction, involving a low-energy D-galactosyl donor, will require high concentrations of both donor and acceptor. Thus, this route may not be operational under normal physiological conditions. Kandler and

Table X

The Levels of Planteose and its Higher Homologs during
the Development of *Sesamum indicum* Seeds[a]

Weeks after anthesis	Oligosaccharide (mg/g fresh weight seeds)			
	Sucrose	Pl	Pl_1	Pl_2
2	4.63	0.66	—	—
4	4.63	3.19	0.61	—
8	4.43	8.95	1.30	—
10	4.83	11.96	1.56	0.15
Mature, resting	6.00	18.48	1.86	0.51

[a] Planteose is indicated as Pl; Pl_1 (sesamose) and Pl_2 are the higher homologues (tetra- and pentasaccharides, respectively), which contain additional α-D-galactosyl groups attached to C-6 of the existing D-galactosyl group of the molecule; —, not detected. After Dey (1980b).

Hopf (1982) have reported their preliminary observations stating that the enzyme preparations from *Fraxinus excelsior* and *Sesamum indicum* are able to use UDP-D-galactose, but not galactinol, as the D-galactosyl donor for the synthesis of planteose. Table X shows that planteose accumulates in *S. indicum* seeds during development and maturation. The enzyme preparation from 10-week-old seeds (following anthesis) was able to catalyse the formation of labelled planteose from both UDP-D-galactose and galactinol as D-galactosyl donors and [^{14}C] sucrose as the acceptor (Dey, 1980b). The nucleotide sugar gave an approximately threefold higher labelling of planteose relative to the incorporation of galactinol. The enzyme extract from 8-week-old seeds gave a substantially lower incorporation of label; the extracts from immature or fully mature, resting seeds were totally unable to synthesise planteose. Compounds such as D-galactose, D-galactosyl phosphate, melibiose, and raffinose failed to act as D-galactosyl donors. These results indicate that the choice of the physiological state of seeds is important in demonstrating the *in vitro* synthesis of planteose. It is possible that two specific transferases are involved in the synthesis, each requiring a specific D-galactosyl donor.

The *in vivo* role of planteose is most likely to be that of a storage oligosaccharide which is utilized during seed germination with a primary attack of α-galactosidase followed by invertase.

4. Other Trisaccharides

In addition to raffinose and umbelliferose, two other mono-*O*-α-D-galactosylsucrose isomers are possible in which the D-glucosyl group is substituted at C-3 and C-4, respectively (Table V). The latter possibility is known to exist in the form of a tetrasaccharide, 4^G-α-D-galactopyranosylraffinose (Kato *et al.*, 1979). Specific cleavage of the D-galactosyl group from C-6 of the D-glucosyl moiety will evidently yield the required trisaccharide. However, the C-3-substituted isomer, *O*-α-D-galactopyranosyl-(1→3)-α-D-glucopyranosyl-β-D-fructofuranoside, has been isolated from the seeds of *Lolium* and *Festuca* (MacLeod and McCorquodale, 1958a,b). Both species belong to the family Graminae. The structure of the trisaccharide was established by cleaving the D-fructofuranosyl group, followed by comparing the electrophoretic mobility of the resulting disaccharide with that of marker disaccharides. Sömm and Wickström (1965) confirmed the structure of the trisaccharide by chemical methods. It is possible that the trisaccharide was formed by cleavage of the terminal α-D-galactosyl group of a structurally related tetrasaccharide that also occurs in *Festuca*. This tetrasaccharide has, in fact, been isolated and characterized (Morgenlie, 1970) as being *O*-α-D-galactopyranosyl-(1→4)-*O*-α-D-galactopyranosyl-(1→3)-α-D-glucopyranosyl-β-D-fructofuranoside. Raffinose and stachyose coexist with this tetrasaccharide.

Of the possibilities of substitution at C-1, C-3, C-4, and C-6 of the D-fructosyl moiety of sucrose by α-D-galactosyl groups, thus yielding various mono-*O*-α-D-galactosylsucroses, *O*-α-D-galactopyranosyl-(1→4)-β-D-fructofuranosyl-α-D-glucopyranoside is the only isomer that has not been detected in any plant source. Two other isomers, not yet assigned trivial names, and planteose are well characterized.

Davy and Courtois (1965, 1966) detected *O*-α-D-galactopyranosyl-(1→1)-β-D-fructofuranosyl-α-D-glucopyranoside in the roots of *Silene inflata*. The higher homologous tetrasaccharide lychnose (see Section IV,B,2) was also present at a high concentration in this source (Archambault *et al.*, 1956a,b; Davy, 1966). The trisaccharide probably arises from the cleavage of the terminal α-D-galactosyl group of lychnose.

O-α-D-Galactopyranosyl-(1 → 3)-β-D-fructofuranosyl-α-D-glucopyranoside was also isolated from the roots of *Silene inflata* (Davy and Courtois, 1965,1966). This trisaccharide was separated from other associated oligosaccharides by column chromatography (Davy, 1966; Davy and Courtois, 1966). It is derived from its higher homologous tetrasaccharide, isolychnose (see Section IV,B,3), by the action of α-galactosidase (Courtois *et al.*, 1959).

B. Tetrasaccharides

The tetrasaccharides discussed in this section are mainly the higher homologues of the trisaccharides described in Section IV,A with the exception of the higher homologous tetrasaccharide of umbelliferose, which has not been characterised, although its existence in *Aegopodium podograria* has been demonstrated (Kandler and Hopf, 1982).

1. Stachyose

This tetrasaccharide [O-α-D-galactopyranosyl-(1→6)-O-α-D-galactopyran-osyl-(1→6)-α-D-glucopyranosyl-β-D-fructofuranoside] was first isolated in 1890 by von Planta and Schulz from the rhizomes of *Stachys tuberifera*. It coexists with raffinose and other related oligosaccharides and has been detected in various organs of a number of plants (Tables III, IV, VI). It is the major oligosaccharide in several plant species.

The structure of stachyose (**9**) was elucidated by several workers using enzymatic and chemical methods (Onuki, 1932, 1933; Hérissey *et al.*, 1951; French *et al.*, 1953b; Laidlaw and Wylam, 1953). A detailed description of its isolation and structural determination was presented by French (1954) in an earlier review.

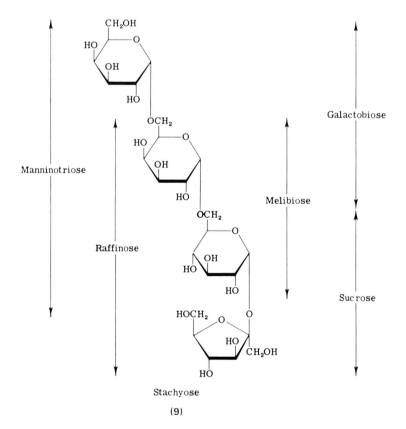

Stachyose

(9)

Stachyose and other oligosaccharides of the raffinose family have been recognised as important transport carbohydrates in a large number of woody plants, cucurbits, and legumes (Zimmermann, 1957; Pristupa, 1959; Webb and Burley, 1964; Webb and Gorham, 1964, 1965; Weidner, 1964; Trip, *et al.*, 1965; Hendrix, 1968; Dickson and Larson, 1975; Dickson and Nelson, 1982; Kandler and Hopf, 1982). As shown in Table IV, stachyose is the main oligosaccharide synthesised during photosynthetic fixation of $^{14}CO_2$ in several plant species. The kinetics of ^{14}C labelling show a rapid saturation of the galactinol pool, followed by saturation of the pools of raffinose and stachyose (Senser and Kandler, 1967b). In *Catalpa* and *Buddleia*, a comparison of $^{14}CO_2$ incorporation into stachyose and raffinose showed that initially (up to 5 min) the radioactivitiy in the latter compound was greater; however, incorporation into stachyose increased with time, and at 60 min it was double that in raffinose. Beyond this time point, if photosynthesis was continued in unlabelled CO_2, the incorporation into raffinose and stachyose still increased while a rapid decrease occurred in the amount of label in galactinol. These results and the shapes of the curves presented in Fig. 10 indicate that stachyose is synthesised at the expense of the pools of raffinose and galactinol, the trisaccharide being derived from galactinol and sucrose (see Section IV,A,1). These inferences are further strengthened by the finding that, during the initial period of $^{14}CO_2$ photoassimilation, the D-galactosyl group of raffinose has a much higher specific radioactivity than the sucrose part of the molecule; the distribution becomes equal after 1 hr. Similarly, the terminal D-galactosyl groups of stachyose initially have the highest specific radioactivity, and the

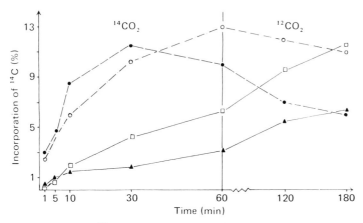

Fig. 10. Proportions of ^{14}C incorporation (%) in different soluble sugars during photoassimilation of $^{14}CO_2$ by the leaves of *Catalpa bignonioides* ○, sucrose (1:4); □, stachyose; ●, galactinol (1:2); ▲, raffinose. After Senser and Kandler (1976b).

distribution becomes equal only after 4 hr. In similar experiments with *Curcurbita pepo*, the kinetics of ^{14}C incorporation into the hexose units of stachyose were comparable to those already mentioned (Hendrix, 1968). However, in this case, a study of time course (15 sec to 5 min) showed that stachyose accounted for $\sim 22\%$ of the total ^{14}C incorporation into the soluble sugars (sucrose, raffinose and stachyose) at 15 sec. The proportion of incorporation in stachyose peaked at 37 sec, the value being 42%, followed by a rapid decline, reaching 9% at 5 min. At 37 sec, the label in raffinose and sucrose was 11 and 2%, respectively. The label in raffinose was $\sim 20\%$ at 1 min and this was unchanged at 5 min; however, the label in sucrose continually increased reaching $\sim 20\%$ by 5 min. At 10 sec, the proportion of label in sucrose was recorded as $\sim 13\%$. Thus, it seems that in *C. pepo* an efficient enzyme system exists that rapidly transforms the initially synthesised pool of sucrose into stachyose via raffinose.

These results also demonstrate that stachyose is the main storage sugar and is rapidly translocated in *C. pepo*. Its concentration per petiole was found to be 0.3–0.4 *M*, and that of sucrose was 0.25–0.35 *M* (Nicholson and Weidner, 1972). Density gradient centrifugation experiments suggested the association of stachyose with particulate fractions. However, this result remains unconfirmed. The translocation of stachyose from its site of synthesis is shown in Fig. 11. In attached leaves, the rate of translocation to other parts of the plant will greatly influence its observed level in this organ. During $^{14}CO_2$ photoassimilation in *C. pepo* it was demonstrated that the specific radioactivity of stachyose decreased linearly with translocation distance (Hendrix, 1968).

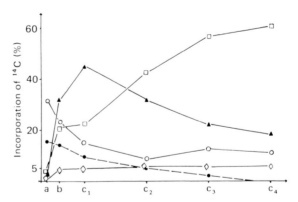

Fig. 11. Proportions of ^{14}C incorporation (%) in different soluble sugars observed in the (a) leaf, (b) leaf stalk, and at varying distances [2 cm (C_1), 7 cm (C_2), 12 cm (C_3), 17 cm (C_4)] from the base of the leaf stalk of *Lamium maculatum*. Photosynthesis in the presence of $^{14}CO_2$ was carried out for 1 hr followed by 5 hr in $^{12}CO_2$. □, stachyose; ▲, raffinose; ○, sucrose; ◇, verbascose; ●, galactinol ($\times 5$). After Senser and Kandler (1976b).

One reason of this observation may be turnover of the oligosaccharide during its transport.

It was apparent from $^{14}CO_2$ photoassimilation experiments using *C. pepo* (Webb and Gorham, 1964; Turgeon and Webb, 1975) that during growth of the leaf to maturity, stachyose is translocated to all parts of the plant except other mature leaves. The rate of translocation was approximately 20 cm/hr and reached a maximum within 10–15 min after synthesis. The sugar was completely translocated from the leaf blade within 45 min. In young leaves and shoots, it was readily metabolised to sucrose and other components; in the stem and roots, the metabolism was slower. A major portion of the translocate passed from the nodal region to the upper and the lower parts of the plant but a small portion was retained by the petiole (see also Hendrix, 1968; Schmitz, 1970; Nicholson and Weidner, 1972). The import of stachyose into the phloem of immature petioles is accompanied by its radial movement into the surrounding tissues (see also Zimmerman, 1957, 1958; Pristupa, 1959; Hartt *et al.*, 1963; Webb and Gorham, 1965). As the young leaf matures, synthesis and export of stachyose begins and import decreases; the fully mature leaf is a net exporter. The synthesis of stachyose and sucrose occurs even in senescent and partly yellowed leaves (Webb and Gorham, 1964; see also Dickson and Larson, 1975). The flow of stachyose from the roots to upper parts of the plant has not been detected.

In experiments with *Cucurbita pepo* and *Cucumis sativa*, the application of labeled D-glucose to small areas of the leaf blade resulted in its conversion to sucrose, raffinose, and stachyose, which were then translocated within the veins and petioles down to internodes of the axis (Schmitz, 1970). Similarly, in *Lamium album*, labelled D-galactose feeding caused labelling of the raffinose family of oligosaccharides. In *Phaseolus vulgaris* leaves, however, in which these oligosaccharides are reported to be absent, sucrose is highly labelled during such experiments (Hoffman *et al.*, 1971). Labelled sucrose, when applied to the leaves of *C. pepo*, is rapidly transported (Hendrix, 1973). These results tend to indicate that the phloem loading system is not specific for stachyose.

It was recently demonstrated that in *C. pepo*, stachyose synthesis takes place in the mesophyll cells. A symplastic pathway of loading of minor veins, rather than an apoplastic pathway, was suggested (Madore and Webb, 1981, 1982). The mechanism of phloem exudation has been studied by several authors (Crafts and Crisp, 1971; Walker and Thaine, 1971; King and Zeevart, 1974; Costello *et al.*, 1982).

The kinetic studies of ^{14}C incorporation into galactinol and stachyose during $^{14}CO_2$ photoassimilation, as discussed earlier, predict the existence of a D-galactosyl transferase responsible for the synthesis of stachyose in a number of plant sources. The enzyme was subsequently isolated by Tanner and Kandler (1966, 1968) from both unripe and ripe seeds of *Phaseolus vulgaris* (see also Kandler, 1967; Tanner, 1969a; Lehle and Tanner, 1973a). The enzyme is able to synthesise labelled stachyose when incubated with raffinose and [^{14}C]galactinol:

$$[^{14}C]\text{Galactinol} + \text{raffinose} \rightleftharpoons [^{14}C]\text{stachyose} + myo\text{-inositol}$$

The entire radioactivity in galactinol was in the D-galactosyl group. In this reaction, the trans-D-galactosylation was found to be approximately 20-fold higher than the hydrolysis of galactinol. The hydrolysis was somewhat higher in the absence of the acceptor molecule; however, this was still one-third of the rate of the transfer reaction. Various properties of the enzyme are summarised in Table XI. The enzyme was purified approximately 25-fold, but the preparation was not free from α-galactosidase activitiy. The relative rates of transferase activity and the ability of the enzyme to hydrolyse galactinol, raffinose, and *p*-nitrophenyl-α-D-galactoside were 37.0:2.2:1.2:32.7, respectively, with all activities measured at pH 7.0. This pH is not far removed from the pH optimum (6.5–6.7) of *P. vulgaris* α-galactosidase (Agrawal and Bahl, 1973). However, the 38-fold purified transferase from *C. pepo* leaves clearly does not possess α-galactosidase activity (Gaudreault and Webb, 1981). The transferase is not inhibited by a high concentration of D-galactose.

The reaction catalysed by the transferase from both sources was found to be freely reversible; however, the synthesis of stachyose is favoured (Table XI). The enzymes are also able to catalyse exchange reactions, as shown in Table XI. The enzymes are quite specific in that they are unable to synthesise raffinose, verbascose, or the higher homologous oligosaccharides that are generally found in the storage organs. On the other hand, Tanner *et al.* (1967) isolated an enzyme from the mature seeds of *V. faba* that is able to transfer the D-galactosyl group of galactinol to raffinose and stachyose, yielding stachyose and verbascose, respectively.

$$[^{14}C]\text{Galactinol} + \text{stachyose} \rightleftharpoons \text{verbascose} + myo\text{-inositol}$$

Both transferase activities were claimed to reside in the same protein. Some hydrolysis of galactinol occurs during these reactions; it was 7% during stachyose synthesis and 20% during verbascose synthesis. The ratio of D-galactosyl transfer to raffinose and stachyose is approximately 2.6:1. Raffinose and stachyose show mutual inhibition of the transfer reactions, with the synthesis of verbascose inhibited to a greater extent by raffinose. However, the kinetics of the mixed inhibition are not available for possible confirmation that the two activities are resident in the same protein.

The synthesised stachyose is transported from the leaves to other organs where it serves various functions. In the storage organs such as roots and seeds, where endogenous synthesis may also occur, stachyose may be deposited as such or be transformed to other α-D-galactosyl-containing oligosaccharides. In this respect, the exchange properties of the transferase may prove to be of much importance. As described for raffinose (Section IV,A,1), stachyose functions in seeds as a storage carbohydrate and is metabolised during germination. Two enzymes, α-galactosidase and β-fructofuranosidase, are involved in the initial degradation of the tetrasaccharide. The former enzyme liberates raffinose and, eventually, sucrose, whereas the latter yields mannino-triose (**9**).

The accumulation of stachyose, along with sucrose and raffinose, in leaves is known to provide frost-hardiness to winter-hardy plants. There is a seasonal variation of the pool size of these oligosaccharides (Kandler and Hopf, 1982). Experiments with the needles of *Picea excelsa* involving $^{14}CO_2$ photoas-similation demonstrated an increase in the specific radioactivity of stachyose, along with that of raffinose, during the winter (see also Trip *et al.*, 1963). This reflects a turnover of the sugar. The translocation is also dependent on the season. Thus, photoperiod and temperature are important factors in cold acclimatization and regulation of the metabolism of the oligosaccharides (Aronsson *et al.*, 1976; Kandler *et al.*, 1979). In *C. pepo*, sugar translocation is completely inhibited at 5°C (Webb, 1970).

Stachyose, synthesised by the mature leaves, is transported to and metabolised by immature leaves (Webb and Gorham, 1964, 1965; see also Bachofen, 1962), thus serving as a ready source of energy. The immature leaves are not able to synthesise the oligosaccharide (Turgeon and Webb, 1975), and the specific D-galactosyl transferase was found to be totally absent from these tissues (Gaudreault and Webb, 1981). However, in the young leaves an active alkaline α-galactosidase is present that more efficiently hydrolyses stachyose than raffinose (Gaudreaut and Webb, 1982, 1983). In this respect it is of interest that although the mature leaves of *C. pepo* contain α-galactosidase (Thomas and Webb, 1977), inhibition of sugar export by blocking the vascular system results in accumulation of stachyose and raffinose in the blade and no free D-galactose is detectable (Webb, 1971). The presence of α-galactosidase in the phloem of some plants—for example, in *Populus tremuloides*—has also been reported (Pridham, 1960). The compartmentation of the enzyme is a possible explanation for the occurrence of high levels of α-D-galactosyl oligosaccharides in plant tissues (see Fig. 9).

2. Lychnose

The basic structure of this tetrasaccharide [O-α-D-galactopyranosyl-(1→6)-α-D-glucopyranosyl-1-O-α-D-galactopyranosyl-β-D-fructofuranoside] resem-bles that of raffinose in which a second α-D-galactosyl group is attached to the C-1 of the D-fructose moiety (**10**). The structure was deduced by methods

Table XI

Comparison of Some of the Properties of Galactinol: raffinose 6-α-D-galactosyltransferase from Various Sources

Properties	Source of the enzyme		
	Cucurbita pepo[a]	*Phaseolus vulgaris*[b]	*Vicia faba*[c]
K_m (mM)			
Raffinose	4.6	0.84	0.85
Melibiose	5.2	12.00	3.90
Stachyose	—	—	3.30
Galactinol	7.7	7.30	11.00
V_{max} (%)			
Raffinose	—	100	100
Melibiose	—	46	80
Stachyose	—	—	35
Galactinol	—	—	—
pH Optimum	6.5–6.9	6.0–7.0	6.3
Inhibitors	Melibiose, *myo*-inositol, Tris, Zn^{2+}, Mn^{2+}, Ni^{2+}, Mg^{2+}, Ca^{2+}, Cu^{2+}, Ag^+	Sulphydryl compounds	Raffinose, stachyose
Specificity			
Donor	Galactinol and p-nitrophenyl-α-D-galactoside; UDP-D-galactose and melibiose do not act as donors	Galactinol; UDP-D-galactose and ADP-D-galactose do not act as donors	Galactinol

Acceptor	Raffinose, melibiose (poor, forms manninotriose), D-galactose (poor, two unidentified products); D-fructose, D-glucose, cellobiose, gentiobiose, maltose, melizitose, sucrose, trehalose, maltotriose, manninotriose, and stachyose do not act as acceptors	Raffinose, D-glucose (poor, forms melibiose), D-galactose (poor), lactose (poor); glycerol, D-fructose, sucrose, maltose, cellobiose, trehalose, gentiobiose, melizitose, and stachyose do not act as acceptors	Stachyose, raffinose, melibiose, D-galactose (poor), D-glucose (poor), lactose (poor); glycerol, D-fructose D-glucose 1-phosphate, D-glucose 6-phosphate, sucrose, maltose, cellobiose, trehalose, gentiobiose, melizitose, and verbascose do not act as acceptors
Reversibility of the reaction	Freely reversible	Freely reversible; $\dfrac{[\text{stachyose}][\textit{myo}\text{-inositol}]}{[\text{raffinose}][\text{galactinol}]} = 4$	—
Exchange reactions	$\textit{myo}$-[^{14}C]Inositol + galactinol $\rightleftharpoons$ [^{14}C]galactinol + $\textit{myo}$-inositol; [^{14}C]Galactinol + stachyose $\rightleftharpoons$ [^{14}C]stachyose + galactinol [^{14}C]Raffinose + stachyose $\rightleftharpoons$ [^{14}C]stachyose + raffinose	$\textit{myo}$-[^{14}C]Inositol + galactinol $\rightleftharpoons$ [^{14}C]galactinol + $\textit{myo}$-inositol [^{14}C]Raffinose + stachyose $\rightleftharpoons$ [^{14}C]stachyose + raffinose	—
Stability	For 1 month at 4°C in the presence of the 20 mM 2-mercaptoethanol	For several months at 0°C	—

[a] Gaudreault and Webb (1981).
[b] Tanner and Kandler (1968), Lehle and Tanner (1973a).
[c] Tanner et al. (1967), Lehle and Tanner (1973a).

involving methylation and partial hydrolysis (Archambault *et al.*, 1956a,b; Wickström *et al.*, 1958a,b). The oligosaccharide is generally found in the vegetative storage organs and leaves of plants belonging to the Caryophyllaceae (Schwarzmaier, 1973); thus, it has chemotaxonomic importance. It was first isolated from the roots of *Lychnis dioica* (Archambault *et al.*, 1956a,b; Courtois, 1957; Wickström *et al.*, 1958a,b), followed by its isolation from other species, for example, *Cucubalus baccifera* (Courtois and Ariyoshi, 1960a), *Dianthus caryophyllus* (Courtois and Ariyoshi, 1962a,b), *D. lumnitzeri* (Königshofer *et al.*, 1979), *Lychnis alba* (Paquin, 1958), and *Silene inflata* (Davy and Courtois, 1965). Its presence was also demonstrated in the seeds of *Sesamum indicum* (Hatanaka, 1959). The members of the raffinose family of oligosaccharides coexist in the vegetative parts and those the planteose family coexist in the seeds of *S. indicum*.

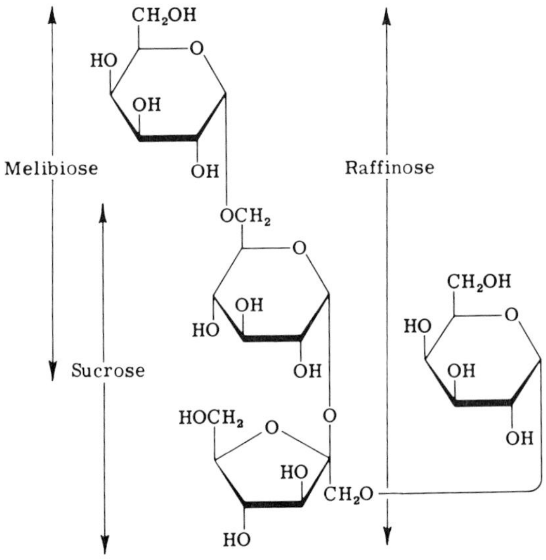

Lychnose

(10)

Schwarzmaier (1973) demonstrated that in several species of Caryophyllaceae, foliar photoassimilation of $^{14}CO_2$ gave rise to labelled lychnose in addition to D-galactose, sucrose, raffinose, and other oligosaccharides. Sucrose and raffinose were labelled at an early stage, whereas longer photosynthetic periods were required for the labelling of lychnose. This indicates that raffinose is probably the precursor of lychnose. However, in *in vitro* experiments, an enzyme preparation from the leaves of *Cerastium arvense* was unable to synthesise lychnose from raffinose and a labelled D-galactosyl donor such as UDP-D-galactose or galactinol (Schwarzmaier, 1973). On the other hand, during $^{14}CO_2$ assimilation when no net synthesis of lychnose was shown to occur, there was appreciable labelling of lychnose. One way of explaining this observation is via an exchange reaction:

Lychnose + [^{14}C]raffinose $\rightleftharpoons$ [^{14}C]lychnose + raffinose

The mechanism of the initial synthesis of lychnose could be similar to that of the D-fructosylkestose series of oligosaccharides (Kandler and Hopf, 1982), for example:

Raffinose + α-D-galactosylsucrose(s) $\rightleftharpoons$ lychnose + sucrose

However, such a pathway for lychnose, catalysed by a transferase, has not been thus far demonstrated.

3. Isolychnose

This tetrasaccharide [*O*-α-D-galactopyranosyl-(1→6)-α-D-glucopyranosyl-3-*O*-α-D-galactopyranosyl-β-D-fructofuranoside] is an isomer of lynchose in which the α-D-galactosyl group is present at C-3 of the D-fructose moiety (**11**) instead of at C-1 (**10**). The structure of isolychnose has been determined by enzymatic and chemical methods (Wickström *et al.*, 1959).

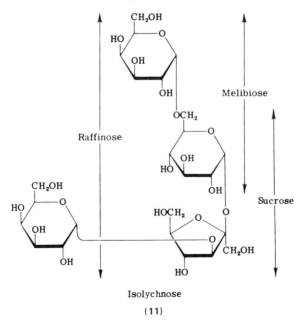

Isolychnose

(11)

This oligosaccharide was first isolated from the roots of *Lychnis dioica* (Wickström *et al.*, 1959) and was further shown to co-occur with lychnose in various roots and leaves as described earlier (see Section IV,B,2).

The discussions on the metabolism of lynchose also apply for isolychnose, except that in the latter case different specific enzymes may be involved in the exchange and the transfer reactions. However, α-galactosidase and β-fructofuranosidase are most likely to be involved in the breakdown of both tetrasaccharides.

4. Sesamose

Sesamose [O-α-D-galactopyranosyl-(1→6)-O-α-D-galactopyranosyl-(1→6)-β-D-fructofuranosyl-α-D-glucopyranoside] was first isolated from the seeds of *Sesamum indicum* by Hatanaka (1959), and this was later confirmed by other workers (Wankhede and Tharanathan, 1976; Amuti and Pollard, 1977; Kandler and Hopf, 1982). The structure of sesamose (**12**), elucidated by applying enzymatic and chemical methods, indicated that the oligosaccharide was a higher homologue of planteose (Hatanaka, 1959). Thus, analogous to the co-occurrence of stachose with raffinose in plants, sesamose probably also occurs in those plant organs where planteose exists (see Section IV,A,3).

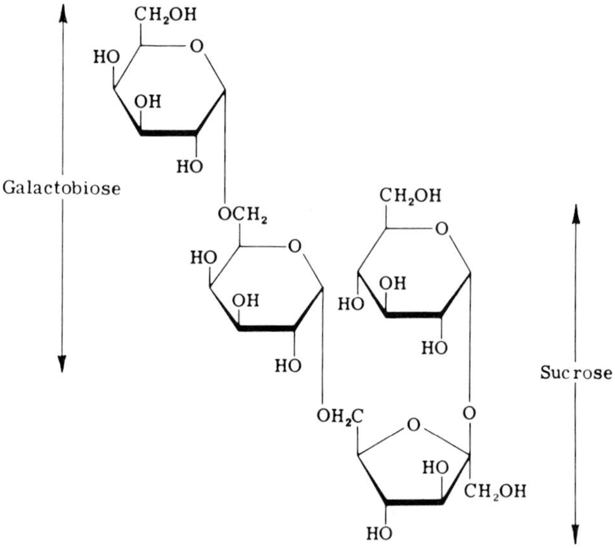

Sesamose

(12)

The formation of sesamose in *S. indicum* seeds occurs during the maturation period (Dey, 1980b). In *in vitro* experiments, it was shown that UDP-D-galactose, rather than galactinol, serves as the D-galactosyl donor for its synthesis (Dey, 1980b).

The *in vivo* utilization of sesamose probably involves the sequential action of α-galactosidase and invertase (see also Section IV,A,3).

5 Other Tetrasaccharides

Several α-D-galactosyl-containing tetrasaccharides may be included in this section; however, these are largely secondary oligosaccharides. For example, the cleavage of the terminal D-fructosyl group of verbascose (see Section IV,C,1) by invertase yields verbascotetraose [O-α-D-galactopyranosyl-(1→6)-O-α-D-galactopyranosyl-(1→6)-O-α-D-galactopyranosyl-(1→6)-D-glucopyranose]. This tetrasaccharide has been shown to occur in the free form in *Betula papyrifera* (Haq and Adams, 1962), *Theobroma cacao* (Cerbulis, 1955a), and *Vicia sativa* (Courtois *et al.*, 1956a). The structural elucidation of this oligosaccharide has been discussed by French (1954).

Kato *et al.* (1979) isolated two tetrasaccharides, I and II, from cotton seeds and elucidated their structures as being O-α-D-galactopyranosyl-(1→6)-α-D-glucopyranosyl-(1→2)-[O-β-D-fructofuranosyl-(2→1)]-β-D-fructofuranoside

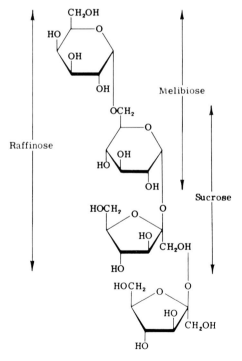

Tetrasaccharide I

(13)

(**13**) and *O*-α-D-galactopyranosyl-(1→6)-[*O*-α-D-galactopyranosyl-(1→4)]-*O*-α-D-glucopyranosyl-(1→2)-β-D-fructofuranoside (**14**), respectively. The tetrasaccharide I was earlier detected in *Triticum* species (White and Secor, 1953; Saunders and Walker, 1969; Saunders *et al.*, 1975) and its structure was determined (Saunders, 1971). It is likely that this oligosaccharide is synthesised in the seeds via the transferase activity of invertase. Plant invertases are known to transfer a D-fructosyl group from one sucrose molecule to C-1 of the D-fructose and/or C-6 of the D-glucose moieties of another sucrose molecule, yielding isokeotose and/or neokeotose (Allen and Bacon, 1956; Kandler and Hopf, 1982). A transfer of the D-fructosyl group to raffinose, instead of sucrose, will thus yield tetrasaccharide I. In this context, it should be remembered that sucrose and the raffinose family of oligosaccharides also co-occur in the seeds mentioned above.

Tetrasaccharide II

(**14**)

Morgenlie (1970) reported the isolation of a tetrasaccharide, *O*-α-D-galactopyranosyl-(1→4)-*O*-α-D-galactopyranosyl-(1→3)-α-D-glucopyranosyl-β-D-fructofuranoside, from *Festuca rubra*. Members of the raffinose family of oligosaccharides coexist with this tetrasaccharide. Higher homologues have not thus far been detected; however, the related trisaccharide that also occurs in this source is described in Section IV,A,4.

A homologous tetrasaccharide of umbelliferose has been reported to occur in *Aegopodium podagraria* (Kandler and Hopf, 1982).

A tetrasaccharide composed of L-arabinose and D-galactose was isolated by Iskenderov (1971) from the fruits of *Hedera pastuchovi*. The oligosaccharide has a linear structure: O-α-D-galactopyranosyl-(1→3)-O-α-D-galacto-pyranosyl-(1→3)-α-D-galactopyranosyl-β-L-arabinopyranoside.

C. Higher Homologous Oligosaccharides

The D-galactose-containing tetrasaccharides have been discussed in Section IV,B and in this section higher homologs are described which are formed by subsequent linking of new α-D-galactosyl groups to the already existing α-D-galactosyl moieties. Thus, four homologous series are considered here: stachyose, sesamose, lychnose and isolychnose series. As regards the series based upon the tetrasaccharide, α-D-galactosylumbelliferose, its existence in *Aegopodium podagraria* has only been predicted but thus far not demonstrated.

1. Stachyose Series

The first member of this series (**15**) is the pentasaccharide verbascose, which was first isolated from the roots of *Verbascum thapsus* (Bourquelot and Bridel, 1910). Other members of the series were also present in this source and were separated by column chromatography (Hérissey *et al.*, 1954a). The structure was subsequently elucidated (Murakami, 1940, 1943; Courtois *et al.*, 1955). The oligosaccharide occurs in most leguminous plants along with raffinose and stachyose (Kandler and Hopf, 1982), the highest concentrations being in the storage organs (see Tables III and VI). Compared to other oligosaccharides, its relative concentration was high in the seeds of *Cajanus cajan*, *Cicer arietinum*, *Dolichos uniflorus*, *Lens esculenta*, *Lupinus angustifolius*, *L. luteus*, *Phaseolus vulgaris*, *Pisum sativum*, *Vicia faba*, *V. sativa*, and *Vigna radiata* (Cristofaro *et al.*, 1974; Cerning-Béroard and Filiartre, 1977, 1979, 1980; Holl and Vose, 1980; Kandler and Hopf, 1982; Sosulski *et al.*, 1982; see also Amuti and Pollard, 1977).

Although the formation of verbascose was demonstrated in leaves during $^{14}CO_2$ photoassimilation (Kandler and Hopf, 1982), *in vitro* biosynthesis by leaf extracts has not so far, been shown. It is most likely that galactinol-mediated synthesis occurs here, similar to that established for raffinose and stachyose (see Sections, IV,A,1 and IV,B,1). This pathway has, in fact, been shown to operate in *Vicia faba* seed; the isolated enzyme displays the dual property of synthesizing both stachyose and verbascose (Tanner *et al.*, 1967).

The hexasaccharide of the stachyose series, ajugose, was first detected in the roots of *Ajuga nipponensis* (Murakami, 1942). This was followed by its isolation from the roots of *Salvia pratensis* (Courtois *et al.*, 1956c) and *V. thapsus* (Hérissey *et al.*, 1954a). The seeds of various leguminous plants also contain ajugose along with the oligosaccharides of the raffinose family, albeit in much smaller quantities (Courtois *et al.*, 1956c); Tharanathan *et al.*, 1976;

Amuti and Pollard, 1977; Cerning-Béroard and Filiatre, 1977, 1980). The higher homologous oligosaccharides with a degree of polymerization up to 15 have also been detected in some plant species (Hérissey *et al.*, 1954b; Hattori and Hatanaka, 1958; Cerning-Béroard and Filiatre, 1977, 1980). Courtois *et al.*, (1960b)) elucidated the structure of ajugose.

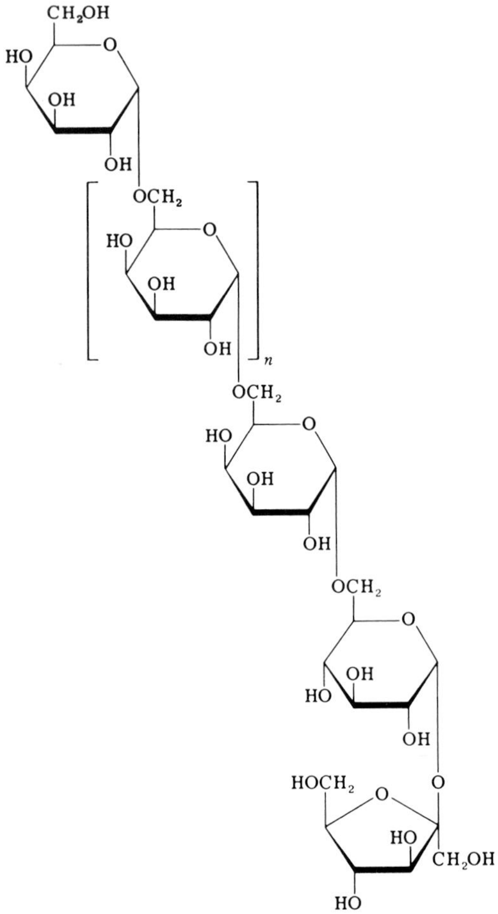

Stachyose series

(15)

The pathway of ajugose synthesis presumably involves galactinol as the D-galactosyl donor. The enzyme from *V. faba* seeds was unable to synthesise ajugose from verbascose and galactinol (Tanner *et al.*, 1967); however, the enzyme preparations from *P. sativum* and *V. sativa* were able to synthesise the hexasaccharide (Kandler and Hopf, 1982). The physiological state of the seeds and the level of natural occurrence of the oligosaccharide in the source are probably important factors in demonstrating the *in vitro* synthesis. Figure 12 illustrates the generally accepted pathway for the biosynthesis of the raffinose family of oligosaccharides.

The degradation of this series of oligosaccharides involves the participation of α-galactosidase and invertase as described for raffinose and stachyose. Likewise, the function of these oligosaccharides is mainly as storage carbohydrates. Since their concentration in the leaves is not very high, they are relatively unimportant in providing frost hardiness to the plant.

2. Sesamose Series

The pentasaccharide of this series (**16**) was first isolated by Hatanaka (1959) from the seeds of *Sesamum indicum*. The members of this series (up to hexasaccharide) were isolated from this source by using a charcoal–Celite column (Wankhede and Tharanathan, 1976). Paper-chromatographic separation was also demonstrated (Dey, 1980b). The presence of the oligosaccharides of this series has not yet been detected in any other plant source. It is important to select the seeds at the right stage of maturity as only fully mature seeds are likely to contain the higher homologues of the series (Dey, 1980b).

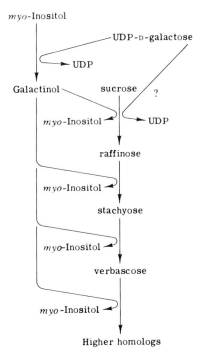

Fig. 12. Postulated pathway for the biosynthesis of the raffinose family of oligosaccharides.

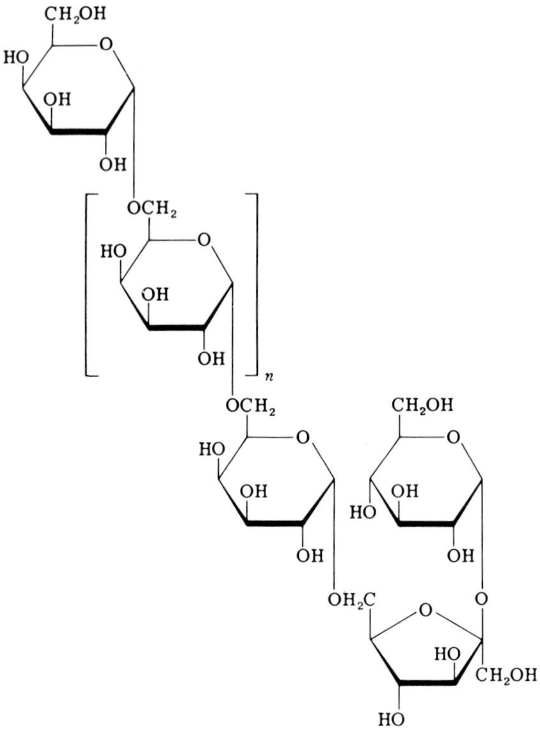

Sesamose series

(16)

In mature seeds of *S. indicum*, the concentration of the oligosaccharides of this series decreases with increasing degree of polymerization (Dey, 1980b). The biosynthetic pathway of the pentasaccharide was shown to involve UDP-D-galactose as the D-galactosyl donor. Whereas the enzyme preparation from 8-week-old seeds (following anthesis) displayed no apparent synthesis of the pentasaccharide, an extract from 10-week-old seeds gave positive indication of synthesis (Dey, 1980b).

The *in vivo* degradation of the sesamose series of oligosaccharides requires sequential action of α-galactosidase and invertase; the levels of these enzymes generally increase during germination.

3. Lychnose Series

The oligosaccharides of this series (17) occur in the vegetative storage organs of plant species belonging to Caryophyllaceae—for example, in the roots of *Cucubalus baccifera*, *Dianthus caryophyllus*, and *Lynchis dioica* (Courtois *et al.*, 1958, 1960a; Wickström *et al.*, 1959; Courtois and Ariyoshi,

1960a, 1962a,b). In *L. dioica,* the oligosaccharides occur mainly during the autumn season. Members of both lychnose and isolychnose series coexist in the same plant sources (Courtois and Davy, 1965; Courtois and Percheron, 1965), and oligosaccharides containing 8 to 12 D-galactose groups have been detected (Courtois *et al.,* 1960a). The penta- and hexasaccharides are more abundant than the higher homologues. The lower homologues were extracted from the roots of *L. dioica* with 70% ethanol; however, the higher homologues were extractable from the residual tissue using water (Courtois *et al.,* 1960a).

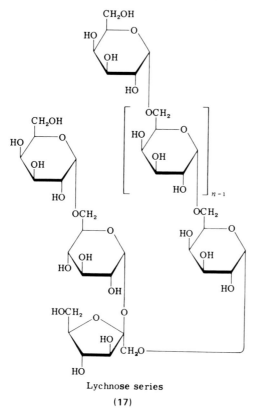

Lychnose series

(17)

4. *Isolychnose Series*

The members of this series (18) have been found in the roots of some plant species belonging to the family Sileneae (Courtois and Percheron, 1965; Davy and Courtois, 1966). As mentioned earlier, the oligosaccharides of both series (lychnose and isolychnose) often coexist in the same plant source. However, in *L. dioica,* the oligosaccharides of isolychnose series preponderate in the spring season (Courtois *et al.,* 1958; Wickström *et al.,* 1959). Various members of the isolychnose series have been separated by column chromatography using either cellulose or activated charcoal. On the other hand, if the members of both series are present in the extract, it is difficult to resolve the isomeric oligosaccharides of identical molecular weights.

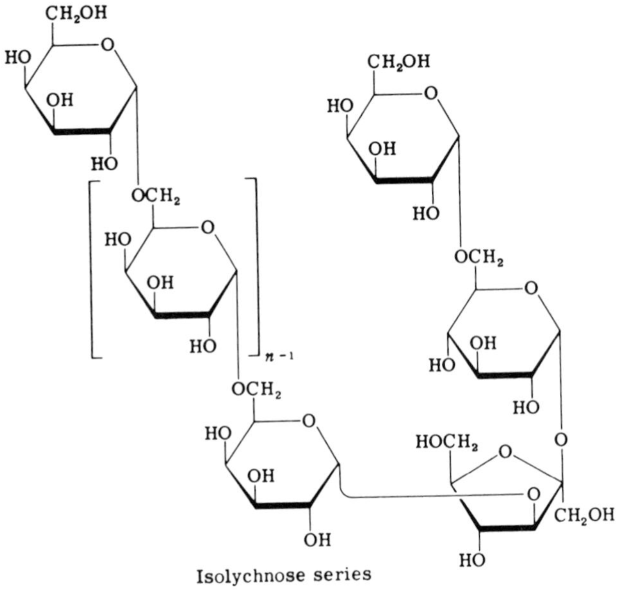

Isolychnose series

(18)

In $^{14}CO_2$ photoassimilation experiments using leaves of plant species belonging to the Caryophyllacea, Schwarzmair (1973) demonstrated the formation of labelled oligosaccharides of the lychnose and isolychnose series. However, they became labelled only after prolonged photosynthesis (~ 18 hr), while raffinose was labelled at an earlier stage. Both exchange and transfer reactions, catalysed by specific enzymes, are probably involved in *in vivo* biosynthesis (see also Sections IV,B,2 and 3); no *in vitro* synthesis of any of the oligosaccharides has been demonstrated.

VI. CONCLUDING REMARKS

Although raffinose occurs ubiquitously in higher plants, the distribution of some of its isomers is not so wide. The chemotaxonomic importance of these oligosaccharides is not well defined and thus wider and more systematic survey of plant species will provide much needed information. The higher homologs of several of these trisaccharides—for example, those of umbelliferose—have not been isolated; on the other hand, the sesamose series of oligosaccharides are known to occur only in the seeds of *Sesamum indicum*. Here, it should be remembered that variations in the levels of the oligosaccharides occur with the season of the year as well as the physiological state of the plant organ. For example, the lychnose series of oligosaccharides occurs in the roots of *Lychnis dioica* mainly in the autumn and the isolychnose series occurs in the spring (Courtois *et al.*, 1958; Wickström *et al.*, 1959).

The D-galactosylsucrose derivatives rank next to sucrose with respect to their distribution in the plant kingdom. However, our knowledge of their cellular location is incomplete. In several instances, vacuoles are recognised as the site of sucrose accumulation (Matile, 1978; Nishimura and Beevers, 1978); thus, it may be speculated that the D-galactosyloligosaccharides are also located in this organelle (see also Kandler and Hopf, 1982). In this respect, much work is required along the lines that have been followed in the study of *in vivo* sucrose accumulation. A generalised knowledge of the subcellular and cellular sites of the biosynthesis of the oligosaccharides will be of much importance. In one instance (i.e., *Cucurbita pepo*), isolated mesophyll cells have been shown to synthesis stachyose (Madore and Webb, 1982). The mechanism(s) involved in the initial transport of the newly synthesised oligosaccharide is yet to be generally established. Preliminary observations on the vein loading in *C. pepo* (Madore and Webb, 1981) indicates a symplastic pathway (see Giaquinta, 1983, for phloem loading of sucrose). It is, however, not known how the activities of the hydrolytic enzymes that act on these oligosaccharides are controlled in various tissues. Are the enzymes somehow separated, by means of compartmentation, from the site of the sugar synthesis and the route of transport? Are these enzymes present in inactive forms? Are there endogenous inhibitors of these enzymes? Does the *in vivo* pH, the specificity of the enzymes, or other factors play any role? Answers to these questions would enrich our understanding of the *in vivo* metabolism of the oligosaccharides. The glycoprotein nature of several of these enzymes (Dey and Del Campillo, 1984; Dey *et al.*, 1983) suggest a possibility for their existence as glycoconjugates, formed with reserve lectins. This view is probably true in the case of storage organs such as seeds (Basha and Roberts, 1981), where protein bodies are the sites of storage proteins, lectins, and several glycosidases (Matile, 1975; Van der Wilden *et al.*, 1980; Larkins, 1981; Van der Wilden and Chrispeels, 1983; Dey, 1984a). Glycan chains of the glycoproteins are considered important in targeting the glycoproteins to the protein bodies (Neufeld and Ashwell, 1980). Some α-galactosidases display lectin activity (Dey *et al.*, 1982b) and "lectin-like" activity (Shannon and Hankins, 1981; Dey, 1984c). This additional property of the enzyme is likely to enable its bonding to specific sites. These findings are only recent, and further exploratory research in this field will be of much value.

As regards *in vivo* synthesis, the most widely studied D-galactosyloligosaccharides are raffinose, umbelliferose, and stachyose. Here, biosynthesis in the leaves, during photoassimilation of labelled CO_2, has been the main subject of investigation. However, it is not clear what percentage of the oligosaccharides from the leaves finally becomes deposited in the storage organs. Evidently, the oligosaccharides are also synthesised independently in the storage organs (see Handley *et al.*, 1983). At this site, the exchange reaction catalysed by the specific transferases must also contribute to the total synthesis. It is possible that this property of these enzymes plays an important role in determining which of the oligosaccharides is finally deposited and stored in highest concentration in the tissue. Reports of *in vivo* synthesis of other oligosaccharides, such as planteose, sesamose, lychnose, and isolychnose, are scant.

With respect to the *in vitro* synthesis of the D-galactosyloligosaccharides, not a single D-galactosyl transferase has yet been purified to homogenity. There is much scope for study in this regard, and a comparative assessment of the physical and kinetic properties of these enzymes from various plant sources, including foliar and storage tissues, is well warranted. The enzymes for the synthesis of lychnose and isolychnose are yet to be isolated.

REFERENCES

Aaman, P. (1979). *J. Sci. Food Agric.* **30**, 869–875.
Agrawal, K. M. L. and Bahl, O. P. (1973). *In* "Methods in Enzymology" (V. Ginsburg, ed.), Vol. 28, pp. 720–728. Academic Press, New York.
Albrecht, G. J. and Kauss, H. (1971). *Phytochemistry* **10**, 1293–1298.
Alden, J. and Hermann, R. K. (1971). *Bot. Rev.* **37**, 37–142.
Allen, P. J. and Bacon, J. S. D. (1956). *Biochem. J.* **63**, 200–206.
Amuti, K. S. and Pollard, C. J. (1977). *Phytochemistry* **16**, 533–537.
Anderson, L. and Wolter, K. (1966). *Annu. Rev. Plant Physiol.* **17**, 209–222.
Angyal, S. J. and Anderson, L. (1959). *Adv. Carbohydr. Chem.* **14**, 135–212.
Anker, L. (1974). *Acta Bot.* **23**, 705–714.
Anker, L., de Bruyn, M. A. A. and Wierex, M. A. C. I. (1973). *Acta Bot. Neerl.* **22**, 75–76.
Aplin, R. T., Durham, L. J., Kanazawa, Y and Safe, S. (1967). *J. Chem. Soc.* pp. 1346–1347.
Archambault, A., Courtois, J. E., Wickström, A. and Le Dizet, P. (1956a) *Bull. Soc. Chim. Biol.* **38**, 1121–1131.
Archambault, A., Courtois, J. E., Wickström, A. and Le Dizet, P. (1956b). *C. R. Hebd. Seances Acad. Sci.* **242**, 2875–2877.
Aronsson, A., Ingestad, T. and Lööf, L.G. (1976). *Physiol. Plant.* **36**, 127–132.
Arora, S. K. (1983). *In* "Chemistry and Biochemistry of Legumes" (S. K. Arora, ed.), pp. 1–50. Arnold, London.
Augier, J. (1947). *C. R. Hebd. Seances Acad. Sci.* **224**, 1654–1656.
Avigad, G. (1982). *In* "Encyclopedia of Plant Physiology, New Series" (F. H. Loewus and W. Tanner, eds.), Vol. 13A, pp. 217–347. Springer-Verlag, Berlin and New York.

Bachofen, R. (1962). *Vierteljahrsschr. Naturforsch. Ges.* Züerich **107**, 41–47.
Baker, D. B. and Ray, P. M. (1965). *Plant Physiol.* **40**, 360–368.
Basha, S. M. M. and Roberts, R. M. (1981). *Plant Physiol.* **67**, 936–939.
Bean, R. C. and Hassid, W. Z. (1955). *J. Biol. Chem.* **212**, 411–425.
Beck, G. (1969). *Z. Pflanzenphysiol.* **61**, 360–366.
Beveridge, R. J., Dekker, R. F. H., Richards, G. N. and Towsey, M. (1972). *Aust. J. Chem.* **25**, 677–678.
Beveridge, R. J., Ford, C. W., and Richards, G. N. (1977). *Aust. J. Chem.* **30**, 1583–1590.
Bidwell, R. G. S. (1958). *Can. J. Bot.* **36**, 337–349.
Bieleski, R. L. (1982). In "Encyclopedia of Plant Physiology, New Series" (F. A. Loewus and W. Tanner, eds.), Vol. 13A, pp. 158–192. Springer Verlag, Berlin and New York.
Bisson, M. A. and Kirst, G. O. (1979). *Aust. J. Plant Physiol.* **6**, 523–538.
Boerheim-Svendsen, A. (1954). Ph. D. Thesis, University of Oslo.
Boerheim-Svendsen, A. (1956). *Acta Chem. Scand.* **10**, 1500–1501.
Boerheim-Svendsen, A. (1958). *Medd. Nor. Farm. Selsk.* **20**, 1–18.
Bouchardat, M. (1871). *Bull. Soc. Chim. Fr.* **16**, 36–38.
Bourdu, R., Cartier, D. and Gorenflot, R. (1963). *Bull. Soc. Bot. Fr.* **110**, 107–109.
Bourne, E. J., Pridham, J. B. and Walter, M. W. (1962). *Biochem. J.* **82**, 44P.
Bourne, E. J., Walter, M. W. and Pridham, J. B. (1965). *Biochem. J.* **97**, 802–806.
Bourquelot, E. and Bridel, M. (1910). *C. R. Hebd. Seances Acad. Sci.* **151**, 760–762.
Bowles, D. (1979). *FEBS Lett.* **102**, 1–3.
Brown, R. J. and Serro, F. R. (1953). *J. Am. Chem. Soc.* **75**, 1040–1042.
Carruthers, A., Dutton, J. V., Oldfield, J. F. T., Elliott, C. W., Heaney, R. K. and Teague, H. J. (1963). *Int. Sugar J.* **65**, 266–270.
Cerbulis, J. (1955a). *Arch. Biochem. Biophys.* **58**, 406–413.
Cerbulis, J. (1955b). *J. Am. Chem. Soc.* **77**, 6054–6056.
Cerning-Béroard, J. and Filiartre, A. (1977) *Comm. Eur. Commun.* [Rep.] EUR **EUR5686**, 65–79.
Cerning-Béroard, J. and Filiartre, A. (1979). *Lebensm.-Wiss. Technol.* **12**, 273–280.
Cerning-Béroard, J. and Filiartre, A. (1980). *Z. Lebensm.-Unters.-Forsch.* **171**, 281–285.
Chan, P. H. and Hassid, W. Z. (1975). *Anal. Biochem.* **64**, 272–279.
Clermont, S. and Percherson, F (1979). *Phytochemistry* **18**, 1963–1965.
Colclasure, G. C. and Yapp, J. H. (1976). *Physiol Plant.* **37**, 298–302.
Colin, H. and Guéguen, E. (1930). *C. R. Hebd. Seances Acad. Sci.* **191**, 163–164.
Cooper, R. A. and Greenshields, R. N. (1961). *Biochem. J.* **81**, 6.
Costello, R. L., Bassham, J. A. and Calvin, M. (1982). *Plant Physiol.* **69**, 77–82.
Court, G. J. (1980). *J. Phycol.* **16**, 270–279.
Courtois, J. E. (1957). *Biokhimiya* **22**, 248–258.
Courtois, J. E. (1968). *Bull. Soc. Bot. Fr.* **115**, 309–344.
Courtois, J. E. and Ariyoshi, U. (1960a). *Bull. Soc. Chim. Biol.* **42**, 737–751.
Courtois, J. E. and Ariyoshi, U. (1960b). *C. R. Hebd. Seances Acad. Sci.* **250**, 1369–1371.
Courtois, J. E. and Ariyoshi, U. (1962a). *Bull. Soc. Chim. Biol.* **44**, 23–30.
Courtois, J. E. and Ariyoshi, U. (1962b). *Bull. Soc. Chim. Biol.* **44**, 31–37.
Courtois, J. E. and Davy, J. (1965). *C. R. Hebd. Seances Acad. Sci.* **261**, 3483–3485.
Courtois, J. E. and Percheron, F. (1965). *Mem. Soc. Bot. Fr.* pp. 29–39.
Courtois, J. E. and Percheron, F. (1971) In "Chemotaxonomy of Leguminosae" (J. B. Harborne, D. Boulter and B. L. Turner, eds.), pp. 207–229. Academic Press, New York.
Courtois, J. E., Wickström, A., Fleury, P. and Le Dizet, P. (1955). *Bull. Soc. Chim. Biol.* **37**, 1009–1021.
Courtois, J. E., Archambault, A. and Le Dizet P. (1956a). *Bull. Soc. Chim. Biol.* **38**, 359–363.
Courtois, J. E., Wickström, A. W. and Le Dizet, P. (1956b). *Bull. Soc. Chim. Biol.* **38**, 863–870.

Courtois, J. E., Archambault, A. and Le Dizet, P. (1956c). *Bull. Soc. Chim. Biol.* **38**, 1117–1119.

Courtois, J. E., Le Dizet, P. and Wickström, A. (1958). *Bull Soc. Chim. Biol.* **40**, 1059–1065.

Courtois, J. E., Le Dizet, P. and Petek, F. (1959). *Bull. Soc. Chim. Biol.* **38**, 1121–1131.

Courtois, J. E., Le Dizet, P. and Davy, J. (1960a). *Bull. Soc. Chim. Biol.* **42**, 351–364.

Courtois, J. E., Dillemann, G. and Le Dizet, P. (1960b). *Ann. Pharm. Fr.* **18**, 17–28.

Courtois, J. E., Petek, F. and Douy, T. (1961). *Bull. Soc. Chim. Biol.* **43**, 1189–1196.

Crafts, A. S. and Crisp, C. E. (1971). "Phloem Transport in Plants." Freeman, San Francisco, California.

Craigie, J. S. (1974). *In* "Algal Physiology and Biochemistry" (W. D. P. Stewart, ed.), pp. 206–235. Univ. of California Press, Berkeley.

Craigie, J. S., McLachlan, J. and Tocher, R. D. (1968). *Can. J. Bot.* **46**, 605–611.

Cristofaro, E., Mottu, F. and Wuhrmann, J. J. (1974). *In* "Sugars in Nutrition" (H. L. Sipple and K. W. McNutt, eds.), pp. 313–336. Academic Press, New York.

Cronquist, A. (1968). "The Evolution and Classification of Flowering Plants," Nelson, London.

Crowden, R. K., Harborne, J. B. and Heywood, V. H. (1969). *Phytochemistry* **8**, 1963–1984.

Culberson, C. F. (1969). "Chemical and Botanical Guide to Lichen Products." Univ. of North Carolina Press, Chapel Hill.

Culberson, C. F., Culberson, W. L. and Johnson, A. (1977). "Chemical and Botanical Guide to Lichen Products," Suppl. II. Am. Biol. Lichnol Soc., St. Louis, Missouri.

Dahlgren, R. (1975). *Bot. Not.* **129**, 119–147.

Dalessandro, G. and Northcote, D. H. (1977a). *Biochem. J.* **162**, 267–279.

Dalessandro, G. and Northcote, D. H. (1977b). *Biochem. J.* **162**, 281–288.

Dalessandro, G. and Northcote, D. H. (1977c). *Planta* **134**, 39–44.

Davy, J. (1966). *Ann. Farm. Fr.* **24**, 703–709.

Davy, J. and Courtois, J. E. (1965). *C. R. Hebd. Seances Acad. Sci.* **261**, 3483–3485.

Davy, J. and Courtois, J. E. (1966). *Medd. Nor. Farm. Selsk.* **28**, 197–210; *Chem. Abstr.* **67**, 91053S (1967).

Dey, P. M. (1969). *Biochim. Biophys. Acta.* **191**, 644–652.

Dey, P. M. (1978). *Adv. Carbohydr. Chem. Biochem.* **35**, 341–376.

Dey, P. M. (1979). *Phytochemistry* **18**, 35–38.

Dey, P. M. (1980a). *FEBS Lett.* **112**, 60–62.

Dey, P. M. (1980b). *FEBS Lett* **114**, 153–156.

Dey, P. M. (1980c). *Adv. Carbohydr. Chem. Biochem.* **37**, 283–372.

Dey, P. M. (1981). *Phytochemistry* **20**, 1493–1496.

Dey, P. M. (1983). *Eur. J. Biochem.* **136**, 155–159.

Dey, P. M. (1984a). *Phytochemistry* **23**, 257–260.

Dey, P. M. (1984b). *Phytochemistry* **23**, 729–732.

Dey, P. M. (1984c). *Eur. J. Biochem.* **140**, 385–390.

Dey, P. M. and Del Campillo, E. (1984). *Adv. Enzymol.* **56**, 141–249.

Dey, P. M. and Dixon, M. (1974). *Biochim. Biophys. Acta* **370**, 269–275.

Dey, P. M. and Kauss, H. (1981). *Phytochemistry* **20**, 45–48.

Dey, P. M. and Pridham, J. B. (1969). *Biochem. J.* **115**, 47–54.

Dey, P. M. and Pridham, J. B. (1972). *Adv. Enzymol.* **36**, 91–130.

Dey, P. M., Pridham, J. B. and Sumar, N. (1982a). *Phytochemistry* **21**, 2195–2199.

Dey, P. M., Naik, S. and Pridham, J. B. (1982b). *FEBS Lett.* **150**, 233–237.

Dey, P. M., Del Campillo, E. and Pont Lezica, R. (1983). *J. Biol. Chem.* **258**, 923–929.

Dickinson, B. D., Hopper, J. E. and Davies, M. D. (1973). *In* "Biogenesis of Plant Cell Wall Polysaccharides" (F. A. Loewus, ed.), pp. 29–48. Academic Press, New York.

Dickson, R. E. and Larson, P. R. (1975). *Plant Physiol.* **56**, 185–193.

Dickson, R. E. and Nelson, E. A. (1982). *Physiol. Plant.* **54**, 393–401.

Doll, S. and Willenbrink, J. (1980). *Dev. Plant Biol.* **4**, 437–438.

Dupéron, R. (1964). *C. R. Hebd. Seances Acad. Sci.* **258**, 5960–5963.
Dutton, J. V. (1966). *Int. Sugar J.* **68**, 261–264.
El-Shimi, N. M., Luh, B. S. and Ahmed, A. E. T. (1980). *J. Food Sci.* **45**, 1652–1657.
Ernst, R. (1967). *Am. Orchid Soc. Bull.* **36**, 1068–1073.
Ernst, R., Arditti, J. and Healey, P. L. (1971). *Am. J. Bot.* **58**, 827–835.
Eskin, N. A. M., Johnson, S., Vaisey-Genser, M. and McDonald, B. E. (1980). *Can. Inst. Food Sci. Technol. J.* **13**, 40–42.
Fan, D. F. and Feingold, D. S. (1969). *Plant Physiol.* **44**, 599–604.
Fan, D. F. and Feingold, D. S. (1970). *Plant Physiol.* **46**, 592–597.
Feige, G. B. (1973). *Z. Pflanzenphysiol.* **69**, 290–292.
Feige, G. B. (1974). *Z. Pflanzenphysiol.* **72**, 272–275.
Feingold, D. S. (1982). *In* "Encyclopedia of Plant Physiology, New Series" (F. A. Loewus and W. Tanner, eds.), Vol. 13A, pp. 3–76. Springer-Verlag, Berlin and New York.
Feingold, D. S. and Avigad, G. (1980). *In* "The Biochemistry of Plants" (J. Preiss, ed.), Vol. 3, pp. 101–170. Academic Press, New York.
Ferguson, J. D. and Street, H. G. (1957). *Ann. Bot. (London)* [N. S.] **22**, 525–538.
Foglietti, M. J. and Percheron, F. (1976). *Biochimie* **58**, 499–504.
French, D. (1954). *Adv. Carbohydr. Chem.* **9**, 149–184.
French, D. (1955). *J. Am Chem. Soc.* **77**, 1024–1025.
French, D., Wild, G. M., Young, B. and W. J. (1953a). *J. Am. Chem. Soc.* **75**, 709–712.
French, D., Wild, G. M. and James, W. J. (1953b). *J. Am Chem. Soc.* **75**, 3664–3666.
French, D., Youngquist, R. W. and Lee, A. (1959). *Arch Biochem. Biophys.* **85**, 471–473.
Frydman, R. B. and Neufeld, E. F. (1963). *Biochem. Biophys. Res. Commun.* **12**, 121–125.
Gander, J. E. (1976). *In* "Plant Biochemistry" (J. Bonner and J. E. Varner, eds.), 3rd ed., pp. 337–379. Academic Press, New York.
Gaudréault, P.-R. and Webb, A. J. (1981). *Phytochemistry* **20**, 2629–2633.
Gaudréault, P.-R. and Webb, J. A. (1982). *Plant Sci. Lett.* **24**, 281–288.
Gaudréault, P.-R. and Webb, J. A. (1983). *Plant Physiol.* **71**, 662–668.
Gendraud, M. and Cloux, D. (1979). *Phytochemistry* **18**, 1631–1633.
Giaquinta, R. T. (1983). *Annu. Rev. Plant Physiol.* **34**, 347–387.
Ginsberg, V. and Neufeld, E. T. (1957). *Am. Chem. Sci. Abstr.* **132**, 27.
Ginsberg, V. Stumpf, P. K. and Hassid, W. Z. (1956). *J. Biol. Chem.* **223**, 977–982.
Gitzelmann, R. and Auricchio, S. (1965). *Pediatrics* **36**, 231–236.
Glaser, L. (1972). *In* "The Enzymes" (P. D. Boyer, ed.), 3rd ed., Vol. 6, pp. 355–380. Academic Press, New York.
Göering, H., Recklin, E. and Kaiser, R. (1968). *Flora (Jena)* **159**, 82–103.
Gohl, B., Larsson, K., Nilsson, M., Theander, O. and Thomke, S. (1977). *Cereal Chem.* **54**, 690–697.
Gomyo, T. and Nakamura, M. (1966). *Agric. Biol. Chem.* **30**, 425–427.
Gorenflot, R. and Bourdu, R. (1962). *Rev. Cytol. Biol. Veg.* **25**, 349–360.
Grisebach, H. (1980). *In* "The Biochemistry of Plants" (J. Preiss, ed.), Vol. 3, pp. 171–197. Academic Press, New York.
Gross, D. (1974). *Int. Comm. Unif. Methods Sugar Anal.* Subject 15.
Gross, K. C. (1981). Ph.D. Thesis, pp. 1–68. North Carolina State University at Raleigh.
Gross, K. C. and Pharr, D. M. (1982). *Plant Physiol.* **69**, 117–121.
Gustafson, G. L. and Gander, J. E. (1972). *J. Biol. Chem.* **274**, 1387–1397.
Handley, L. W., Pharr, D. M. and McFeeters, R. F. (1983). *Plant Physiol.* **72**, 498–502.
Hansen, P. (1967). *Physiol. Plant.* **20**, 302–391.
Hansen, R. G. and Gitzelmann, R. (1975). *In* "Physiological Effects of Food Carbohydrates" (A. Jeanes and J. E. Hodge, eds.), pp. 100–122. Am. Chem. Soc., Washington, D.C.

Haq, S. and Adams, G. A. (1962). *Can. J. Biochem. Physiol.* **40**, 987–997.
Hartt, C. E., Kortschak, H. P., Forbes, A. J. and Burr, G. O. (1963). *Plant Physiol.* **38**, 305–318.
Hassid, W. Z., Putman, E. W. and Ginsberg, V. (1956). *Biochim. Biophys. Acta* **20**, 17–22.
Hatanaka, S. (1959). *Arch. Biochem. Biophys.* **82**, 188–194.
Hattori, S. and Hatankana, S. (1958). *Bot. Mag.* **71**, 417–423.
Hattori, S. and Shiroya, T. (1951). *Arch. Biochem. Biophys.* **34**, 121–134.
Hegnauer, R. (1964). "Chemotaxonomie der Pflanzen," Vol. 3, pp. 173–184. Birkhaeuser, Basel.
Hegnauer, R. (1969). "Chemotaxonomie der Pflanzen," Vol. 5, pp. 329–330. Birkhaeuser, Basel.
Hegnauer, R. (1973). "Chemotaxonomie der Pflanzen," Vol. 6, pp. 554–629. Birkhaeuser, Basel.
Hendrix, J. E. (1968). *Plant Physiol.* **43**, 1631–1636.
Hendrix, J. E. (1973). *Plant Physiol.* **52**, 688–689.
Hendrix, J. E. (1982). *Plant Sci. Lett.* **25**, 1–7.
Hérissey, H. (1957). *Bull. Soc. Chim. Biol.* **39**, 1553–1555.
Hérissey, H., Wickström, A. and Courtois, J. E. (1951). *Bull. Soc. Chim. Biol.* **33**, 642–648.
Hérissey, H., Fleury, P., Wickström, A., Courtois, J. E. and Le Dizet, P. (1954a). *Bull. Soc. Chim. Biol.* **36**, 1507–1518.
Hérissey, H., Fleury, P., Wickström, A., Courtois, J. E. and Le Dizet, P. (1954b). *Bull. Soc. Chim. Biol.* **36**, 1519–1524.
Heywood, V. H. (1972). *Bot. J. Linn. Soc.* **64**, S–1.
Hiller, K. (1972). *Bot. J. Linn. Soc.* **64**, 369–384.
Hiller, K. and Kothe, N. (1969). *Planta Med.* **17**, 79–86.
Hoffman, F., Kull, U. and Jeremias, K. (1971). *Z. Pflanzenphysiol.* **64**, 223–231.
Hoffmann-Ostenhof, O. (1969). *Ann. N.Y. Acad. Sci.* **165**, 815–819.
Holl, F. B. and Vose, J. R. (1980). *Can. J. Plant Sci.* **60**, 1109–1114.
Hopf, H. (1973). Ph.D. Thesis, University of Munich.
Hopf, H. and Kandler, O. (1974). *Plant Physiol.* **54**, 13–14.
Hopf, H. and Kandler, O. (1976). *Biochem. Physiol. Pflanz.* **169**, 5–36.
Hopf, H. and Kandler, O. (1977). *Phytochemistry* **16**, 1715–1717.
Hopper, J. E. and Dickinson, D. H. (1972). *Arch. Biochem. Biophys.* **148**, 523–535.
Hughes, R. and Street, H. E. (1974). *Ann. Bot. (London)* [*N.S.*] **38**, 555–564.
Huller, T. L. and Smith, F. (1966). *Arch. Biochem. Biophys.* **115**, 505–509.
Imhoff, V. (1970). *C. R. Hebd. Seances, Acad. Sci.* **270**, 2441–2443.
Imhoff, V. (1973). *Hoppe-Seyler's Z. Physiol. Chem.* **354**, 1550–1554.
Impellizzeri, G., Mangiafico, S., Oriente, G., Piattelli, M., Sciuto, S., Fattorusso, E., Magno, S., Santacroce, C. and Sica, D. (1975). *Phytochemistry* **14**, 1549–1557.
Iskenderov, G. B. (1971). *Khim. Prir. Soedin.* **7**, 514–515; *Chem. Abstr.* **76** 43971d (1972).
Jeremias, K. (1958). *Planta* **52**, 195–205.
Jeremias, K. (1962). *Ber. Dtsch. Bot. Ges.* **75**, 313–332.
Johnston, J. F. W. (1843). *Philos. Mag.* [4] **23**, 14–18.
Jukes, C. and Lewis, D. H. (1974). *Phytochemistry* **13**, 1519–1521.
Kabat, E. A., MacDonald, D. L., Ballou, C. E. and Fischer, H. O. L. (1953). *J. Am. Chem. Soc.* **75**, 4507–4509.
Kandler, O. (1967). *In* "Harvesting the Sun: Photosynthesis in Plant Life" (A. S. Pietro, F. A. Greer and T. J. Army, eds.), pp. 131–152. Academic Press, New York.
Kandler, O. and Hopf. H. (1980). *In* "The Biochemistry of Plants" (J. Preiss, ed.), Vol. 3, pp. 221–270. Academic Press, New York.
Kandler, O. and Hopf. H. (1982). *In* "Encyclopedia of Plant Physiology, New Series" (F. A. Loewus and W. Tanner, eds.), Vol. 13A, pp. 348–383. Springer-Verlag, Berlin and New York.
Kandler, O., Dover, C. and Ziegler, P. (1979). *Ber Dtsch. Bot. Ges.* **92**, 225–241.

Kapitsa, O. S. and Evdonina, L. V. (1981). Genetika (Moscow) 17, 424–436.
Kato, K. Abe, M., Ishiguro, K. and Ueno, Y. (1979). Agric. Biol. Chem. 43, 293–297.
Kauss, H. (1967). Z. Pflanzenphysiol. 56, 453–465.
Kauss, H. (1968). Z. Pflanzenphysiol. 58, 428–433.
Kauss, H. (1969). Ber. Dtsch. Bot. Ges. 82, 115–125.
Kauss, H. (1973). Plant Physiol. 52, 613–615.
Kauss, H. (1974). In "Membrane Transport in Plants" (U. Zimmerman and J. Dainty, eds.), pp. 90–94. Springer-Verlag, Berlin, and New York.
Kauss, H. (1977). In "Plant Biochemistry II" 'D. H. Northcote, ed.), Vol. 13, pp. 119–140. University Park Press, Baltimore, Maryland.
Kauss, H. (1979). Prog. Phytochem. 5, 1–27.
Kauss, H. (1981). Plant Physiol. 68, 420–424.
Kauss, H. (1983). Plant Physiol. 71, 169–172.
Kauss, H. and Quader, H. (1976). Plant Physiol. 58, 295–298.
Kauss, H. and Schobert, B. (1971). FEBS Lett. 19, 131–135.
Kauss, H., Thomson, K. S., Tetour, M. and Jeblick, W. (1978). Plant Physiol. 61, 35–37.
Kauss, H,, Thomson, K. S., Thomson, M. and Jeblick, W. (1979). Plant Physiol. 63, 455–459.
King, R. W. and Zeevart, J. A. D. (1974). Plant Physiol. 53, 96–103.
Kirst, G. O. (1980). Phytochemistry 19, 1107–1110.
Kirst, G. O. and Bisson, M. A. (1979). Aust. J. Plant Physiol. 6, 539–556.
Kirst, G. O. and Bisson, M. A. (1980). Dev. Plant Biol. 4, 485–486.
Königshofer, H., Albert, R. and Kinzel, H. (1979). Z. Pflanzenphysiol. 92, 449–453.
Korytnyk, W. and Metzler, E. (1962). Nature (London) 915, 616–617.
Kremer, B. P. (1978a). Can. J. Bot. 56, 1655–1659.
Kremer, B. P. (1978b). Mar. Biol. 48, 47–55.
Kremer, B. P. (1978c). Phycologia 17, 430–434.
Kremer, B. P. (1979). Z.Pflanzenphysiol. 93, 139–147.
Kremer, B. P. and Feige, G. B. (1979). Z. Naturforsch., C: Biosci. 34C, 1209–1314.
Kremer, B. P. and Kirst, G. O. (1981). Plant Sci. Lett. 23, 349–357.
Kremer, B. P. and Vogl, R. (1975). Phytochemistry 14, 1309–1314.
Kreuzer, H. P. and Kauss, H. (1980). Planta 147, 435–438.
Kuhn, R. and Löw, I. (1949). Chem. Ber. 82, 479–481.
Kylin, H. (1918). Hoppe-Seyler's Z. Physiol. Chem. 101, 236–247.
Labaneiah, M. E. O. and Luh, B. S. (1981). Cereal Chem. 58, 135–138.
Laidlaw, R. A. and Wylam, C. B. (1953). J. Chem. Soc. pp. 567–571.
Larcher, W., Heber, U. and Santarius (1973). In "Temperature and Life" (H. J. Precht, H. Christopherson, H. Hensel and W. Larcher, eds.), pp. 195–292. Springer-Verlag, Berlin and New York.
Larkins, B. A. (1981). In "The Biochemistry of Plants" (A. Marcus, ed.)., Vol. 6, pp. 449–489. Academic Press, New York.
Lee, L., Kimusa, A. and Tochikura, T. (1978). Biochim. Biophys. Acta 527, 301–304.
Lehle, L. and Tanner, W. (1973a). In "Methods in Enzymology" (V. Ginsburg, ed.), Vol. 28, pp. 522–530.
Lehle, L. and Tanner, W. (1973b). Eur. J. Biochem. 38, 103–110.
Lehle, L., Tanner, W. and Kandler, O. (1970). Hoppe-Seyler's Z. Physiol. Chem. 351, 1494–1498.
Leigh, R. A., apRees, T., Fuller, W. A. and Banfield, J. (1979). Biochem. J. 178, 539–547.
Levitt, J. (1972). In "Responses of Plants to Environmental Stresses". Academic Press, New York.
Lindberg, B. (1955a). Acta Chem. Scand. 9, 1093–1096.
Lindberg, B. (1955b). Acta Chem. Scand. 9, 1097–1099.
Linden, J. C. (1982). Enzyme Microb. Technol. 4, 130–136.

Lineberger, R. D. and Steponkus, P. L. (1980). *Plant Physiol.* **65**, 298–304.
Lobelle-Rich, P. and Reeves, R. G. (1982). *In* "*Methods in Enzymology*" (W. A. Wood. ed.), Vol. 90, pp. 552–555. Academic Press, New York.
Loewus, F. A. and Dickinson, D. B. (1982). *In* "Encyclopedia of Plant Physiology, New Series" (F. A. Loewus and W. Tanner, eds.), Vol. 13A, pp. 193–216. Springer-Verlag, Berlin and New York.
Loewus, F. A. and Jang, R. (1958). *J. Biol. Chem.* **232**, 505–519.
Loewus, F. A. and Loewus, M. W. (1980). *In* "The Biochemistry of Plants" (J. Preiss ed.), Vol. 3, pp. 43–76. Academic Press, New York.
Loewus, F. A., Jang, R. and Seegmiller, C. G. (1958). *J. Biol. Chem.* **232**, 533–541.
Lohmar R. L., Jr. (1957). *In* "The Carbohydrates" (W. Pigman, ed.), pp. 241–298. Academic Press, New York.
MacLeod, A. M. and McCorquodale, H. (1958a). *New Phytol.* **57**, 168–182.
MacLeod, A. M. and McCorquodale, H. (1958b). *Nature (London)* **182**, 815–817.
Madore, M. and Webb, J. A. (1981). *Can. J. Bot.* **59**, 2550–2557.
Madore, M. and Webb, J. A. (1982). *Can. J. Bot.* **60**, 126–130.
Majak, W., Craigie, J. S. and McLachlan, J. (1966). *Can. J. Bot.* **44**, 541–549.
Malca, I., Endo, R. M. and Long, M. R. (1967). *Phytopathology* **57**, 272–278.
Maretzki, A. and Thom, M. (1978). *Plant Physiol.* **61**, 544–548.
Matile, P. (1975). "The Lytic Compartment of Plant Cells," Vol. 1. Springer-Verlag, Berlin and New York.
Matile, P. (1978). *Annu. Rev. Plant Physiol.* **29**, 193–213.
Mayes, J. S., Miller, L. R. and Myers, F. K. (1970). *Biochemi. Biophys. Res. Commun.* **39**, 661–665.
Metcalf, T. N., III, Wang, J. L., Schubert, K. R. and Schneider, M. (1983). *Biochemistry* **22**, 3969–3975.
Miller, L. P. (1973). *In* "Phytochemistry" (L. P. Miller, ed.), Vol. 1, pp. 145–175. Van Nostrand-Reinhold, Princeton, New Jersey.
Moreno, A. and Cardini, C. E. (1966). *Plant Physiol.* **41**, 909–910.
Morgenlie, S. (1970). *Acta Chem. Scand.* **24**, 2149–2155.
Murakami, S. (1940). *Acta Phytochim.* **11**, 213–229.
Murakami, S. (1942). *Acta Phytochim.* **13**, 37–56.
Murakami, S. (1943). *Acta Phytochim.* **14**, 161–184.
Nagashima, H. (1976). *Bull. Jpn. Soc. Phycol.* **24**, 103–110.
Nagashima, H. and Fukuda, I. (1981). *Phytochemistry* **20**, 439–442.
Nagashima, H., Nakamura, S. and Nisizawa, K. (1969). *Shokubutsugaku Zasshi* **82**, 379–386; *Chem. Abstr.* **72**, 97498r (1970).
Nath, R. L. and Rydon, H. N. (1954). *Biochem. J.* **57**, 1–10.
Neufeld, E. F. and Ashwell, G. (1980). *In* "The Biochemistry of Glycoproteins and Proteoglycans" (W. J. Lennarz, ed.), pp. 241–266. Plenum, New York.
Neufeld, E. F., Ginsburg, V., Putman, E. W., Fanshier, D. and Hassid, W. Z. (1957). *Arch. Biochem. Biophys.* **69**, 602–616.
Nicholson, J. F. and Weidner, T. M. (1972). *Trans. Ill. State Acad. Sci.* **65**, 3–6.
Nishimura, M. and Beevers, H. (1978). *Plant Physiol.* **62**, 44–48.
Oesch, F. (1969). *Planta* **86**, 360–380.
O'Kelley, J. C. (1955). *Am. J. Bot.* **42**, 322–327.
Oliver, I. T. (1961). *Biochim. Biophys. Acta* **52**, 75–81.
Olson, A. C., Becker, R., Miers, J. C., Gumbmann, M. R. and Wagner, J. R. (1975). *In* "Protein Nutritional Quality of Foods and Feeds" (M. Friedman, ed.), Part 2, pp. 551–563. Dekker, New York.
Onuki, M. (1932). *Nippon Nogei Kagaku Kaishi* **8**, 445–462; *Chem. Abstr.* **26** 4308 (1932).
Onuki, M. (1933). *Sci. Pap. Inst. Phys. Chem. Res. (Jpn.)* **20**, 201–244.

Ordin, L. and Bonner, J. (1957). *Plant Physiol.* **32**, 212–215.
Ovcharov, K. G. and Koshelev, Y. P. (1974). *Fiziol. Rast.* **21**, 969–974; *Chem. Abstr.* **82**, 14036W (1975).
Paquin, R. (1958). M.Sc. Thesis, University of Montreal.
Pazur, J. H. and Shadaksharaswamy, M. (1961). *Biochem. Biophys. Res. Commun.* **5**, 130–134.
Pazur, J. H., Shadaksharaswamy, M. and Meidell, G. E. (1962). *Arch. Biochem. Biophys.* **99**, 78–85.
Peat, S. and Rees, D. A. (1961). *Biochem. J.* **79**, 7–12.
Peel, A. J. (1975). *In* "Encyclopedia of Plant Physiology, New Series" (M. H. Zimmermann and J. A. Milburn, eds.), Vol. 1, pp. 171–196. Springer-Verlag, Berlin and New York.
Petek, F., Villarroya, G. and Courtois, J. E. (1966). *C. R. Hebd. Seances Acad. Sci.* **263**, 195–197.
Pharr, D. M., Sox, H. N., Locy, R. D. and Huber, S. C. (1981). *Plant Sci. Lett.* **23**, 25–33.
Plant, A. R. and Moore, K. G. (1982). *Phytochemistry* **21**, 985–989.
Pressey, R. and Shaw, R. (1969). *Eur. Potato J.* **12**, 64–66.
Pridham, J. B. (1960). *Biochem. J.* **76**, 13–17.
Pridham, J. B. and Dey, P. M. (1974). *In* "Plant Carbohydrate Biochemistry" (J. B. Pridham, ed.), pp. 83–96. Academic Press, New York.
Pridham, J. B. and Hassid, W. Z. (1965). *Plant Physiol.* **40**, 984–986.
Pristupa, N. A. (1959). *Fiziol. Rast. (Engl. Transl.)* **6**, 26–32.
Putman, E. W. and Hassid, W. Z. (1954). *J. Am. Chem. Soc.* **76**, 2221–2223.
Quednow, K. G. (1930). *Bot. Arch.* **30**, 51–108.
Rackis, J. J. (1975). *In* "Physiological Effects of Food Carbohydrates" (A. Jeanes and J. E. Hodge, eds.), pp. 207–222. Am. Chem. Soc., Washington, D.C.
Rathbone, G. B. (1980). *Dev. Food Carbohydr.* **2**, 145–185.
Reed, R. H., Collins, J. C. and Russel, G. (1980). *J. Exp. Bot.* **31**, 1539–1554.
Roberts, R. M. and Butt, V. S. (1969). *Planta* **84**, 250–262.
Roberts, R. M., Heiseman, A. and Wicklin, A. (1971). *Plant Physiol.* **48**, 36–42.
Ruttloff, H., Taeufel, A., Krause, W. G., Haenel, H. and Taeuful, K. (1967). *Nahrung* **11**, 39–46.
Santarius, K. A. and Milde, H. (1977). *Planta* **136**, 163–166.
Saunders, R. M. (1971). *Phytochemistry* **10**, 491–493.
Saunders, R. M. and Walker, H. B., Jr. (1969). *Cereal Chem.* **47**, 85–92.
Saunders, R. M., Betschart, A. A. and Lorenz, K. (1975). *Cereal Chem.* **52**, 472–478.
Schiweck, H. (1978). *Int. Comm. Unif. Methods Sugar Anal.* Subject 15.
Schiweck, H. (1982). *Int. Comm. Unif. Methods Sugar Anal.* Subject 15.
Schiweck, H. and Buesching, L. (1969). *Zucker* **22**, 377–384.
Schmitz, K. (1970). *Planta* **91**, 96–110.
Schobert, B., Untner, E. and Kauss, H. (1972). *Z. Pflanzenphysiol.* **67**, 385–398.
Schwarz, V., Goldberg, L., Komrower, G. M. and Holzel, A. (1956). *Biochem. J.* **62**, 34–40.
Schwarzmaier, G. (1973). Ph.D. Thesis, University of Munich.
Schweizer, T. F., Horman, I. and Wünsch, P. (1978). *J. Sci. Food Agric.* **28**, 148–154.
Sellmair, J. and Kandler, O. (1970). *Z. Pflanzenphysiol.* **63**, 65–83.
Sellmair, J., Beck, E., and Kandler, O. (1969). *Z. Pflanzenphysiol.* **61**, 338–342.
Sellmair, J., Beck, E., Kandler, O. and Dress, A. (1977). *Phytochemistry* **16**, 1201–1204.
Senser, M. and Kandler, O. (1966). *Ber. Dtsch. Bot. Ges.* **79**, 210.
Senser, M. and Kandler, O. (1967a). *Phytochemistry* **6**, 1533–1540.
Senser, M. and Kandler, O. (1967b). *Z. Pflanzenphysiol.* **57**, 376–388.
Serro, F. R. and Brown, R. J. (1954) *Anal. Chem.* **26**, 890–892.
Seybold, S. (1969). *Flora (Jena), Abt. A* **160**, 561–575.
Shannon, L. M. and Hankins, C. N. (1981). *In* "The Phytochemistry of Cell Recognition and Cell Surface Interactions" (F. A. Loewus and C. A. Ryan, eds.), pp. 93–114, Plenum, New York.
Shibuya, I. (1961). *Diss. Abstr.* **21**, 2463–2464.

Shiroya, T. (1963). *Phytochemistry* **2**, 33–46.

Sidbury, J. B., Jr. (1969). *In* "Galactosemia" (D. Y. Hsia, ed.), p. 13. Thomas, Springfield, Illinois.

Siddiqui, I. R., Wood, P. J. and Khanzada, G. (1973). *Carbohydr. Res.* **29**, 255–258.

Silva, H. C. and Luh, B. S. (1979). *Can. Inst. Food Sci. Technol. J.* **12**, 103–107.

Sioufi, A., Percheron, F. and Courtois, J. E. (1970). *Phytochemistry* **9**, 991–999.

Smart, E. L. and Pharr, D. M. (1981). *Planta* **153**, 370–375.

Sömme, R. and Wickström, A. (1965). *Acta Chem. Scand.* **19**, 537–540.

Sosulski, F. W., Elkowicz, L. and Reichert, R. D. (1982). *J. Food Sci.* **47**, 498–502.

Spang, B., Claude, F. and Kauss, H. (1981). *Plant Physiol.* **67**, 190–191.

Stacey, B. E. (1974). *In* "Plant Carbohydrate Biochemistry" (J. B. Pridham, ed.), pp. 47–59. Academic Press, New York.

Stanek, J., Cerny, M., Pacak, J., and Mayer, K. (1965). "The Oligosaccharides." Academic Press, New York.

Su, C. and Hassid (1962). *Biochemistry* **1**, 468–474.

Takhtajan, A. (1959). "Die Evolution der Angiospermen." Fischer, Jena.

Tanner, W. (1969a). *Ann. N.Y. Acad. Sci.* **165**, 726–742.

Tanner, W. (1969b). *Allg. Prakt. Chem.* **20**, 152.

Tanner, W. and Kandler, O. (1966). *Plant Physiol.* **41**, 1540–1542.

Tanner, W. and Kandler, O. (1968). *Eur. J. Biochem.* **4**, 233–239.

Tanner, W., Lehle, H. and Kandler, O. (1967). *Biochem. Biophys. Res. Commun.* **29**, 166–171.

Tharanathan, R. N., Wankhede, D. B. and Rao, R. R. (1976). *J. Food Res.* **41**, 715–716.

Theander, O. and Aman, P. (1976). *Swed. J. Agric. Res.* **6**, 81–85.

Thomas, B. and Webb, J. A. (1977). *Phytochemistry* **16**, 203–206.

Trip, P., Krotkov. G. and Nelson, C. D. (1963). *Can. J. Bot.* **41**, 1005–1010.

Trip, P., Nelson, C. D. and Krotkov, G. (1965). *Plant Physiol.* **40**, 740–747.

Turgeon, R. and Webb, J. A. (1975). *Planta* **123**, 53–62.

Turnquist, R. L. and Hanson, R. G. (1976). *In* "The Enzymes" (P.D. Boyer, ed.), 3rd ed.) Vol. 13, pp. 51–71. Academic Press, New York.

Turnquist, R. L., Turnquist, M. M., Bachmann, R. C. and Hanson, R. G. (1974). *Biochim. Biophys. Acta* **364**, 59–67.

Van der Wilden, W. and Chrispeels, M. J. (1983). *Plant Physiol.* **71**, 82–87.

Van der Wilden, W., Herman, E. M. and Chrispeels, M. J. (1980). *Proc. Natl. Acad. Sci. U.S.A.* **77**, 428–432.

Veluthambi, K., Mahadevan, S. and Maheshwari, R. (1981). *Plant Physiol.* **68**, 1369–1374.

Veluthambi, K., Mahadevan, S. and Maheshwari, R. (1982a). *Plant Physiol.* **69**, 1247–1251.

Veluthambi, K., Mahadevan, S. and Maheshwari, R. (1982b). *Plant Physiol.* **70**, 686–688.

Venkataraman, R. and Reithel, F. J. (1958). *Arch. Biochem. Biophys.* **75**, 443–452.

Vincent, C. and Delachanal, H. (1889). *C.R. Hebd. Seances Acad. Sci.* **109**, 676–679.

Von Planta, A. and Schulz, E. (1890). *Ber. Dtsch. Chem. Ges.* **23**, 1692–1699.

Wahab, A. H. and Burris, J. S. (1975). *Iowa State J. Res.* **50**, 29–45.

Walker, H. G., Jr., Ricci, B. A. and Goodwin, J. C. (1965). *J. Am. Soc. Sugar Beet Technol.* **13**, 503–508.

Walker, T. S. and Thaine, R. (1971). *Ann. Bot. (London)* [N.S.] **35**, 773–790.

Walter, M. W. (1963). *Ph.D. Thesis*, University of London.

Wankhede, D. B., and Tharanathan, R. N. (1976). *J. Agric. Food Chem.* **24**, 655–659.

Wattiez, N. and Hans, M. (1943). *Bull. Acad. R. Med. Belg.* **8**, 386–396.

Webb, J. A. (1970). *Can. J. Bot.* **48**, 935–942.

Webb, J. A. (1971). *Can. J. Bot.* **49**, 717–733.

Webb, J. A. (1973). *Plant Physiol.* **51**, Suppl. 12.

Webb, J. A. (1982). *Can. J. Bot.* **60**, 1054–1059.

Webb, J. A. and Burley, J. W. A. (1962). *Science* **137**, 776.

Webb, J. A. and Burley, J. W. A. (1964). *Plant Physiol.* **39**, 973–977.

Webb, J. A. and Gorham, P. R. (1964). *Plant Physiol.* **39**, 663–672.

Webb, J. A. and Gorham, P. R. (1965). *Can. J. Bot.* **43**, 97–103.

Weidner, T. M. (1964). Ph.D. Thesis, Ohio State University, Columbus.

White, L. M. and Secor, G. E. (1953). *Arch. Biochem. Biophys.* **44**, 244–245.

Whittingham, C. P., Keys, A. J. and Bird, I. F. (1979). *In* "Encyclopedia of plant Physiology, New Series" (M. Gibbs and E. Latzko, eds.), Vol. 6, pp. 313–326. Springer-Verlag, Berlin and New York.

Wickberg, B. (1958). *Acta Chem. Scand.* **12**, 1187–1201.

Wickström, A. and Boerheim-Svendsen, A. (1956). *Acta Chem. Scand.* **10**, 1199–1207.

Wickström, A., Courtois, J. E., Le Dizet, P. and Archambault, A. (1958a). *Bull. Soc. Chim. Fr.* pp. 1410–1415.

Wickström, A., Courtois, J. E., Le Dizet, P. and Archambault, A. (1958b). *C.R. Hebd. Seances Acad. Sci.* **246**, 1624–1626.

Wickström, A., Courtois, J. E., Le Dizet, P. and Archambault, A. (1959). *Bull. Soc. Chim. Fr.* pp. 871–878.

Wiencke, C. and Laeuchli, A. (1981). *Z. Pflanzenphysiol.* **103**, 247–258.

Willenbrink, J. and Doll, S. (1979). *Planta* **147**, 159–162.

Wynd, F. L. (1933). *Ann. Mo. Bot. Gard.* **20**, 569–581.

Yasui, T. (1980). *Agric. Biol. Chem.* **44**, 2253–2255.

Yasui, T. and Ohno, S. (1982). *Nippon Nogei Kagaku Kaishi* **56**, 1053–1056.

Ziegler, H. (1975). *In* "Encyclopedia of Plant Physiology. New Series" (M. H. Zimmermann and J. A. Milburn, eds.), Vol. 1, pp. 59–100. Springer-Verlag, Berlin and New York.

Zimmermann, M. H. (1957). *Plant Physiol.* **32**, 288–291.

Zimmermann, M. H. (1958). *In* "The Physiology of Forest Trees" (K. V. Thimann, ed.), pp. 381–400. Ronald Press, New York.

Zimmermann, M. H. and Ziegler, H. (1975). *In* "Encyclopedia of Plant Physiology, New Series" (M. H. Zimmermann and J. A. Milburn, eds.), Vol. 1, pp. 480–505. Springer-Verlag, Berlin and New York.

Chapter 3

Glycosides

P. M. DEY[*] and R. A. DIXON

Department of Biochemistry
Royal Holloway College, (University of London)
Egham Hill, Egham
Surrey, England

I. GENERAL

A covalent linkage formed between the hemiacetal hydroxyl group of a reducing sugar and the OH group of a hydroxyl compound, with the elimination of a water molecule, is called an *O*-glycosidic bond. The sugar moiety (termed the glycone residue) may exist in the pyranose or the furanose ring form, and the glycosidic bond with the hydroxyl compound (termed the aglycone residue) may exist in the α or β configuration. If the linkage of the reducing sugar is with the SH group of a thiol compound, the product is termed an *S*-glycoside. Similarly, the formation of *N*-glycosides involves linkage with amino groups. *C*-Glycosides, such as *C*-glycosyl flavonoids, also occur in plants; these compounds are resistant to acid hydrolysis under conditions that bring about hydrolysis of other glycoside classes.

Although several different sugars are known to constitute the glycone moiety of the glycoside classes just described, glucose seems to be the predominant sugar. Other sugar moieties are apiose, arabinose, fructose, fucose, galactose, glucosamine, glucuronic acid, mannose, rhamnose, ribose, sorbose, and xylose (Pridham, 1965; Barz and Köster, 1981). Some of the more complex glycones are shown in Table I. The structures of these may give an indication of the nature of the glycosidases involved in their degradation. The

[*] To whom all enquiries should be addressed.

Table I

Oligosaccharide Components of Some Phenolic Glycosides[a]

Trivial name	Structure
Disaccharides	
Sophorose	2-*O*-β-D-Glucosyl-D-glucose
Laminaribiose	3-*O*-β-D-Glucosyl-D-glucose
Maltose	4-*O*-α-D-Glucosyl-D-glucose
Gentiobiose	6-*O*-β-D-Glucosyl-D-glucose
Unnamed	*O*-Galactosylgalactose
Unnamed	2-*O*-β-D-Glucosyl-D-mannose (acetylated)
Unnamed	*O*-Glucosylrhamnose
Neohesperidose	2-*O*-α-L-Rhamnosyl-D-glucose
Rungiose	3-*O*-α-L-Rhamnosyl-D-glucose
Rutinose	6-*O*-α-L-Rhamnosyl-D-glucose
Robinobiose	6-*O*-α-L-Rhamnosyl-D-galactose
Unnamed	4-*O*-α-L-Rhamnosyl-D-xylose
Unnamed	*O*-Rhamnosylrhamnose
Sambubiose	2-*O*-β-D-Xylosyl-D-glucose
Lathyrose	2-*O*-β-D-Xylosyl-D-galactose
Primverose	6-*O*-β-D-Xylosyl-D-glucose
Vicianose	6-*O*-α-L-Arabinosyl-D-glucose
Unnamed	2-*O*-Apiofuranosyl-D-glucose
Unnamed	6-*O*-Apiofuranosyl-D-glucose
Trisaccharides	
Linear:	
Gentiotriose	*O*-β-D-Glucosyl-(1→6)-*O*-β-D-glucosyl-(1→6)-D-glucose
Soroborose	*O*-β-D-Glucosyl-(1→6)-*O*-α-D-glucosyl-(1→4)-D-glucose
Sophorotriose	*O*-β-D-Glucosyl-(1→2)-*O*-β-D-glucosyl-(1→2)-D-glucose
Unnamed	*O*-β-D-Glucosyl-(1→2)-*O*-α-L-rhamnosyl-(1→6)-D-glucose
Rhamninose	*O*-α-L-Rhamnosyl-(1→4)-*O*-α-L-rhamnosyl-(1→6)-D-galactose
Unnamed	*O*-α-L-Rhamnosyl-(1→3)-*O*-α-L-rhamnosyl-(1→6)-D-galactose
Branched:	
2[G]-Glucosylrutinose	*O*-β-D-Glucosyl-(1→2)-*O*-[α-L-rhamnosyl-(1→6)]-β-D-glucose
3[G]-Glucosylneohesperidose	*O*-β-D-Glucosyl-(1→3)-*O*-[α-L-rhamnosyl-(1→2)]-β-D-glucose
Unnamed	*O*-L-Rhamnosyl-(1→4)-*O*-[D-glucosyl-(1→6)]-D-glucose
Unnamed	*O*-β-D-Xylosyl-(1→2)-*O*-[α-L-rhamnosyl-(1→6)]-β-D-glucose

[a] From Pridham (1965) and Barz and Köster (1981).

main physiological functions of the conjugating sugar moieties include solubilisation, detoxification, and storage of their respective aglycones. However, the fate and the mode of utilization of the sugar moieties upon degradation of the conjugates are largely unexplored; some glycones are presumably channeled into catabolic routes.

The possibility of a reserve function for glycosides is largely speculative; however, data presented in Table II show accumulation of considerable amounts of certain glycosides in different plant species. Furthermore, experimental evidence that has appeared in the literature from time to time may enable one to propose a possible reserve role for some glycosides. For example, some flavonoid and alkaloid glycosides have been shown to be utilised in those plant organs that display rapid growth (Robinson, 1974; Barz, 1975; Barz and Hösel, 1975). Seasonal variation of the levels of these compounds has also been recorded (Ellis, 1974; Barz and Hösel, 1975; Waller and Nowacki, 1978). Other notable examples of seasonal variation include that of salicin in willow and arbutin in *Viccinium* leaves (Thieme, 1965b). Diurnal variations also occur in the levels of some glycosides (Barz and Köster, 1981).

Metabolic studies indicate that some complex phenols and their glycosides can undergo turnover and degradation in plants (Barz and Köster, 1981). The sites of accumulation and degradation of the glycosides seem to be different; for example, dhurrin (a cyanogenic glycoside) is located in the leaf epidermal layer in *Sorghum*, while the specific β-glucosidase for its degradation is found in mesophyll tissue (Kojima *et al.*, 1979). Many plant secondary products are stored in vacuoles; thus, they must be translocated to other sites for catabolic processes to occur (see Macleod and Pridham, 1966; Robinson, 1974; Barz and Hösel, 1975). In this connection it is important to note that specific glucosidases are not likely to coexist with the glycosides in vacuoles (Hösel, 1981).

Phenolic glucosides may undergo degradation by the catabolic routes shown in Fig. 1. Three routes of utilization seem to be possible: (a) cleavage of the glycosidic linkage by the action of either specific or nonspecific glycosidases (Hösel and Conn, 1982; Dey and Del Campillo, 1984), liberating glucose for further utilization; (b) phosphorylation of the primary alcoholic group of the glycone moiety followed by hydrolysis (Pridham and Saltmarsh, 1962; Nanayakkara *et al.*, 1970); (c) transfer of the glycosyl group to a

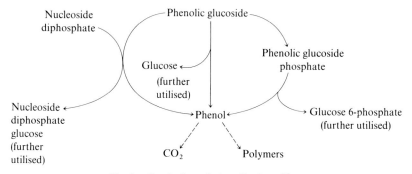

Fig. 1. Catabolism of phenolic glucosides.

TABLE II
Distribution of Some Glycosides in Plants

Glycoside	Plant	Amount (% dry weight)	Reference
Cyanogenic glycosides			
Amygdalin	*Prunus communis* (bitter almond)	1.8	Webster (1913)
	P. persica (bark of peach)	2.0	Rabaté (1933)
Gynocardin	*Gynocardia odorata* (seed)	5.0	Rosenthaler (1932)
Prulaurasin	*P. laurocerasus* (leaf)	1.4	Rosenthaler (1932)
Glucosinolates			
Benzylglucosinolate	*Carica papaya* (fruit latex)	7.3–11.6	Tang (1973)
	Lepidium sativum (leaf)	0.1[a]	Gil and MacLeod (1980)
Glucobrassicin	*Isatis tinctoria* (seeds)	0.23[a]	Elliot and Stowe (1971)
Total glucosinolates	*Brassica oleracea*	Maximum of 0.13[a] (depending on variety)	Van Etten *et al.*, (1979)
1-Methoxy-3-indolylmethyl glucosinolate	*Brassica oleracea* (Kohlrabi seedlings)	0.01[a]	Goetz and Schraudolf (1983)
Lactone glycosides			
Aesculin	*Bursaria spinosa* (leaf)	5.0	Dick (1943)
Cichoriin	*Cichorium intibus* (flower)	4.5	Merz (1932)

Compound	Plant	Amount[a]	Reference
Daphnin	*Daphne mezereum* (leaf)	22.0	Asai (1930)
Vellein	*Velleia discophora* (stem)	5.0	Bottomley and White (1951)
Phenolic glycosides			
Arbutin	*Pyrus communis* (leaf, seed)	5.0	Miller (1973)
Phloridzin	*Saxifraga crassifolia*	18.0	Miller (1973)
	Malus spp. (root)	12.0	Williams (1966)
Salicin	Members of Salicaceae (various organs)	5.0	Iwamoto et al., (1945); Thieme (1965a)

[a] Percentage of fresh weight.

nucleoside diphosphate, yielding a high-energy glycosyl donor (see Sutter and Grisebach, 1975).

The distribution of various glycosides with reference to plant systematics has been described by Seigler (1981). In the following sections, discussion is limited to the metabolism of only the glycone moieties of glycosides. For a full account of the biochemistry of these secondary products, the reader is directed to Volume 7 of *The Biochemistry of Plants: A Comprehensive Treatise* (P. K. Stumpf and E. E. Conn, Eds., Academic Press, New York, 1981).

II. CYANOGENIC GLYCOSIDES

Glycosides producing HCN on hydrolysis have been known for many years to exist in plants. The occurrence of these cyanogenic glycosides has been documented in over 2000 plant species belonging to 110 families (Gibbs, 1974; Hegnauer, 1977). Their general structure is represented as follows,

Glycone   Aglycone

where the glycone moiety is an O-β-D-glucosyl group in most cases and R^1 and R^2 of the aglycone may be H, aliphatic, or aromatic substituents (in some, a cyclopentenyl ring is present) (Conn, 1981; Nahrstedt, 1982; Seigler, 1982). Some cyanogenic glycosides may contain an additional glucose unit linked by an O-β-$(1\rightarrow6)$ bond to the existing glycone moiety, as in amygdalin, vicianin, lucumin, linustatin, and neolinustatin.

Glycosylation of the α-hydroxynitrile is the last step in the biosynthesis of cyanogenic glycosides. UDP-Glucose is the glycosyl donor, and the reaction is catalysed by a transferase.

α-Hydroxylnitrile   Cyanogenic glucoside

An enzyme preparation from the seedlings of flax was shown to synthesise linamarin ($R^1 = R^2 = CH_3$, a glycoside of the cyanohydrin of acetone) (Hahlbrock and Conn, 1970). A glycoside of the cyanohydrin of 2-butanone

[(*R*)-lotaustralin] is also found in flax. The same glycosyltransferase seems to be responsible for the synthesis of both compounds. The enzyme is specific for the aliphatic side chains of acetone and 2-butanone, the latter being somewhat preferred. On the other hand, a transferase from the seedlings of *Sorghum bicolor* requires the presence of an aromatic group in the acceptor molecule (Reay and Conn, 1974). This enzyme exhibited strict stereospecificity and glycosylated *p*-hydroxy-(*S*)-mandelonitrile from a racemic mixture (*R*,*S*), forming (*S*)-dhurrin. Similarly, a glycosyltransferase from *Triglochin maritima* formed the corresponding enantiomer, (*R*)-taxiphyllin, from *p*-hydroxy-(*R*)-mandelonitrile (Hösel and Nahrstedt, 1980). The major site of dhurrin accumulation in *Sorghum* seedlings is the vacuole (Saunders and Conn, 1978; see also Matile, 1976, 1978; Leigh and Barton, 1976; Sasse *et al.*, 1979; Saunders, 1979).

D-Glucose is liberated from cyanogenic glycosides by the action of β-glucosidase; the aglycone further decomposes to produce HCN:

$$\begin{array}{c} R^1 \\ \diagdown \\ C \\ \diagup \diagdown \\ R^2 C{\equiv}N \end{array} \begin{array}{c} O\text{-}\beta\text{-}\text{D-glucose} \end{array} \xrightarrow[\beta\text{-Glucosidase}]{} \begin{array}{c} R^1 \\ \diagdown \\ C \\ \diagup \diagdown \\ R^2 C{\equiv}N \end{array} \begin{array}{c} OH \end{array} + \text{Glucose}$$

α-hydroxynitrile lyase ↓

$$\begin{array}{c} R^1 \\ \diagdown \\ C{=}O + \text{HCN} \\ \diagup \\ R^2 \end{array}$$

Bitter almond β-glucosidase has been studied intensively since its detection nearly 150 years ago. This enzyme hydrolyses the endogenous cyanogenic glucosides amygdalin and prunasin. Two specific forms, A and B, were resolved, and kinetic studies indicated their participation in the following sequence of reactions (Haisman and Knight, 1967; Haisman *et al.*, 1967; Lalégérie, 1974):

$$\begin{array}{c} H \\ \diagdown \\ C \\ \diagup \diagdown \\ C{\equiv}N \end{array} \begin{array}{c} O\text{-}\beta\text{-}\text{D-Glucose-}(1{\rightarrow}6)\text{-}O\text{-}\beta\text{-}\text{D-Glucose} \end{array} \xrightarrow[\beta\text{-Glucosidase A}]{} \begin{array}{c} H \\ \diagdown \\ C \\ \diagup \diagdown \\ C{\equiv}N \end{array} \begin{array}{c} O\text{-}\beta\text{-}\text{D-Glucose} \end{array}$$

Amygdalin   Prunasin

β-Glucosidase B ↓

$$\begin{array}{c} H \\ \diagdown \\ C \\ \diagup \diagdown \\ C{\equiv}N \end{array} \begin{array}{c} OH \end{array} + \text{Glucose}$$

Some β-glucosidases are not only specific for the glucose moiety of glycosides and its anomeric configuration, but also display much specificity for the aglycone residue (Hösel and Conn, 1982; Dey and Del Campillo, 1984). β-Glucosidases from plants that contain cyanogenic glucosides appear to be specific for these particular substrates (Butler et al., 1965). Thus, β-glucosidases highly specific for dhurrin and triglochinin have been reported (see Table III). These glucosidases display very little activity towards the synthetic substrate p-nitrophenyl β-D-glucoside (generally used for routine β-glucosidase assay). However, this substrate is utilised by the β-glucosidase of almond in preference to the endogenous glucosides.

The well-acknowledged physiological role of cyanogenic glycosides in plants is mainly as a deterrent to herbivores. This is attributed to their ability to produce toxic HCN when macerated or crushed. How far the seed cyanogenic glycosides may contribute as carbohydrate reserves is not known.

III. PHENOLIC AND FLAVONOID GLYCOSIDES

Plant phenolics and flavonoids have been the subject of several previous reviews (e.g., Harborne, 1964; Pridham, 1965; Courtois and Percheron, 1970; Miller, 1973; Hahlbrock, 1981). The occurrence of their glycosides is very widespread in higher plants. Discussion in this section is limited only to aspects of metabolism concerning the glycone moieties.

UDP-D-Glucose is generally regarded as the glucosyl donor for the formation of β-glucosides of most phenolic compounds. Thus the synthesis of arbutin (4-hydroxyphenyl β-D-glucoside) was achieved by using wheat germ glucosyltransferase (Cardini and Leloir, 1957). However, another transferase was found in this source that transferred a second β-glucosyl group to the primary alcoholic group of the existing glycone moiety yielding 4-hydroxy-phenyl β-D-gentiobioside (Yamaha and Cardini, 1960a,b). ADP-D-Glucose

Table III

Relative Activities of β-Glucosidases with Respect to Some Cyanogenic β-Glucosides

Source of β-glucosidase (references)	PNPG[a]	Dhurrin	Taxiphyllin	Linamarin	Triglochinin
Sorghum bicolor (Eklund, 1981)	100	25,000	25	10	625
Triglochin maritima (Nahrstedt et al., 1979)	100	91	—	—	3900
Alocasia macrorrhiza (Hösel and Nahrstedt, 1975)	100	14	1	1	1000

[a] p-Nitrophenyl β-D-glucoside.

served as a better glucosyl donor in the second glycosylation (Trivelloni *et al.*, 1962). TDP-D-Glucose was found to be the glycosyl donor for the formation of the immediate precursor of rutin [O-α-L-rhamnosyl-(1→6)-β-D-glucosyl-(1→3)-quercetin], and TDP-L-rhamnose served as the final glycosyl donor (Jacobelli *et al.*, 1958). UDP-D-Glucuronate was a glycosyl donor for the glucuronosylation of quercetin (Marsh, 1960). Glucose esters of aromatic acids are also formed with UDP-D-glucose as the glycosyl donor (Corner and Swain, 1965; Macheix, 1977; Schlepphorst and Barz, 1979).

In the glycosylation of flavonoid derived compounds (e.g. chalcones, flavones, flavonols, isoflavones, and anthocyanins), glycosyltransferases show much specificity for the position of the phenolic hydroxyl groups (Bajaj *et al.*, 1983). There also exist specific transferases that glycosylate the hydroxyl groups of sugars of flavonoid glycosides; UDP-activated sugar derivatives of apiose, arabinose, glucose, glucuronic acid, rhamnose, and xylose are generally known to act as glycosyl donors (Hösel, 1981). It seems that glycosylation is the last step in the biosynthesis of flavonoid glycosides (see Sutter and Grisebach, 1973; Poulton and Kauer, 1977).

It is difficult to say with certainty whether the glycosyl acceptor specificity of the transferases is restricted by the type of flavonoid (flavonol, flavone, anthocyanin, etc.) on which these enzymes act. For example, a flavonol glucosyltransferase from soybean cell cultures is unable to glucosylate anthocyanins (Poulton and Kauer, 1977), whereas cyanidin-specific trans-ferases from various sources can also use flavonols as substrates (Saleh *et al.*, 1976a,b). However, transferase specificity with respect to the position of the hydroxyl group of the glucosyl acceptor is known to be narrow (Sutter and Grisebach, 1973; Van Brederode *et al.*, 1975). Although, as already noted, separate transferases exist for further glycosylation of the existing sugars of flavonoid glycosides (Sutter *et al.*, 1972; Ortmann *et al.*, 1972), separation of the transferases involved in the synthesis of a flavonol 3-triglucoside has not been achieved (Shute *et al.*, 1979).

A high degree of specificity is displayed by many β-glucosidases with respect to their action on phenolic and flavonoid glucosides (see Table IV). Multiple forms of the enzyme (displaying varying specificity) may occur in a single source (Podstolski and Lewak, 1970; Marcinowski and Grisebach, 1978; Dey and Del Campillo, 1984). It is evident from the data presented in Table IV that not only the glycone moiety and the anomeric nature of the glucoside determine the specificity but also the aglycone moiety may play an important role (Hösel and Conn, 1982). Thus, use of a chromogenic glucoside, instead of the natural substrate(s), as a general substrate for assaying the enzymic activity may lead to misinformation.

An alternative to glucosyl hydrolysis is the glucosyltransferase reaction. In pulse-labelling experiments with plant cell suspension cultures using ^{14}C-labelled 3-O-β- and 7-O-β-glucosides of kaempferol, the 3-O-glucoside was

Table IV

Relative Activities of β-Glucosidases with Respect to Some Phenolic and Flavonoid Glucosides

Source of β Glucosidase	PNPG[a]	Biochanin 7-β-glucoside	Coniferin	Formononetin 7-β-glucoside	Phloridzin	Syringin
Almond (Hösel et al., 1978)	100	—	3.7	—	—	0.01
Apple (Podstolski and Lewak, 1970)						
Seed						
I	100	—	—	—	756	—
II	100	—	—	—	182	—
III	100	—	—	—	0.9	—
Leaf						
I	100	—	—	—	3700	—
II	100	—	—	—	150	—
III	100	—	—	—	0.8	—
Chick pea tissue (Hösel and Barz, 1975)	100	218	2.7	137	—	—
Suspension cultured chick pea (Hösel et al., 1978)	100	—	284	280	—	40
Soybean (Hösel and Todenhagen, 1980)	100	—	20,000	1	—	18,200
Spruce (Marcinowski and Grisebach, 1978)						
I	100	—	650	—	—	412
II	100	—	68	—	—	55

[a] p-Nitrophenyl β-D-glucoside.

metabolised faster than the 7-*O* isomer (Muhle *et al.*, 1976). Similar observations were made by Dittrich and Kandler (1971) using intact plant systems. However, no specific β-glucosidase could be detected (Surholt and Hösel, 1978). In this respect the findings of Sutter and Grisebach (1975) may be of relevance, as some phenolic glucosides are energy-rich and are involved in enzyme-catalysed transfer reactions, e.g.,

Kaempferol 3-*O*-β-D-glucoside + UDP $\longrightarrow$ UDP-D-glucose + kaempferol

The ΔG° for the hydrolysis of the glucoside was approximately -25 kJ/mol.

The β-glucosidase specific for isoflavone 7-*O*-β-D-glucoside in *Cicer arietinum* roots is clearly involved in the metabolism of isoflavones in this organ (Hösel, 1976). In *Phaseolus aureus* and *C. arietinum* tissues, results of pulse-labelling experiments demonstrated that although the level of cinnamoyl glucose esters remains constant, there is considerable turnover (Molderez *et al.*, 1978). Similarly, there is turnover of some flavonol (isorhamnetin, kaempferol, and quercetin) glucosides and their biosides and triosides during development of *Cucurbita maxima* seedlings (Strack, 1973; Strack and Reznik, 1976). The monosides displayed a half-life of 30–36 hr, the biosides a half-life of 48 hr, but the triosides were not turned over. The fates of the released glycone moieties remain relatively unexplored.

IV. GLUCOSINOLATES

Glucosinolates are thio-β-D-glucosides with the general structure

$$R-C\begin{matrix} \diagup NOSO_3^- \ X^+ \\ \diagdown S\text{-}\beta\text{-}D\text{-Glucose} \end{matrix}$$

where R may be an aliphatic, aromatic, or heterocyclic group and X usually represents a potassium ion. These compounds are water soluble and nonvolatile. Their occurrence is restricted to dicotyledenous plants, especially in the order Capporales and families Capparidaceae, Cruciferae, Gyrostemonaceae, Limnanthaceae, Moringaceae, Resedaceae, Salvadoracea, Tovariaceae, and Tropaeolaceae. Thus, glucosinolates are useful chemotaxonomic markers (see Kjaer, 1974; Rodman, 1978). Several recent general reviews on glucosinolates are available (Van Etten and Tookey, 1979; Underhill, 1980; Tookey *et al.*, 1980; Larsen, 1981).

Glucosinolates may occur in all parts of the plant, generally in low concentration (0.1% or less with respect to fresh weight) (Cole, 1976; Van Etten *et al.*, 1976); however, in seeds they may account for up to 10% of the dry weight (Josefsson, 1972; Wetter and Dyck, 1973; Van Etten *et al.*, 1974). Thus, glucosinolates could possibly play a storage role in seeds. They have been found to accumulate in seed endosperm (Kjaer, 1960), and to be located in vacuoles in horseradish root cells (Grob and Matile, 1979; Matile, 1980).

Mahadevan and Stowe (1972) have demonstrated that indole glucosinolates in *Isatis tinctoria* undergo a rapid turnover with a half-life of approximately 48 hr. Sinapoylglucoraphanin in *Raphanus sativus* seeds rapidly disappears during germination (Linschied *et al.*, 1980).

Glucosinolate biosynthesis has been previously reviewed (Underhill *et al.*, 1973; Kjaer and Larsen, 1973, 1976, 1977, 1980; Larsen, 1981). Glucosylation is the penultimate step producing disulphoglucosinolate; this intermediate is sulphonated to form the final glucosinolate molecule:

The glucosyltransferase is highly specific for UDP-D-glucose (Matsuo and Underhill, 1971); however, little specificity is shown towards the side chain of the thiohydroximate.

The cleavage of the glucose residue of glucosinolates is catalysed by specific enzymes known as myrosinases (thioglucoside glucohydrolases; EC 3.2.3.1). Plants containing glucosinolates display a high activity of this enzyme. Myrosinases act at a slower rate on desulphated substrates (Ettlinger and Lundeen, 1957; Reese *et al.*, 1958; Nagashima and Uchiyama, 1959a,b). Simple O-β-D-glucosides are very poor substrates (Reese *et al.*, 1958; Tsuruo and Hata, 1968). Multiple forms of the enzyme often occur, differing mainly in isoelectric point and carbohydrate content (Björkman and Janson, 1972; Björkman and Lönnerdal, 1973; Lönnerdal and Janson, 1973) but generally displaying similar substrate specificities. In many cases ascorbate has been shown to activate myrosinase, bringing about an increase in V_{max} and an increase in K_m. The properties of myrosinase from three different sources are outlined in Table V. It is clear that the situation with respect to reports of different multiple forms of the enzyme, even from the same source, is very complex.

V. *C-* AND *N*-GLYCOSIDES

The *C*-glycoside vitexin (8-*C*-D-glucopyranosylapigenin, a *C*-glycosyl-flavone) was first isolated in 1900 (Perkin, 1900); however, the *C*-glycosidic nature of the linkage was recognised only in 1957 (Evans *et al.*, 1957). Since then, a number of *C*-glycosyl derivatives of flavones, xanthones, and gallic acid have been shown to exist in plants (Wallace and Aston, 1966; McCarthy, 1969; Courtois and Percheron, 1970; Miller, 1973; Wallace and Grisebach, 1973; Grün and Franz, 1979, 1980; Tameyama and Yoshida, 1979; Auterhoff *et al.*, 1980; Fujita and Inoue, 1980; Young and Siegler, 1982. *C*-Glycosylflavones are widespread in the superorder Chenopodiiflora (Gornal *et al.*, 1979; Seigler, 1981; see also Chopin and Bouillant, 1975).

C-Glycosylflavones accumulate during the early development of primary leaves of *Avena sativa*, but then rapidly decrease at a later stage (Popovici and Wissenböck, 1976, 1977). The flavone derivatives involved were isoorientin and isovitexin (both being *O*-2″-arabinosides) and isoswertisin and vitexin (both being *O*-2″-rhamnosides). Similarly, Grün and Franz (1979) showed by pulse-labelling experiments that in *Aloe* species the *C*-glycosides are not end products, since they are turned over at a considerable rate.

Some indication as to the rate of biosynthesis of the *C*-glycosides comes from studies on flavonol, xanthone, and gallic acid derivatives (Wallace and Alston, 1966; Wallace and Grisebach, 1973; Tameyama and Yoshida, 1979; Fujita and Inoue, 1980). Grün and Franz (1981) have recently shown that a crude enzyme extract from the leaves of *Aloe arborescens* can utilise UDP-D-glucose for the synthesis of the *C*-glycoside aloin. Whereas ADP-glucose, GDP-glucose, glucose 1-P, and D-glucose failed to act as glycosyl donors, UDP- galactose acted as a poor donor for the transfer of D-galactose to aloe emodin anthrone. No other cofactor was required for the *C*-glycosylation. The molecular mechanism of this reaction remains unknown.

Aloe emodin anthrone   Aloin

The glucosyl moieties of *C*-glyosides may be further substituted by *O*-glycosyl groups. This type of glycosylation at the sugar residue also proceeds

Table V

Properties of Myrosinase from Higher Plant Sources

Species	Source	Form[a]	Molecular weight	Subunit molecular weight	% Carbohydrate		pI
Sinapis alba	Mature seed	I					5.00
		II					5.55
Sinapis alba	Seed	A	≥ 151,000				5.90
		B	≥ 151,000				5.45
		C	151,000	62,000	18		5.08
Sinapis alba	Seed powder	I	152,000	40,000	Ia 15.8	⎱	
					Ib 17.8	⎰ 4.60	
					Ic 22.5		
		II	125,000	30,000	8.6		4.80
Brassica napus		A					6.20
		B					5.60
		C	135,000	65,000	C_1 9.3		4.96
					C_2 15.2		4.99
					C_3 17.4		5.06
		D					4.90
Crambe abyssinica	Seed	I	> 200,000				
		II	110,000				

[a] Nomenclature of authors.

via nucleotide-activated sugar donors, as, for example, in the substitution of arabinosyl or glucosyl groups at C-2 of apigenin 6-C-glucoside (Van Brederode and Van Nigtevecht, 1974a,b; see also Van Nigtevecht and Van Brederode, 1975); however, different transferases are involved.

The most important *N*-glycosides are the nucleosides. Others include cobalamins (vitamin B_{12}), which are D-ribofuranosyl derivatives of dimethylbenzimidazole, and vicine, which is an *N*-β-D-glucosyl derivative of a diaminopyrimidinedione (Hérissey and Cheymol, 1931).

N-Glycosylated glycosyl conjugates seem to occur essentially as detoxification products for both exogenous and endogenous compounds. Some metabolically active compounds can also be stored as *N*-glycosylated conjugates, which thus serve a reservoir function. For example, nicotinic acid conjugates (*N*-glycosides and *N*-methyl derivatives) are used for pyridine nucleotide biosynthesis (Barz, 1977). Some fungicides and herbicides may be converted by the plant into *N*-glucoside conjugates (Fredrick and Genlete, 1961; Kamimura *et al.*, 1974; Edwards *et al.*, 1982). Little is known about the enzymology of *N*-glycoside synthesis or hydrolysis.

pH Optimum	Temperature optimum	K_m (M)		Preferred substrates	References
		+ Ascorbate	− Ascorbate		
6.50	75°C		8×10^{-5}	p-Hydroxy benzyl glucosinolate	Vose (1972)
6.50	45°C	8×10^{-5}	Very little activity	p-Hydroxy benzyl glucosinolate	
4.20	67°C	2.5×10^{-4}		Glucotropaeolin	Björkman and
5.50	60°C	3.0×10^{-4}	6×10^{-5}	Glucotropaeolin	Janson (1972);
5.80	60°C	4.0×10^{-4}	1.7×10^{-4}	Glucotropaeolin	Björkman and Lönnerdal (1973)
					Ohtsura and Hata (1972)
5.10	60°C	4.0×10^{-4}	5×10^{-4}	Glucotropaeolin	Lönnerdal and
5.10	60°C	4.0×10^{-4}	6×10^{-5}	Glucotropaeolin	Janson (1973); Börkman and
4.25	60°C	3.0×10^{-4}	3×10^{-5}	Glucotropaeolin	Lönnerdal (1973)
					Tookey (1973)
8.9		5.0×10^{-3}		epi-Progoitrin	

REFERENCES

Asai, T. (1930). *Chem. Zentralbl.* **II,** 408.
Auterhoff, H., Graf, E., Eurisch, G. and Alexa, M. (1980). *Arch Pharmacol.* **313,** 113–120.
Bajaj, K. L., de Luca, V., Khouri, H. and Ibrahim, R. K. (1983). *Plant Physiol.* **72,** 891–896.
Barz, W. (1975). *Planta Med., Suppl.* pp. 117–133.
Barz, W. (1977). *In* "Plant Cell Culture and its Biotechnological Application" (W. Barz, E. Reinhard and M. H. Zenk, eds.), pp. 153–171. Springer-Verlag, Berlin and New York.
Barz, W. and Hösel, W. (1975). *In* "The Flavonoids" (J. B. Harborne, T. J. Mabry and H. Mabry, eds.), pp. 916–969. Chapman & Hall, London.
Barz, W. and Köster, J. (1981). *In* "The Biochemistry of Plants" (E. E. Conn, ed.), Vol. 7, pp. 35–84. Academic Press, New York.
Björkman, R. and Janson, J. C. (1972). *Biochim. Biophys. Acta* **276,** 508–518.
Björkman, R. and Lönnerdal, B. (1973). *Biochim. Biophys. Acta* **327,** 121–131.
Bottomley, W. and White, D. E. (1951). *Aust. J. Sci. Res., Ser. A* **4,** 107–111.
Butler, G. W., and Bailey, R. W. and Kennedy, L. D. (1965). *Phytochemistry* **4,** 369–381.
Cardini, C. E. and Leloir, L F. (1957). *Cienc. Invest.* **13,** 514.
Chopin, J. and Bouillant, M. L. (1975). *In* "The Flavonoids" (J. B. Harborne, T. J. Mabry and H. Mabry, eds.), pp. 692–742. Chapman & Hall, London.
Cole, R. A. (1976). *Phytochemistry* **15,** 759–762.

Conn, C. E. (1981). *In* "The Biochemistry of Plants" (E. E. Conns, ed.), Vol. 7, pp. 479–500. Academic Press, New York.

Corner, J. J. and Swain, T. (1965). *Nature (London)* **207**, 635.

Courtois, J. E. and Percheron, F. (1970). *In* "The Carbohydrates" (W. Pigman and D. Horton, eds.), 2nd ed., Vol. 2A, pp. 213–240. Academic Press, New York.

Dey, P. M. and Del Campillo, E. (1984). *Adv. Enzymol.* **56**, 141–249.

Dick, A. T. (1943). *J. Counc. Sci. Ind. Res. (Aust.)* **16**, 11–14; *Chem Abstr.* **37**, 4204 (1943).

Dittrich, P. and Kandler, O. (1971). *Ber Dtsch. Bot. Ges.* **84**, 465–472.

Edwards, V. T., McMinn, A. L. and Wright, A. N. (1982). *Prog. Pestic. Biochem.* **2**, 71–125.

Eklund, S. H. (1981). M. S. Thesis, University of California, Davis.

Elliot, M. C. and Stowe, B. B. (1971) *Plant Physiol.* **48**, 498–503.

Ellis, B. E. (1974). *Lloydia* **37**, 168–184.

Ettlinger, M. G. and Lundeen, A. J. (1957). *J. Am. Chem. Soc.* **79**, 1764–1765.

Evans, W. H., McGoskin, A., Jurd, L., Robertson, A. and Williamson, W. R. N. (1957). *J. Chem. Soc.* pp. 3510–3523.

Fredrick, J. F. and Genlete, A. C. (1961). *Arch Biochem. Biophys.* **92**, 356–361.

Fujita, M. and Inoue, T. (1980). *Chem. Pharmacol. Bull.* **28**, 2482–2486.

Gibbs, R. D. (1974). "Chemotaxonomy of Flowering Plants," Vol. 4. McGill-Queen's Univ. Press, Montreal.

Gil, V. and MacLeod, A. J. (1980). *Phytochemistry* **19**, 1365–1368.

Goetz, J. K. and Schraudolf, H. (1983). *Phytochemistry* **22**, 905–907.

Gornal, R. J., Bohm, B. A. and Dahlgren, R. (1979). *Bot. Not.* **132**, 1–30.

Grob, K. and Matile, P. (1979). *Plant Sci. Lett.* **14**, 327–335.

Grün, M. and Franz, G. (1979). *Pharmazie* **34**, 669–670.

Grün, M. and Franz, G. (1980). *Planta Med.* **39**, 288.

Grün, M. and Franz, G. (1981). *Planta* **152**, 562–564.

Hahlbrock, K. (1981). *In* "The Biochemistry of Plants" (E. E. Conn, ed.), Vol. 7, pp. 425–456. Academic Press, New York.

Hahlbrock, K. and Conn, E. E. (1970). *J. Biol. Chem.* **245**, 917–922.

Haisman, D. R. and Knight, D. J. (1967). *Biochem. J.* **103**, 528–534.

Haisman, D. R., Knight D. J. and Ellis, M. J. (1967). *Phytochemistry* **6**, 1501–1505.

Harborne, J. B., ed. (1964). "Biochemistry of Phenolic Compounds." Academic Press, New York.

Hegnauer, R. (1977). *Plant Syst. Evol., Suppl.* **1**, 191–207.

Hérissey, H. and Cheymol, J. (1931). *Bull. Soc. Chim. Biol.* **13**, 29.

Hösel, W. (1976). *Planta Med.* **30**, 97–103.

Hösel, W. (1981). *In* "The Biochemistry of Plants" (E. E. Conn, ed.), Vol. 7, pp. 725–753. Academic Press, New York.

Hösel, W. and Barz, W. (1975). *Eur. J. Biochem.* **57**, 607–616.

Hösel, W. and Conn, E. E. (1982) *Trends Biochem. Sci.* **7**, 219–221.

Hösel, W. and Nahrstedt, A. (1975). *Hoppe-Seyler's Z. Physiol. Chem.* **356**, 1265–1275.

Hösel, W. and Nahrstedt, A. (1980). *Arch. Biochem. Biophys.* **203**, 753–757.

Hösel, W. and Todenhagen, R. (1980). *Phytochemistry* **19**, 1349–1353.

Hösel, W., Surholt, E. and Borgman, E. (1978). *Eur. J. Biochem.* **84**, 487–492.

Iwamoto, H. K., Evans, W. E. and Krantz, J. C. (1945). *J. Am. Pharm. Assoc.* **34**, 205–209.

Jacobelli, G., Tabone, M. J. and Tabone, D. (1958). *Bull. Soc. Chim. Biol.* **40**. 955–961.

Josefsson, E. (1972). *In* "Rapeseed" (L. Appelqvist and R. Ohlson, eds.), pp. 354–377. Elsevier, Amsterdam.

Kamimura, S., Nishikawa, M., Saeki, H. and Takahi, Y. (1974). *Phytopathology* **64**, 1273–1281.

Kjaer, A. (1960). *Fortschr. Chem. Org. Naturst.* **18**, 122–176.

Kjaer, A. (1974). *In* "Chemistry in Botanical Classification" (G. Bendz and J. Santesson, eds.), pp. 229–234. Academic Press, New York.

Kjaer, A. and Larsen, P. O. (1973). *Biosynthesis* **2**, 71–107.
Kjaer, A. and Larsen, P. O. (1976). *Biosynthesis* **4**, 179–203.
Kjar, A. and Larsen, P. O. (1977). *Biosynthesis* **5**, 120–135.
Kjar, A. and Larsen, P. O. (1980). *Biosynthesis* **6**, 155–180.
Kojima, M., Poulton, J. E., Thayer, S. S. and Conn, E. F. (1979). *Plant Physiol.* **63**, 1022–1028.
Lalégérie, P. (1974). *Biochimie* **57**, 1163–1172.
Larsen, P. O. (1981). *In* "The Biochemistry of Plants" (E. E. Conn, ed.), Vol. 7, pp. 501–525. Academic Press, New York.
Leigh, R. A. and Barton, D. (1976). *Plant Physiol.* **58**, 656–662.
Linscheid, M., Wendisch, D. and Strack, D. (1980). *Z. Naturforsch., C: Biosci.* **35C**, 907–914.
Lönnerdal, B. and Janson, J. C. (1973). *Biochim. Biophys. Acta* **315**, 421–429.
McCarthy, T. J. (1969). *Planta Med.* **17**, 1–7.
Macheix, J.-J. (1977). *C.R. Hebd. Seances Acad. Sci.* **284**, 33–36.
Macleod, N. J. and Pridham, J. B. (1966). *Phytochemistry* **5**, 777–781.
Mahadevan, S. and Stowe, B. B. (1972). *Plant Physiol.* **50**, 43–50.
Marcinowski, S. and Grisebach, H. (1978). *Eur. J. Biochem.* **87**, 37–44.
Marsh, C. A. (1960) *Biochim. Biophys. Acta* **44**, 359–361.
Matile, P. (1976). *Nova Acta Leopold., Suppl.* **7**, 139–156.
Matile, P. (1978). *Annu. Rev. Plant Physiol.* **29**, 193–213.
Matile, P. (1980). *Biochem. Physiol. Pflanz.* **175**, 722–731.
Matsuo, M. and Underhill, E. W. (1971). *Phytochemistry* **10**, 2279–2286.
Merz, K. W. (1932). *Arch. Pharm. (Weinheim Ger.)* **270**, 476–495.
Miller, L. P. (1973). *In* "Phytochemistry" (L. P. Miller, ed.), Vol. 1, pp. 297–375. Van Nostrand-Reinhold, Princeton, New Jersey.
Molderez, M., Nagels, L. and Parmentier, F. (1978) *Phytochemistry* **17**, 1747–1750.
Muhle, E. Hösel, W. and Barz, W. (1976). *Phytochemistry* **15**, 1669–1672.
Nagashima, Z. and Uchiyama, M. (1959a). *Nippon Nogei Kagaku Kaishi* **33**, 1068–1071.
Nagashima, Z. and Uchiyama, M. (1959b). *Nippon Nogei Kagaku Kaishi* **33**, 1144–1149.
Nahrstedt, A. (1982). *In* "Cyanide in Biology" (B. Vennsland, E. E. Conn, C. J. Knowles, J. Westley and F. Wissing, eds.), pp. 145–181. Academic Press, New York.
Nahrstedt, A., Hösel, W. and Walther, A. (1979). *Phytochemistry* **18**, 1137–1141.
Nanayakkara, S., Pridham, J. B. and Young M. A. (1970). *Biochem. J.* **121**, 12P–13P.
Ohtsura, M. and Hata, K. (1972). *Agric. Biol. Chem.* **36**, 2495–2501.
Ortmann, R., Sutter, A. and Grisebach, H. (1972). *Biochim. Biophys. Acta* **258**, 71–87.
Perkin, A. (1900). *J. Chem. Soc., Trans.* **77**, 416.
Podstolski, A. and Lewak, S. (1970). *Phytochemistry* **9**, 289–296.
Popovici, G. and Wissenböck, G. (1976). *Ber. Dtsch. Bot. Ges.* **87**, 483–489.
Popovici, G. and Wissenböck, G. (1977). *Z. Pflanzenphysiol.* **82**, 450–454.
Poulton, J. E. and Kauer, M. (1977). *Planta* **136**, 53–59.
Pridham, J. B. (1965). *Adv. Carbohydr. Chem.* **20**, 371–408.
Pridham, J. B. and Saltmarsh, M. J. (1962). *Biochem. J.* **82**, 44P–45P.
Rabaté, J. (1933). *Bull. Soc. Chim. Biol.* **15**, 385–394.
Reay, P. F. and Conn, E. E. (1974). *J. Biol. Chem.* **249**, 5826–5830.
Reese, E. T., Clapp, R. C. and Mandels, M. (1958). Arch. Biochem. Biophys. **75**, 228–242.
Robinson, T. (1974). *Science* **184**, 430–435.
Rodman, J. E. (1978). *Phytochem. Bull.* **11**, No. 1–2 (*Phytochem. Sect. Bot. Soc. Am.*), 6–31.
Rosenthaler, M. (1932). *In* "Handbuch der Pflanzen Analyse" (G. Klein, ed.), Vol. 3, Part 2, pp. 1036–1057. Springer-Verlag, Vienna.
Saleh, N. A. M., Poulton, J. G. and Grisebach, H. (1976a). *Phytochemistry* **15**, 1865–1868.
Saleh, N. A. M., Fritsch, H., Witkop, P. and Grisebach, H. (1976b), *Planta* **133**, 41–45.
Sasse, F., Backs-Hüsemann, D. and Barz, W. (1979). *Z. Naturforsch., C: Biosci.* **34C**, 848–853.

Saunders, J. A. (1979). *Plant Physiol.* **64**, 74–78.
Saunders, J. A. and Conn and E. E. (1978). *Plant Physiol.* **61**, 154–157.
Schlepphorst, R. and Barz, W. (1979). *Planta Med.* **36**. 333–342.
Seigler, D. S. (1981) *In* "The Biochemistry of Plants" (E. E. Conn, ed.), Vol. 7., pp. 139–176. Academic Press, New York.
Seigler, D. S. (1982). *In* "Cyanide in Biology" (B. Vennesland, E. E. Conn, C. J. Knowles, J. Westley and F. Wissing, eds.), pp. 133–143. Academic Press, New York.
Shute, J. L., Jourdan, P. S. and Mansell, R. L. (1979). *Z. Naturforsch., C: Biosci.* **34C**, 738–741.
Strack, D. (1973). Ph.D. Thesis, University of Cologne.
Strack, D. and Reznik, H. (1976). *Z. Pflanzenphysiol.* **79**, 95–108.
Surholt, E. and Hösel, W. (1978). *Phytochemistry* **17**, 873–877.
Sutter, A. and Grisebach, H. (1973). *Biochim. Biophys. Acta* **309**, 289–295.
Sutter, A. and Grisebach, H. (1975). *Arch. Biochem. Biophys.* **167**, 444–447.
Sutter, A., and Ortmann, R. and Grisebach, H. (1972). *Biochim. Biophys. Acta* **258**, 71–87.
Tameyama, M. and Yoshida, S. (1979). *Bot. Mag* **92**, 69–73.
Tang, C.-S. (1973). *Phytochemistry* **12**, 769–773.
Thieme, H. (1965a). *Pharmazie* **20**, 436–440.
Thieme, H. (1965b). *Planta Med.* **13**, 431–438.
Tookey, H. L. (1973). *Can. J. Biochem.* **51**, 1305–1310.
Tookey, H. L., Van Etten, C. H. and Daxenbichler, M. E. (1980). *In* "Toxic Constituents of Plant Foodstuffs" (I. E. Liener, ed.), pp. 103–142. Academic Press, New York.
Trivelloni, J. C., Recondo, E. and Cardini, C. E. (1962). *Nature (London)* **195**, 1202.
Tsurno. I. and Hata, T. (1968). *Agric. Biol. Chem.* **32**, 1425–1431.
Underhill, E. W. (1980). *In* "Encyclopedia of Plant Physiology, New Series" (A. Pirson and M. H. Zimmermann, eds.), Vol. 8, pp. 493–511. Springer-Verlag, Berlin and New York.
Underhill, E. W. Wetter, L. R. and Chisholm, M. D. (1973). *Biochem. Soc. Symp.* **38**, 303–326.
Van Brederode, J. and Van Nigtevecht, G. (1974a). *Biochem. Genet.* **11**, 6581.
Van Brederode, J. and Van Nigtevecht, G. (1974b). *Phytochemistry*, **13**, 2763–2766.
Van Brederode, J., Van Nigtevecht, G. and Kamsteeg, J. (1975). *Heredity* **35**, 429–430.
Van Etten, C. H. and Tookey, H. L. (1979). *In* "Herbivores: Their Interaction with Secondary Plant Metabolites" (G. A. Rosenthal and D. H. Janzen, eds.), pp. 471–500. Academic Press, New York.
Van Etten, C. H., McGrew, C. E. and Daxenbichler, M. E. (1974). *J. Agric. Food Chem.* **22**, 483–487.
Van Etten, C. H., Daxenbichler, M. E., Williams, P. H. and Kwolek, W. F. (1976). *J. Agric. Food Chem.* **24**, 452–455.
Van Nigtevecht, G. and Van Brederode, J. (1975). *Heredity* **35**, 429.
Vose, J. R. (1972). *Phytochemistry* **11**, 1649–1653.
Wallace, J. W. and Alston, R. E. (1966). *Plant Cell Physiol.* **7**, 699–700.
Wallace, J. W. and Grisebach, H. (1973). *Biochim. Biophys. Acta* **304**, 837–841.
Waller, G. R. and Nowacki, E. K. (1978). *In* "Alkaloid Biology and Metabolism in Plants" (G. R. Waller and E. K. Nowacki, eds.), Chapters 1 and 6. Plenum, New York.
Wester, D. H. (1913). "Anleitung zur Darstellung Phytochemischer Übungspräparat für Pharmazenten Chemiker, Technologen u.a." Springer-Verlag, Berlin and New York.
Wetter, L. R. and Dyck, J. (1973). *Can. J. Anim. Sci.* **53**, 625–626.
Williams, A. H. (1966). *In* "Comparative Phytochemistry" (T. Swain, ed.), pp. 297–307. Academic Press, New York.
Yamaha, T. and Cardini, C. E. (1960a). *Arch. Biochem. Biophys.* **86**, 127–132.
Yamaha, T. and Cardini, C. E. (1960b). *Arch. Biochem. Biophys.* **86**, 133–137.
Young, D. A. and Siegler, D. S. (1982). *In* "CRC Hand Book of Biosolar Resources" (A. Mitsui and C. C. Black, eds.), Vol. 1, Part 1. CRC Press, Boca Raton, Florida.

Chapter **4**

Starch

D. J. MANNERS

Department of Brewing and Biological Sciences
Heriot-Watt University
Edinburgh, Scotland

I. INTRODUCTION

Starch, the reserve carbohydrate of the majority of higher plants, occurs as water-insoluble granules that vary in size and shape, depending on the species and maturity of the plant. Within these granules, there is normally a mixture of two polysaccharides: amylose and amylopectin. The former amounts to ~20–30% of most starches and is largely composed of long linear chains of

BIOCHEMISTRY OF STORAGE
CARBOHYDRATES IN GREEN PLANTS

$(1 \rightarrow 4)$-linked α-D-glucopyranose residues. By contrast, amylopectin, which is the major starch component, is a macromolecule containing literally thousands of shorter branched chains of $(1 \rightarrow 4)$-linked α-D-glucose residues. The average chain length of the constituent chains in amylose and amylopectin is $\sim 10^3$ and ~ 20–25, respectively, although individual chains may vary considerably in length from these average values. Some starch granules, e.g. from wheat and amylomaize, contain significant amounts (up to 20%) of a third polysaccharide, often referred to as the intermediate fraction. This also contains chains of $(1 \rightarrow 4)$-linked α-D-glucose residues, but the number of chains per molecule and the average length of these chains is different from that in either amylose or amylopectin.

In this chapter the structure and properties of the starch components and the starch granule will be reviewed. This will be followed by a discussion of various enzymes (Table I) that catalyze the synthesis and degradation of starch. Many of these enzymes exist in multiple forms, which have been incorrectly described as isoenzymes or isozymes by various authors. It should be emphasized that the term isoenzyme or isozyme should apply only to those multiple forms of an enzyme arising from genetically determined differences in primary structure, and not to those derived by modification from the same primary sequence (Enzyme Nomenclature Recommendations of IUPAC and IUB, 1972). In many instances, "multiple form" is the correct term, covering all proteins possessing the *same* catalytic activity and occurring naturally in a single species.

Throughout this discussion of the structure and metabolism of starch, work published during the last decade is emphasized, since much of the earlier work has already been reviewed (Manners, 1974a,b,c). References are also given to a number of other recent reviews that describe the biochemistry of starch in certain selected plant tissues in more detail than is possible in a general review.

II. THE STRUCTURE AND PROPERTIES OF THE STARCH COMPONENTS

A. Introduction

The classical studies on the fractionation of starch by Schoch, Haworth, and Meyer have been well documented (Schoch, 1945; Greenwood, 1956). The addition of *n*-butanol, cyclohexanol, or thymol to an aqueous dispersion of starch resulted in the precipitation of an insoluble amylose complex, and the amylopectin could then be isolated from the supernatant solution. The first stage of this fractionation depends on the fact that when starch granules are heated in water, they swell, gelatinize, and are then ruptured to give an

Table I

List of Major Starch-Metabolising Enzymes

Enzyme commission (EC) number	Recommended name	Systematic name
2.4.1.1	Phosphorylase	1,4-α-D-Glucan : orthophosphate α-D-glucosyltransferase
2.4.1.13	Sucrose synthase	UDPglucose : D-fructose 2-α-D-glucosyltransferase
2.4.1.14	Sucrose-phosphate synthase	UDPglucose : D-fructose 6-phosphate 2-α-D-glucosyltransferase
2.4.1.18	Q-enzyme or 1,4-α-glucan branching enzyme	1,4-α-D-Glucan : 1,4-α-D-glucan 6-α-D-(1,4-α-D-glucano)transferase
2.4.1.21	Starch synthase	ADPglucose : 1,4-α-D-glucan 4-α-D-glucosyltransferase
2.4.1.24	T-enzyme or 1,4-α-D-glucan 6-α-D-glucosyltransferase	1,4-α-D-glucan : 1,4-α-D-glucan (D-glucose) 6-α-D-glucosyltransferase
2.4.1.25	D-enzyme or 4-α-D-Glucanotransferase	1,4-α-D-Glucan : 1,4-α-D-glucan 4-α-D-glucosyltransferase
2.7.7.27	ADPglucose pyrophosphorylase	ATP : α-D-glucose 1-phosphate adenyltransferase
3.2.1.1	α-Amylase	1,4-α-D-Glucan glucanohydrolase
3.2.1.2	β-Amylase	1,4-α-D-Glucan maltohydrolase
3.2.1.3	Glucoamylase or amyloglucosidase	1,4-α-D-Glucan glucohydrolase
3.2.1.41	Pullulanase or limit dextrinase or amylopectin 6-glucanohydrolase	Pullulan 6-glucanohydrolase
3.2.1.68	Isoamylase	Glycogen 6-glucanohydrolase

aqueous dispersion or paste, depending on the actual concentration of starch. The gelatinization temperature is characteristic of the plant species, and is usually in the range 55–80°C. During this process, the internal organization of the starch granule is destroyed, and the component polysaccharides are released as individual molecules. At this stage, the essentially linear amylose has a high affinity for polar organic molecules, and forms an insoluble complex. This fractionation effectively negates the so-called unitary theory, which suggested that a starch granule consisted of a single giant macromolecule.

The structural analysis of the isolated amylose and amylopectin components has been carried out by standard methods based on methylation, periodate oxidation, and partial acid hydrolysis studies (for details, see Williams, 1968). Both components consist of chains of $(1\rightarrow4)$-linked α-D-glucose residues, but there are large differences in the relative lengths and number of chains per molecule (Table II). This results in significant differences in molecular size and shape and in their physicochemical properties, particularly solubility in water, and in their capacity to interact with iodine. In both components, it is now clear that the linear chains are interlinked by means of $(1\rightarrow6)$-α-D-glucosidic linkages.

The overall properties of amylose, of amylopectin, and of phytoglycogen, a related $(1\rightarrow4)$-α-D-glucan present in sweet corn, are summarized in Table II. The average chain length (CL), defined as the number of glucose residues per

Table II

Properties of Starch-Type Polysaccharides

Property	Amylose	Amylopectin	Phytoglycogen
General structure	Essentially unbranched	Branched	Highly branched
Average chain length (CL)[a]	$\sim 10^3$	20–25	10–14
Degree of polymerization (DP)[a]	$\sim 10^3$	10^4–10^5	$\sim 10^5$
Iodine coloration	Deep blue	Purple	Brown
λ_{max}(nm)	~ 660	530–550	430–450
Stability of aqueous solution	Retrogrades	Stable	Stable
Conversion into maltose (%)			
With β-amylase	~ 70	~ 55	~ 45
With a debranching enzyme[b]			
and then β-amylase	~ 100	~ 75	—
With α-amylase[c]	~ 100	~ 90	~ 80

[a] D-Glucose residues.
[b] Yeast isoamylase.
[c] Some glucose is also produced.

nonreducing end group, can be measured by various chemical and enzymic methods. However, these analyses give statistically average values; individual chains, especially in amylopectin, vary considerably in length. This variation can be studied by the enzymic hydrolysis of the $(1\rightarrow6)$-α-D-glucosidic interchain linkages (i.e. by "debranching"), followed by fractionation of the resulting mixture of linear chains by gel filtration (Lee et al., 1968). This technique is now standard for the determination of the chain profile of branched $(1\rightarrow4)$-α-D-glucans.

The degree of polymerization (DP), defined as the number of glucose residues per reducing end group, is normally calculated from physical measurements of the molecular weight. In completely linear amylose molecules, which are now known to be relatively uncommon, the DP and CL values are the same, whereas in most samples of amylose, DP > CL, and in amylopectins, DP > > > CL.

B. The Structure of Amylose

In much of the work described below, enzymic degradation methods have been widely used. Details of the action patterns of the enzymes are given in Section IV. In summary, α-amylase hydrolyses nonterminal $(1\rightarrow4)$-α-D-glucosidic linkages in both amylose and amylopectin by endo action, giving a mixture of products, while β-amylase catalyzes a stepwise (exo) hydrolysis of alternate linkages, giving maltose as the only low-molecular-weight product. Neither enzyme has any action on $(1\rightarrow6)$-α-D-glucosidic interchain linkages; these latter are selectively hydrolysed by debranching enzymes (e.g. iso-amylase, limit dextrinase, and pullulanase), which, in turn, have no action on $(1\rightarrow4)$-α-D-glucosidic linkages.

The first evidence that amylose was not simply a linear $(1\rightarrow4)$-α-D-glucan came from β-amylolysis studies (Peat et al., 1952a,b). Using crystalline sweet potato β-amylase, only about 70% conversion into maltose occurred, while with amorphous soybean β-amylase, β-amylolysis was complete. These observations were confirmed and extended in many laboratories, and during the next decade there was considerable speculation as to the nature of the barriers to β-amylase. Ester phosphate groups, oxidised glucose residues, and side chains consisting of single β-glucosyl residues were all postulated. However, since debranching enzymes such as yeast isoamylase and bacterial pullulanase effectively removed the barrier (Kjolberg and Manners, 1963; Banks and Greenwood, 1966), it became clear that amylose had a lightly branched structure with some $(1\rightarrow6)$-α-D-glucosidic interchain linkages. Moreover, physicochemical studies indicated that any side chains were of a substantial length. Further work also showed that most samples of amylose were, in fact, heterogeneous, and consisted of a mixture of linear and branched molecules. In one preparation of potato amylose, having a DP of 3200 and a β-amylolysis limit of 77%, the linear fraction (about 40% by weight) had a DP of 1800, while the branched fraction had a DP of $\sim$6000 and a β-amylolysis limit of about 50% (Cowie et al., 1957).

The branched nature of amylose is not confined to potato starch. Table III summarizes some results from this laboratory on amylose samples isolated from starches from various plant sources. The low degree of branching in amylose has been further studied by Hizukuri et al. (1981), who determined the CL and DP of various samples of amylose. Their results (Table IV) show the presence of several branch points in the different samples of amylose, although the overall degree of branching was still extremely low. In all of this work, it is

Table III

Properties of Some Amyloses

Source	β-Amylolysis limit(%)		Degree of polymerisation
	β-Amylase[a]	β-Amylase and isoamylase[b]	
Potato	77	98	3200
Oats	82	101	2550
Malted oats	90	100	1900
Dunaliella bioculata	73	93	—
Haematococcus pluvialis	88	103	—

[a] With purified β-amylase alone.
[b] Combined action of β-amylase and isoamylase.

Table IV

The Branched Nature of Some Amyloses[a]

Source	β-Amylolysis limit(%)	Iodine affinity (g/100 g)	DP[b]	CL[c]	Number of branches per molecule
Potato	68	19.6	6340	520	12.2
Potato (Kenebec)	88	20.1	4850	510	9.5
Tapioca	64	19.0	3390	170	20.0
Maize	76	19.9	1590	180	9.1

[a] Data from Hizukuri et al. (1981).
[b] DP, number of glucose residues per reducing group.
[c] CL, number of glucose residues per nonreducing end group.

assumed that the low degree of branching arose by the limited action of Q-enzyme (see Section III,D) on a proportion of the amylose molecules. The factors that control the action of Q-enzyme on amylose *in vivo* are unknown.

The original work on soybean β-amylase suggested the presence of an additional amylolytic enzyme, which was termed Z-enzyme (Peat *et al.*, 1952b) and was believed to hydrolyse specifically the barrier to β-amylase. It is now known (Banks *et al.*, 1960; Cunningham *et al.*, 1960) that Z-enzyme was a minute trace of α-amylase that enabled the barrier to be bypassed, rather than hydrolysed or removed.

C. The Structure of Amylopectin

1. Introduction

Although the main structural features of amylopectin have been known for many years, details of the fine structure are still uncertain. The realisation that many samples of amylopectin had CL values of 20–25 and DP values of $\sim 10^3$ led, in the period 1937–1940, to three different diagrammatic representations of the macromolecule (Fig. 1). The various models contain three different types of chain (Peat *et al.*, 1956). The A chains are linked to the molecule only by the potential reducing group, while B chains are similarly linked but also carry one or more A chains. The C chain is terminated by the sole reducing group in the molecule. The ratio of A chains to B chains, which is also referred to as the degree of multiple branching, is an important parameter. Enzymic methods for the measurement of this ratio have been devised; the various techniques are not easy, and in some experiments, misleading results have been obtained (see below).

The first experiments of this type showed the presence of approximately equal proportions of both A and B chains. This result is incompatible with either the Haworth or Staudinger structure type (Fig. 1, a and b).

The individual chains in amylopectin vary substantially in length, from about 8 to perhaps 100 or more. A method for examining the distribution of chain lengths has already been described (Section II,A). In the earliest experiments using pullulanase as the debranching enzyme, a trimodal distribution of chains was observed (Fig. 2). However, in later experiments using bacterial isoamylase, a largely bimodal distribution has been obtained, with chain length peaks at about 20, and in excess of 50 (Gunja-Smith *et al.*, 1970). This difference was due to minor differences in the specificity of the two debranching enzymes, rather than to differences in the fine structure of the amylopectin samples, This example provides a cautionary note on the use of enzymes for the structural analysis of branched $(1\rightarrow4)$-α-D-glucans.

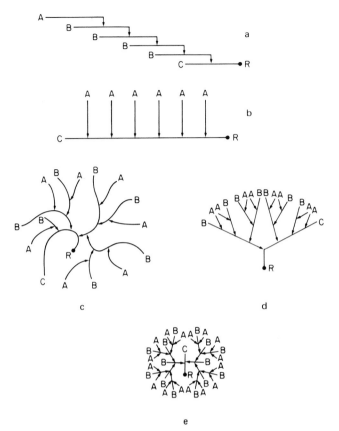

Fig. 1. Diagrammatic representations of the molecular structure of amylopectin as proposed by (a) Haworth, (b) Staudinger, (c) Meyer, (d) Meyer with structure redrawn as a regularly rebranched structure, (e) revised Meyer structure. For definition of A, B, and C chains, see the text. R = reducing group. Note that in structure (e), only half the B chains carry A chains, and that half of the B chains have their nonreducing end groups inside the molecule, and not at the surface as in structures (c) and (d). This figure is reproduced from Manners (1974a).

2. The A:B Chain Ratio in Amylopectin

A survey of the literature (Table V) shows a considerable range of A:B chain ratios. Whereas the first measurements (Peat *et al.*, 1956) indicated that waxy maize starch contained slightly more A chains than B chains, later results (Marshall and Whelan, 1974) suggested that more than twice as many A chains as B chains were present. This surprising result had direct implications on diagrammatic representations of the molecular structure (Fig. 1). With one exception (Marshall, 1975), the consequences of the results of Marshall and Whelan (1974) do not appear to have been incorporated into such diagrams.

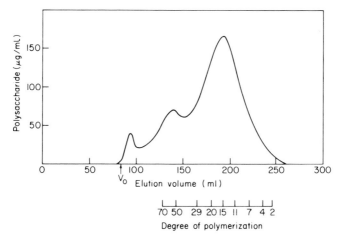

Fig. 2. Profile of chains in broad bean amylopectin as revealed by debranching with bacterial pullulanase followed by gel filtration on Sephadex G-50. An apparently trimodal distribution of chains is shown. Experimental results by courtesy of R. W. Gordon.

Table V

Ratio of A Chains to B Chains in Amylopectins

Source of amylopectin	Ratio	Reference
Waxy maize starch	1.5:1[a]	Peat et al. (1956)
Potato amylopectin	1.3:1	Bathgate and Manners (1966)
Waxy sorghum starch	1.2:1	Bathgate and Manners (1966)
Potato amylopectin	2.0:1	Marshall and Whelan (1974)
Waxy maize starch	2.6:1	Marshall and Whelan (1974)
Waxy sorghum starch	2.6:1	Marshall and Whelan (1974)
Rice amylopectins	1.2–1.5:1	Enevoldsen (1980)
Waxy rice amylopectins	1.1–1.5:1	Enevoldsen (1980)
Waxy maize starch	1.0:1	Manners and Matheson (1981)
Maize amylopectin	1.2:1	Bender et al. (1982)
Potato amylopectin	1.1:1	Bender et al. (1982)

[a]Calculated from the yield of maltose and maltotriose liberated by a debranching enzyme from the β-dextrin.

A critical reexamination of the method used by Marshall and Whelan (1974) led to the conclusion that it was unsatisfactory in several respects (Manners and Matheson, 1981). When appropriate corrections were devised, the A:B ratio in amylopectin then approached 1:1. Independent experiments by Enevoldsen (1980) and by Bender et al. (1982) support the concept of a macromolecule containing only slightly more A chains than B chains.

3. The Overall Molecular Structure of Amylopectin

Before reviewing the more recent structural models for amylopectin, it is necessary to consider the molecular dimensions of the macromolecule. Molecular weight estimations based on light scattering give values in the range of 10^7-10^8 (Banks and Greenwood, 1975), so that a typical molecule could have a DP of 2.5×10^6. Since amylopectins have a CL of ~ 25, individual molecules could contain $\sim 100{,}000$ component chains and interchain linkages (i.e., branch points). A realistic diagrammatic representation of the macromolecule is not therefore practical. It should be emphasized that any diagram can only show a relatively small number of chains and branch points, and simply expresses possible arrangements of the constituent chains.

In 1970, a revision of the hitherto accepted Meyer–Bernfeld model for amylopectin, and also for glycogen, was published (Gunja-Smith et al., 1970). This was based on enzymic degradation studies using β-amylase, muscle phosphorylase, isoamylase, and pullulanase (Fig. 1). Unfortunately, the interpretation of the results is ambiguous (Banks and Greenwood, 1975; Palmer et al., 1983).

An alternative structure based on the assumption that the natural substrate for Q-enzyme was a double helix of amylose was then proposed (Borovsky et al., 1979). The action pattern of this enzyme, as deduced from model experiments using a low-molecular-weight amylodextrin as substrate (see Section III,D,2), resulted in the formation of a branched macromolecule which when unfolded, had a highly extended structure (Fig. 3a). This structure is not compatible with the results of the partial debranching of amylopectin with pullulanase (Manners and Matheson, 1981) or with the pattern of degradation during the initial stages of α-amylolysis (Bertoft and Henriksnäs, 1982).

A cluster-type structure for amylopectin was proposed by French (1973) to explain the physical properties of the molecule, particularly the high viscosity (Fig. 3b). A somewhat similar type of cluster structure was proposed by Robin et al. (1974), based on enzymic degradation studies of acid-treated (lintnerized) potato starch (Fig. 3c). Unfortunately, our understanding of this structure is lessened since these authors describe A and B chains that are defined differently from those used by all other workers. Hence, comparisons of their calculated ratios of A chains to B chains with those in Table V are not possible.

The overall concept of a cluster model was further refined by Manners and Matheson (1981) (Fig. 3d). This model is compatible with the physical properties and with various enzymic degradation studies that show that the branch points are arranged in "tiers" or clusters and are not distributed randomly throughout the macromolecule. While a consensus view of the

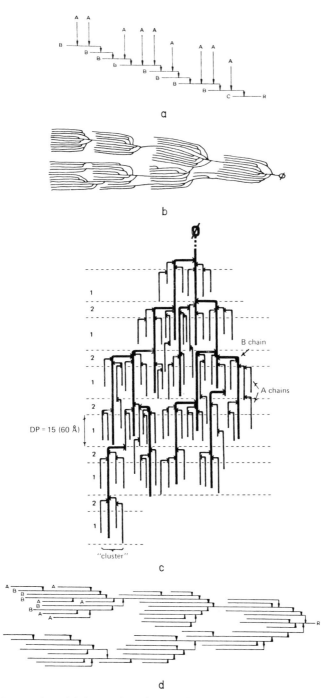

Fig. 3. Structural models for amylopectin: (a) Whelan elongated model (Borovsky *et al.*, 1979); (b) cluster model (French, 1973); (c) cluster model (Robin *et al.*, 1974); (d) cluster model (Manners and Matheson, 1981). In model (d), a proportion of the B chains carry more than one A chain, to account for the change in the A : B chain ratio of the partly debranched material. R or Ø are reducing group. A, B, and C chains are defined in the text. Diagrams are reproduced from the original papers by permission of Carbohydrate Research, Butterworth & Co. Ltd., and Cereal Chemistry.

detailed molecular structure of amylopectin seems therefore to have emerged, the arrangement of the amylopectin molecules within a starch granule remains to be determined.

D. The Intermediate Fraction

The presence of a third starch component has been reported by several workers, but their results tend to be conflicting. This component has been variously described as an anomalous amylose or an anomalous amylopectin.

It is possible that in some starches this material has originated because the initial fractionation of the starch dispersion has not been successful. For example, amylopectin samples from wrinkled-seeded peas and from amylomaize have been reported to have CL values of about 36; however, differential centrifugation of dilute aqueous solutions of these samples showed that they were a mixture of a normal amylopectin and short-chain amylose (Adkins and Greenwood, 1966). The presence of short-chain amylose in an amylopectin sample can be directly shown by enzymic analysis (Banks and Greenwood, 1968). The effective fractionation of amylomaize starch is difficult and has been considered in detail elsewhere (Greenwood and Mackenzie, 1966).

In other work, the presence of an abnormal amylopectin having a higher degree of branching than usual has been reported. In potato and rubber seed starch, approximate CL values of 14 and 16, respectively, have been obtained (Banks and Greenwood, 1959). In wheat starch, ^{13}C-NMR spectroscopy indicated that the abnormal amylopectin had a CL value about 25% shorter than the major amylopectin component (Dais and Perlin, 1982).

The location of the intermediate fraction within the granule and the detailed mechanism of its biosynthesis are not known. Chemically, it contains the same types of glucosidic linkages as amylose and amylopectin, and presumably arises from some variations in the normal activities of starch synthase and Q-enzyme (see Section III).

E. The Starch Granule

Although a detailed discussion of the structure of starch granules would merit a chapter by itself, this section is confined to only three aspects.

Firstly, the heterogeneous nature of the amylose component was readily established by aqueous leaching experiments on intact granules. Using warm water (about 70°C), amylose of relatively low DP and high β-amylolysis limit can be extracted from potato starch granules (Cowie and Greenwood, 1957).

The residual amylose, being of higher molecular weight and having a branched structure, is unable to diffuse out of the granule.

Secondly, it is possible to isolate a relatively insoluble "sac" after the rupture of starch granules. This residue has the iodine-staining properties of amylopectin rather than amylose (Stark *et al.*, 1983) and indicates an amylopectin-rich zone near the surface of the granule.

Thirdly, it has long been known that there are amorphous and crystalline regions within a granule. The former are more susceptible to acidic and enzymic hydrolysis than the latter, whose resistance may be due to double helical chains formed between adjacent amylose molecules, or adjacent clusters of chains in either the same or neighbouring amylopectin molecules. It might be anticipated that the crystalline regions were amylose-rich, but electron microscopy has shown that waxy maize starch granules—which are virtually free of amylose—show both amorphous and crystalline regions, and growth rings (Yamaguchi *et al.*, 1979). A possible relationship between the dimensions of the growth rings and of other structures that are visible in the electron microscope is shown in Fig. 4.

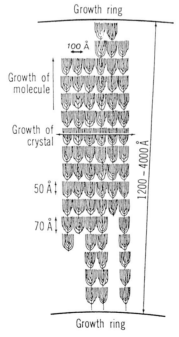

Fig. 4. Schematic representation of the arrangement of amylopectin molecules in a waxy maize starch granule. Diagram by courtesy of K. Kainuma.

It is clear that while substantial progress has been made recently on the determination of the fine structure of the starch components, much further work is still required on the organization of the amylose and amylopectin molecules within starch granules. Additional details of the structure and chemistry of the starch granule are given by Banks and Muir (1980).

III.   THE ENZYMIC SYNTHESIS OF STARCH

A.   Introduction

This section describes the properties of two main classes of enzyme that can synthesize chains of $(1\rightarrow4)$-linked α-D-glucose residues, and of two enzymes that can introduce $(1\rightarrow6)$-α-D-glucosidic interchain linkages into linear or lightly branched $(1\rightarrow4)$-α-D-glucans. While it is now possible to synthesize polysaccharides strongly resembling amylose and amylopectin in the laboratory, the *in vitro* synthesis of a two-component starch granule has not been achieved. A feature of recent work is the demonstration that many of the enzymes involved exist in multiple forms. The *in vivo* significance of this is currently under investigation.

B.   Phosphorylases

These enzymes are transglucosylases that catalyse the reversible transfer of glucose residues between a $(1\rightarrow4)$-α-D-glucan and inorganic phosphate

$$[G]_x + \text{G.I.P.} \rightleftharpoons G(1\rightarrow4)\text{-}[G]_x + P_i$$

where $[G]_x$ is a chain of $(1\rightarrow4)$-linked α-D-glucose residues. Since the reaction is freely reversible, the chain $[G]_x$ may be lengthened or shortened, depending on the relative concentrations of glucose 1-phosphate and inorganic phosphate. The enzyme does not create new chains and is unable to effect *de novo* synthesis. The smallest effective primer or acceptor substrate is maltotetraose (Parrish *et al.*, 1970).

Plant phosphorylases were first described by Hanes in extracts of peas and potatoes, and were used for the first *in vitro* synthesis of an amylose-type molecule. The product gave a blue stain with idoine and was completely degraded by β-amylase. Phosphorylase activity has been detected in virtually all starch-containing plant tissues, and phosphorylase has been purified from many of these sources. Full details of this work can be found in the monograph by Hollo *et al.*, (1971). The present discussion is mainly limited to more recent work.

1. The Purification and Properties of Plant Phosphorylases

Potato phosphorylase, purified about 55-fold, has a molecular weight of 207,000 and unlike animal phosphorylases, is active in the absence of AMP (Lee, 1960a). However, like the animal phosphorylases, it contains 2 molecules of bound pyridoxal 5'-phosphate per molecule (Lee, 1960b). Kinetic studies have shown that enzyme action can involve binary complexes with amylopectin, glucose 1-phosphate, and inorganic phosphate, and also ternary complexes with amylopectin and either glucose 1-phosphate or inorganic phosphate (Gold et al., 1971).

There is evidence for the presence in potatoes of one form of phosphorylase that apparently can catalyse the *de novo* formation of polysaccharide from glucose 1-phosphate (Slabnik and Frydman, 1970). In later work (Frydman and Slabnik, 1973), this activity could not be separated from the classical phosphorylase by gel filtration or ion exchange chromatography, but could be differentiated from it by heat stability and sensitivity to SH reagents, urea, and guanidine hydrochloride. It is possible that this enzyme could be responsible for the synthesis of primers for starch synthase, although its overall distribution in other plants has not yet been established.

Extracts of commercial sweet corn contain a phosphorylase that has been purified 190-fold to a nearly homogeneous state (Lee and Braun, 1973a,b). It had a molecular weight of 315,000 and contained two molecules of pyridoxal 5'-phosphate per molecule, suggesting that the enzyme was dimeric. The primer requirements were similar to those of potato phosphorylase: maltose was inactive, maltotriose was a poor primer, but higher maltosaccharides were effective. These authors emphasised the difficulty in deciding whether or not maltotriose was a primer, since partly purified phosphorylase preparations contained D-enzyme (see Section III,E), which could partly convert malto-triose into higher maltosaccharides. Sweet corn phosphorylase could readily degrade amylopectin, but action on glycogen was limited, so that the enzyme has not been adapted for the *in vivo* degradation of phytoglycogen.

In related studies on an isogenic line of sweet corn, a variety Su_1 contained multiple forms of phosphorylase on gel electrophoresis (Lee, 1972; Lee and Braun, 1973a). Of the total phosphorylase, 90% was in the endosperm, which yielded one major band, two minor bands, and five trace bands of enzyme activity. The embyro contained five bands, two of which were common to the endosperm extracts. Further analysis of these results indicated that three multiple forms, a, b and c, which could form dimers were present. The distribution of the various monomeric and dimeric forms in the endosperm and embyro is shown in Table VI and provides clear evidence of a monomer–dimer interaction with a plant phosphorylase. The results also show the

Table VI

The Relationship of the Multiple Forms of
Sweet Corn Phosphorylase[a]

Proposed form[b]	Endosperm	Embryo
Dimer		
cc	+	−
bc	+	−
bb	+	+
ab	−	+
aa	−	+
Monomer		
c	+	−
b	+	+
a	−	+

[a] Data from Lee (1972).
[b] Monomers a, b, c are arranged in the observed order of increasing electrophoretic mobility, while the dimeric forms are arranged in the expected order of mobility.

variation in the distribution of this enzyme in *different* organs of the *same* plant tissue.

In the normal maize seeds, four different forms of phosphorylase exist (Tsai and Nelson, 1968, 1969). Three of these were present in the endosperm, but the fourth form was present only in the embryo during development and germination. The properties of the four forms are summarised in Table VII. Phosphorylase I was present in all stages of seed germination and endosperm

Table VII

Properties of Multiple Forms of Maize Phosphorylase[a]

Property	Form			
	I	II	III	IV
First detected in development (days)	6	12–16	8–12	?
Present during germination	Yes	No	No	Yes
pH Optimum	5.8	5.9	6.5	6.3
K_m (Glucose 1-phosphate, mM)	3.3	4.0	2.0	1.0
Primer requirement	Yes	No	No	No
Relative activity 22 days after pollination	100	1000	500	20

[a] Data from Tsai and Nelson (1969).

development and required a primer for synthetic activity. By contrast, phosphorylase II appeared only during the stage of rapid starch synthesis and could synthesize polysaccharide in an apparently primer-free system. It is possible that the four forms have different physiological roles in the maize seed. These forms are under separate genetic control, since in the mutant *shrunken-4*, the phosphorylase I activity was only 8% of normal, that of phosphorylase II and III was reduced, while phosphorylase IV was only slightly affected.

Multiple forms of phosphorylase have also been isolated from leaf tissues. The mature banana leaf contained two forms, A and B, with approximate molecular weights of 450,000 and 220,000, respectively (Kumar and Sanwal, 1977). Form A was present in young leaves and decreased during leaf-development, while B increased and was the only form found in mature leaves.

In spinach leaves, the situation is even more complex, in that different phosphorylases are present in the chloroplast and the cytoplasm (Preiss *et al.*, 1980). The enzymes differ in kinetic properties, although their molecular weights (194,000 and 204,000) are not dissimilar. The chloroplastic enzyme has a subunit of molecular weight 92,000, further evidence of a monomer–dimer relationship.

2. The Role of Phosphorylase

With the discovery of starch synthase in 1960, the synthetic role of phosphorylase has had to be reassessed. Many workers have suggested that phosphorylase is primarily a degradative enzyme, and this aspect is considered later in this chapter. In particular, the *in vivo* ratio of inorganic phosphate to glucose 1-phosphate in many plant tissues is quite unfavourable for starch synthesis. However, other workers, in particular Badenhuizen (1969, 1973), have been reluctant to abandon the synthetic role for phosphorylase. For example, in certain plant tissues where starch is being actively synthesized, little or no starch synthase activity could be detected, while phosphorylase was readily demonstrated. Studies of this type require that the assay methods for the two enzymes should be of comparable sensitivity, and this may not always have been the case (Preiss and Levi, 1979).

The previous section showed that multiple forms of phosphorylase exist in certain plants, and that some of these are most active during periods of starch synthesis. It is possible that such enzymes are involved in the synthesis of primers for starch synthase (see Section III,C). This would be a key role in the initiation of starch synthesis. For further discussion and speculation of the *in vivo* role of phosphorylase, the papers and reviews by Badenhuizen (1969, 1973), de Fekete and Vieweg (1973, 1974), and Frydman and Slabnik (1973) should be consulted.

C.   Starch Synthase

The discovery of glycogen synthase in animal tissues by Leloir and his collaborators, in 1957, prompted a search for a related enzyme in the plant kingdom. This was successful, and in 1960, a bean starch granule preparation that catalysed the transfer of glucose residues from UDPG to the granules was described by Leloir (for a review, see Leloir, 1964).

$$UDPG + [G]_x \xrightarrow{\text{synthase}} G\text{-}(1{\to}4)\text{-}[G]_x + UDP$$

Soon afterwards, it was shown that ADPG was a much more active donor substrate:

$$ADPG + [G]_x \xrightarrow{\text{synthase}} G\text{-}(1{\to}4)\text{-}[G]_x + ADP$$

Other nucleoside diphosphate sugar compounds were inactive, and the involvement of ADPG rather than UDPG in the starch synthase reaction is now firmly established. The enzymic activity may be conveniently assayed by measurement of the incorporation of radioactivity from ADP-$[^{14}C]$glucose into $(1{\to}4)$-α-D-glucan.

Although ADPG occurs only in trace quantities in plant tissues, both it and UDPG can be readily synthesized by the corresponding pyrophosphorylase reactions:

$$ATP + \text{glucose 1-phosphate} \rightleftharpoons ADPG + PP_i$$
$$UTP + \text{glucose 1-phosphate} \rightleftharpoons UDPG + PP_i$$

and also from sucrose, by the related sucrose synthase reactions

$$\text{Sucrose} + ADP \rightleftharpoons ADPG + \text{fructose}$$
$$\text{Sucrose} + UDP \rightleftharpoons UDPG + \text{fructose}$$

The latter reactions could provide a metabolic route for the synthesis of starch from sucrose (see Section III,E), and the enzymes that catalyse these reactions have been demonstrated in maize endosperm (de Fekete and Cardini, 1964).

The starch synthase transfers glucose residues from ADPG to an acceptor substrate $[G]_x$, and therefore, like phosphorylase, it does not create new polysaccharide chains. In the laboratory, maltose and higher maltosaccharides could serve as acceptors, although glucose was inactive. The *in vivo* primer has not been clearly identified, and there are reports (see Section III,C,4) of *de novo* synthesis. This latter involves a difficult area for experimental work, since the presence of even a minute trace of maltosaccharide in the enzyme preparation, e.g. by the slight action of α-amylase on a starch granule, could give misleading results. Further discussion of possible *de novo* synthesis will be given in Section III,C,4.

The current literature contains an impressive amount of information on the purification and properties of starch synthases from a wide variety of plant tissues. However, not all of this is consistent, and it is possible that enzyme preparations from *different* organs of the *same* plant may have different catalytic activities. This has already been shown for maize endosperm and embryo preparations (Akatsuka and Nelson, 1966). The earlier studies were hampered by the insoluble nature of the enzyme, which was firmly bound to starch granules. Later work has shown the presence of soluble forms of the enzyme in some plant tissues. It is appropriate therefore to consider the properties of the two forms of starch synthase separately, and to then discuss their *in vivo* role and the metabolic regulation of their activity.

1. The Purification and Properties of Particulate Starch Synthase Preparations

The classic work of Leloir *et al.* (1961) was carried out on carefully prepared starch granules, from young potatoes, sweet corn, and beans, which were stored in a vacuum in a dry state. Enzyme activity was lost on grinding the enzyme in the dry state or in buffer solutions, or with detergents or digitonin.

In later studies (Frydman and Cardini, 1967), grinding the granules led to marked changes in the properties of the enzyme. With potato and waxy maize preparations, a threefold to fourfold increase in activity with ADPG was observed but UDPG was no longer an acceptable substrate. Treatment of the granules led to swelling, followed by loss of the interior of the granules, leaving sacs, which showed an increased level of enzymic activity. This experiment showed the location of some starch synthase on or near to the granule surface. The same workers also examined the donor specificity of various particulate preparations; in all cases, ADPG was the most effective donor, but the relative activity with other sugar nucleotides varied considerably (Table VIII).

Further evidence for the nonidentity of granule bound starch synthases from different plant tissues was obtained by Murata and Akazawa (1968, 1969a). Potassium ions have a stimulatory effect on enzyme activity. The effect was most marked with sweet potato roots (six-fold activation by 0.1 M potassium chloride), was substantial with broad bean seeds (3.5-fold activation), and was much smaller (about 1.5-fold activation) with white potato tubers, taro tubers, rice seeds, and barley seeds. In the sweet potato, maltosaccharides also stimulated the incorporation of radioactivity from ADP-[^{14}C]glucose into starch, and this effect was additive to that due to potassium ions (Murata and Akazawa, 1969b).

The insoluble starch synthase is missing from the endosperm of waxy cereal seeds, whose starch does not contain amylose (Nelson and Rines, 1962). This was first considered (Whelan, 1963) as evidence that amylose was formed from

Table VIII

Donor Specificity of Particulate Starch Synthase Preparations[a]

Source of enzyme preparation	Sugar nucleotide donor[b]			
	ADPG	UDPG	GDPG	TDPG
Potato	36	11.5	1.4	6
Waxy maize	4	0.7	0.5	1
Wrinkled pea	31	10	2.3	1.8
Soya bean	37	<0.1	<0.1	—
Geranium leaves	19	<0.1	<0.1	—

[a] Data from Frydman and Cardini (1967).
[b] Activity expressed as percentage incorporation of radioactivity into starch granules.

UDPG or ADPG by starch synthase, and amylopectin by the combined action of phosphorylase and Q-enzyme on glucose 1-phosphate. However, later work has established that the insoluble synthase may be largely responsible for the synthesis of amylose, while amylopectin results from the combined action of one or more soluble synthase and Q-enzyme complexes. Laboratory experiments on starch synthesis in growing cultures of *Polytoma uvella* have shown (Mangat and Badenhuizen, 1970) that the changes in the insoluble starch synthase follow the percentage of amylose very closely, in confirmation of the suggested role of the particulate enzyme. In other studies on waxy maize mutants, a low level of starch synthase activity of starch granules is ascribed to a small proportion (about 3%) that stain blue with iodine and arise from the embryo and maternal tissue of the seeds (Nelson and Tsai, 1964). Granules from the endosperm were devoid of synthase activity.

2. The Purification and Properties of Soluble Starch Synthase Preparations

The existence of a soluble enzyme that could transfer glucose residues from ADPG to a (1→4)-α-D-glucan was first described by Frydman and Cardini (1964a). The enzyme source was sweet corn, and the acceptor was phytoglycogen. This was followed by an examination of tobacco leaves and potato tubers for a similar enzyme (Frydman and Cardini, 1964b). In both plant sources, a soluble enzyme was detected that could use ADPG as the donor and starch granules as an acceptor. With the leaf preparation, amylopectin and phytoglycogen could also serve as acceptors; this is a further demonstration of the variation in the properties of the enzyme with the plant species.

In the following years, soluble starch synthases were purified from plant tissues, including potatoes (Frydman and Cardini, 1966), sweet corn (Frydman and Cardini, 1965), spinach leaves (Doi *et al.*, 1966; Ozbun *et al.*, 1972), *Chlorella pyrenoidosa* (Preiss and Greenberg, 1967), *Polytoma uvella* (McCracken and Badenhuizen, 1970), and waxy maize (Ozbun *et al.*, 1971). In general, all of these enzymes catalysed the irreversible transfer of glucose residues from ADPG to various acceptors such as amylopectin, phytoglycogen, and maltosaccharides, with the formation of new $(1\rightarrow4)$-α-D-glucosidic linkages. The failure of the earlier experiments to show a soluble enzyme in potato juice could be due to the presence of inhibitory phenolic compounds; this effect could be overcome by homogenisation in the presence of polyvinylpyrrolidone (Frydman and Cardini, 1966). The soluble enzyme differs from the insoluble form in being unable to use UDPG as a donor substrate and starch granules as an effective acceptor. There are therefore either two different starch synthase enzymes in the same tissue or two different forms of a single enzyme whose properties are modified by binding to starch.

The binding of a soluble starch synthase from spinach leaves to amylose was examined by Tanaka and Akazawa (1969). The enzyme was very firmly bound to amylose, but the absolute specificity for ADPG was unchanged. This suggests that binding to a $(1\rightarrow4)$-α-D-glucan does not change the properties of the enzyme, in contrast to the earlier conclusion of Frydman and Cardini (1967). Hence, there are at least two starch synthases of different solubility in most plant tissues.

3. Multiple Forms of Starch Synthase

Not only does starch synthase exist in insoluble and soluble forms, but the latter can also occur in multiple forms. Waxy maize kernels contain two soluble forms that can be separated by DEAE-cellulose column chromatography (Ozbun *et al.*, 1971). The two forms differ in their ability to use various $(1\rightarrow4)$-α-D-glucans as substrates, and one of the forms (I) was able to synthesize polysaccharide in the absence of added primer. Multiple forms have also been detected in various maize varieties by disc gel electrophoresis (Schiefer *et al.*, 1973). Two major types of enzyme were present, and the authors suggested that one of these was responsible for the synthesis of amylose and that the other, in a complex with branching enzyme, produced amylopectin. The ratio of amylose to amylopectin could be related to the ratio of the two forms of synthase.

The soluble synthase from spinach leaves was separated into four forms by DEAE–cellulose chromatography (Ozbun *et al.*, 1972), which had different K_m values for various substrates; form III showed unprimed activity, and the highest affinity for $(1\rightarrow4)$-α-D-glucan substrates. The total enzymic activity was compatible with the observed rate of CO_2 fixation in spinach leaves.

With both maize endosperm and spinach leaves, the form of the enzyme catalysing an unprimed activity was most active in the presence of high concentrations of salt, and the product contained both $(1\rightarrow4)$- and $(1\rightarrow6)$-α-D-glucosidic linkages, indicating the presence of branching enzyme (see Section III,D). This form of the starch synthase had a molecular weight of about 70,000, whereas those forms requiring an added primer for activity have molecular weights in the region of 92,000–95,000 (Hawker *et al.*, 1974).

4. De Novo *Synthesis by Starch Synthase*

The recent literature contains several reports of the apparent unprimed synthesis of starch from ADPG by a starch synthase preparation. For example, a partially purified soluble synthase from potatoes synthesized $(1\rightarrow4)$-α-D-glucan in the absence of added primer in a reaction mixture containing both 0.5 M sodium citrate and 0.05% bovine serum albumin (Hawker *et al.*, 1972). The rate of synthesis was about half that when added amylopectin was present as a primer, and the unprimed product was a branched $(1\rightarrow4)$-α-D-glucan. This would indicate the presence of some Q-enzyme in the synthase preparation. Treatment of the synthase preparation with glucoamylase did not affect the rate of unprimed reaction, hence maltosaccharide chains were believed to be absent from the synthase solution.

Generally similar results have been obtained with a spinach leaf synthase preparation (Fox *et al.*, 1973). With sodium citrate concentrations of 0.25–0.5 M the reaction rate was linear over 1 hr, and greater than that observed when amylopectin or glycogen was used as a primer.

The problem has been investigated by Schiefer *et al.* (1978) using a partly purified starch synthase from sweet corn. In solutions of low ionic strength the reaction was primer-dependent, but in 0.5 M sodium citrate and 0.05% bovine serum albumin, *de novo* synthesis apparently occurred. However, treatment of the enzyme preparation with immobilized amylases completely abolished the unprimed reaction, and the addition of glycogen as a primer restored the original high reaction rate in citrate buffer. The amount of primer required to give a certain reaction rate was much smaller in high concentrations of citrate than in lower concentrations. It was therefore suggested that citrate altered the affinity of the enzyme for the primer but did not change the type of reaction. The endogenous primer in the enzyme preparation was characterised by enzymic analysis as a branched $(1\rightarrow4)$-α-D-glucan.

At the present time, the possibility of true *de novo* synthesis is much less than it seemed to be a decade ago.

5. *The Regulation of Starch Synthesis*

The regulation of glycogen synthesis in animal tissues is at the glycogen synthase level, where various enzymic and hormonal factors control the relative activity of the enzyme. In plant tissues, the work of Preiss and his collaborators (for reviews, see Preiss and Levi, 1980; Preiss, 1982), has established that starch biosynthesis is controlled at the level of ADPG formation. Purified ADPG pyrophosphorylase from *Chlorella pyrenoidosa* was activated about 20-fold by 3-phosphoglyceric acid, the first product of the photosynthetic reduction cycle (Sanwal and Preiss, 1967). ADPG and ATP gave sigmoid-shaped rate-versus-concentration curves in the presence or absence of 3-phosphoglyceric acid. The enzyme was inhibited by inorganic phosphate and ADP. 3-Phosphoglyceric acid was therefore an allosteric effector; it had no action on UDPG pyrophosphorylase. Similar results have been obtained with ADPG-pyrophosphorylase preparations from eight different leaf sources (Furlong and Preiss, 1969) and from maize endosperm (Dickinson and Preiss, 1969). The enzyme from the latter tissue is rather different from the ADPG pyrophosphorylase present in maize embryo tissue (Preiss *et al.*, 1971). Both enzymes were activated by 3-phosphoglycerate, but this metabolite was unable to reverse the inhibition of the embryo enzyme by phosphate. Furthermore, their study of maize mutants suggests that there are two different genes coding for ADPG-pyrophosphorylase, one for that in the endosperm and the other for the embryo enzyme.

D.   Branching Enzymes

These enzymes introduce branch points or inter-chain linkages into linear chains of $(1\rightarrow4)$-linked α-D-glucose residues by a transferase action, as shown in Fig. 5. There appear to be two types of plant branching enzyme, the first of which; Q-enzyme, was originally believed to be able to convert amylose into an amylopectin-type molecule but to be unable to introduce additional branch points into amylopectin. About 4% of branch points was the apparent limit of branching. Those plant tissues, e.g. sweet corn, that synthesize both a two-component starch and phytoglycogen contain a second type of branching enzyme, which can readily introduce further branch points into amylopectin giving a phytoglycogen-type polysaccharide. This distinction in specificity between a Q-enzyme and an amylopectin-branching enzyme is not now as clear-cut as was originally believed, but the two activities are discussed here separately, in accord with the literature.

Since the interchain linkages introduced by Q-enzyme or amylopectin-branching enzyme are hydrolysed by debranching enzymes such as iso-amylase, they are characterised as $(1\rightarrow6)$-α-D-glucosidic linkages.

Experimental studies on branching enzymes have been hampered by the lack of a simple and direct method of assay and by their relative instability. Q-Enzyme and amylopectin-branching enzyme have been assayed indirectly by measurement of the decrease in iodine-staining power of amylose or amylopectin under standard conditions, provided that the enzyme prep-

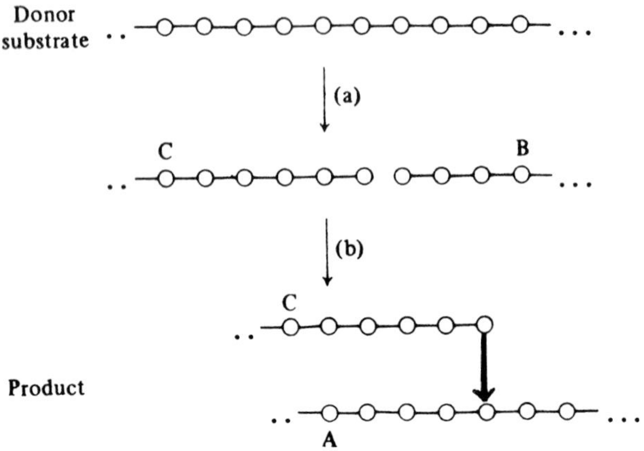

Fig. 5. Formation of $(1\rightarrow 6)$-α-D-glucosidic interchain linkage by Q-enzyme. A short chain of glucose residues (C) is transferred from a donor structure to an acceptor (A), which may be the remaining part of the original chain (B), i.e. intrachain transfer; or it is transferred to an adjacent chain, i.e. interchain transfer. With potato Q-enzyme, interchain transfer appears to predominate (Drummond *et al.*, 1972). Key: $\bigcirc-$, $(1\rightarrow 4)$-linked α-D-glucose residues; arrow shows $(1\rightarrow 6)$-α-D-glucosidic interchain linkage; step (a), scission of nonterminal $(1\rightarrow 4)$ linkage; step (b), formation of $(1\rightarrow 6)$ linkage.

preparations are completely free of α-amylase. This requirement is seldom met, particularly in the early stages of enzyme purification, due to the ubiquitous distribution of α-amylase in plant tissues. Alternatively, the stimulation by a branching enzyme of the synthesis of $(1\rightarrow 4)$-α-D-glucan from glucose 1-phosphate by a phosphorylase can be used as a method of assay. This stimulation arises from the formation of new nonreducing end groups, each of which can act as an acceptor for phosphorylase. Again, it is essential that the phosphorylase preparation must be devoid of α-amylase impurities; for this reason, rabbit muscle phosphorylase rather than a plant phosphorylase has sometimes been used.

Q-Enzyme preparations of varying degrees of purification have been obtained from the potato, broad bean, wrinkled pea, green gram squash, rice, and various varieties of maize (for references, see Manners, 1962). Most of these Q-enzyme preparations show optimum activity at about pH 7.0 and temperatures in the range of 20–25°C (Manners *et al.*, 1968). The following discussion concentrates on potato Q-enzyme, which has been studied in most detail by Whelan and co-workers.

1. The Purification and Properties of Potato Q-Enzyme

Historically, this enzyme was discovered as an impurity in potato phosphorylase (P-enzyme) preparations, and the properties were originally examined using amorphous preparations that were later assessed (Gilbert and Patrick, 1952) as containing only about 5% of active enzyme. Nevertheless, many of the results have been confirmed using more highly purified preparations, a tribute to the value of the original work (Bourne and Peat, 1945; Barker *et al.*, 1949), which was, of course, carried out long before the development of chromatographic techniques for enzyme purification.

Potato Q-enzyme has now been purified, by various chromatographic steps, to near electrophoretic homogeneity (Borovsky *et al.*, 1975). The active enzyme is a monomer with a molecular weight of 85,000, and was free of α-amylase, D-enzyme, limit dextrinase (R-enzyme), and phosphorylase impurities. These authors were able to calibrate the iodine-staining method of assay in terms of the actual number of branch points formed per second, and concluded that their preparation had a specific activity of 6–30 times greater than any previously reported. The molar activity corresponded to the formation of 15 branch points per mole of enzyme per second. In dilute solution, the enzyme was characteristically unstable, unless serum albumin was added. The overall results contrast with another report that Q-enzyme consists of two components, a hydrolase and a transferase, with molecular weights of 70,000 and 20,000, respectively (Griffin and Wu, 1971).

Enzyme action on amyloses of DP 260 and 2050 led to the formation of amylopectin-type polysaccharides, although these were not identical to native amylopectins, as shown by examination of their unit-chain profiles. This is not surprising, since these *in vitro* experiments are not identical to the *in vivo* process, which involves the combined action of Q-enzyme and a starch synthase. In contrast to earlier studies (Peat *et al.*, 1959), the purified Q-enzyme could introduce additional branch points into amylopectin, giving a product with ~6% branch points, but differing significantly from phytoglycogen.

2. The Branching Activity of Q-Enzyme

The mechanism of potato Q-enzyme action on various substrates has been examined in detailed by Borovsky *et al.* (1976, 1979). With amylose, enzyme action involved a random endo-type transglycosylation of the substrate chains, with both interchain and intrachain transfer; i.e. in Fig. 5, the acceptor chain (A) may be part of a different chain to the donor chain or may be part of the original donor chain. Using linear substrates, chains of DP < 40 did not function as acceptors, while those of DP of up to 50 functioned only weakly (Borovsky *et al.*, 1976). However, when a simple branched substrate—amylodextrin—was used, the minimum DP for an effective substrate was reduced to about 25 (Borovsky *et al.*, 1979). Amylodextrin was prepared by the prolonged action of cold acid on waxy maize starch and usually consisted of an A chain and a C chain containing about 10 and 14 glucose residues, respectively (French, 1973). Enzyme action on amylodextrin occurred in two ways, as shown in Fig. 6. The close association of the A and C chains, most probably in a double helical conformation, was believed to provide a favourable environment for the rapid introduction of a second branch point into the amylodextrin molecule. This action of Q-enzyme on a substrate containing two relatively short chains has been extrapolated to amylose, which may also exist as a double helix, and forms the basis of one possible structure for amylopectin (Borovsky *et al.*, 1979). Random enzyme action could give rise to the elongated structure showing some of the features of the older Haworth and Staudinger models, as described in Section II,C (see also Fig. 3a).

Drummond *et al.* (1972) noted that amylopectin that had been incubated with Q-enzyme and then with pullulanase produced maltohexaose and other higher maltosaccharides. However, when amylose was similarly treated, maltohexaose was not formed. This could indicate that maltohexaose chains represent the residual segment of a donor chain after the transfer by Q-enzyme of the outer part of that chain. Hence, Q-enzyme may not be able to act on (1→4) linkages that are less than six glucose residues from a nearby branch point. The length of chain which can be transferred by Q-enzyme has not yet been determined. With the related liver and muscle branching enzymes, there is good evidence that maltoheptaose chains can be readily transferred (Verhue and Hers, 1966; Brown and Brown, 1966).

While the action of Q-enzyme on various branched substrates provides a convenient *in vitro* method for studying the specificity of the enzyme, the *in vivo* situation is rather different. During amylopectin synthesis, a chain-forming enzyme and Q-enzyme react together. The first studies of this type used potato phosphorylase as the chain-forming enzyme, and the product superficially resembled amylopectin (Barker *et al.*, 1950). With the realisation that starch synthase is a more appropriate enzyme, Doi (1969) synthesized a high-molecular-weight polysaccharide by the combined action of spinach starch synthase and potato Q-enzyme on ADPG. The product had an iodine absorption spectrum similar to waxy rice starch, a β-amylolysis limit ranging from 53 to 60%, was partly hydrolysed by a *Pseudomonas* isoamylase, and had a sedimentation constant showing a strong concentration dependence, which is typical of a native amylopectin. Although the individual chain profiles were not examined in this study, there seems little doubt that an amylopectin-type α-glucan had been formed. The synthesis would appear to be a sequential process in which Q-enzyme introduced new branch points into growing chains of (1→4)-linked α-D-glucose residues, which in turn provided more acceptor sites for the glucose residues being transferred from ADPG by starch synthase. The overall process leads to the formation of both A and B chains.

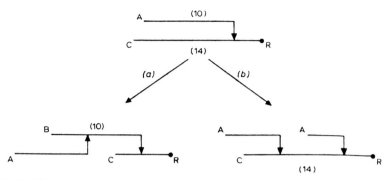

Fig. 6. The action of Q-enzyme on amylodextrin, from Borovsky *et al.* (1979). The approximate number of D-glucose residues in the A and C chains is shown in brackets. Step (a) is transfer of a segment of the C chain to the A chain; step (b) is transfer in the opposite direction.

3. The Purification and Properties of Amylopectin-Branching Enzymes

Since sweet corn differs from most plants in synthesizing both starch and phytoglycogen, it was of interest to examine extracts of this plant for a branching enzyme capable of acting readily on amylopectin. This work was originally carried out independently in three laboratories. Lavintman and Krisman (1964) and Manners and Rowe (1964) first reported the presence in sweet corn of an enzyme that converted amylopectin into a phytoglycogen-type polysaccharide. Later work from these two laboratories (Lavintman, 1966; Manners *et al.*, 1968) described the separation of an amylopectin-branching enzyme from a Q-enzyme and examined the effect of pH, temperature, and various inhibitors on the two enzymes. The branch points introduced into amylopectin, like those in potato amylopectin, were hydrolysed by an isoamylase preparation from sweet corn.

The general conclusions were later confirmed by Hodges *et al.* (1969), although their enzyme preparation had little action on samples of potato and maize amylopectin, having β-amylolysis limits of 46 and 43%. These values are 10% lower than other literature values, hence their samples may not have been typical. It should also be appreciated that the action of a branching enzyme in introducing further branch points into amylopectin may be sterically restricted by the presence of the original branch points. These constraints do not exist if a branching enzyme is acting in concert with starch synthase.

Lavintman (1966) also reported that Q-enzyme preparations from wheat, barley, and normal maize had no action on amylopectin. It seems probable that the existence of the amylopectin-branching enzyme in higher terrestrial plants is confined to those plants, such as sweet corn, that synthesize a phytoglycogen.

4. Multiple Forms of Branching Enzymes

It has been assumed so far that only one Q-enzyme or amylopectin-branching enzyme is present in a particular plant tissue. However, it has been known for some time (Fredrick, 1971) that certain algal branching enzyme preparations exhibited heterogeneity on polyacrylamide gel electrophoresis. There is now good evidence for the existence of multiple forms of branching enzymes in spinach leaves and maize endosperm.

Two multiple forms of spinach leaf Q-enzyme have been reported (Preiss and Levi, 1979), one of which was associated with starch synthase activity. The two forms have molecular weights of $\sim 80,000$ but show different relative activities in buffers of different pH. Both forms could stimulate starch synthase activity, in accord with the previous work of Doi (1969).

By using two different assay methods, Boyer and Preiss (1978a) were able to detect two forms of branching enzyme in developing maize kernels. The two forms were separated by DEAE–cellulose chromatography. One form (enzyme I) was free from starch synthase activity, and the ratio of activity, determined by phosphorylase stimulation and by iodine staining, was ~ 30–60. Enzyme II was associated with starch synthase and the assay ratio was ~ 300–500; this enzyme preparation could be fractionated further on 4-aminobutyl-Sepharose to give fractions IIa and IIb. All three enzyme preparations could introduce further branch points into amylopectin, IIb being more active than IIa in this respect.

In amylomaize, which has a high amylose content and an amylopectin with a low degree of branching, the total content of Q-enzyme was about one-third of that in normal maize (Baba et al., 1982). Amylomaize also contained three Q-enzyme fractions with approximate molecular weights of 90,000, 92,000, and 45,000. These authors concluded that the decreased content of amylopectin in amylomaize starch granules was related to the reduction in the total Q-enzyme content and not to qualitative changes in the properties of the enzymes.

In comparative studies of various maize genotypes, Boyer and Preiss (1978b, 1981) have obtained evidence for independent genetic control of the multiple forms of the branching enzymes and starch synthases. Kernels from an amylose-extender (ae) mutant lacked branching enzyme IIb and had a low activity of citrate-stimulated starch synthase I. By contrast, kernels from a dull (du) mutant did not contain starch synthase II and branching enzyme IIa (see Table IX). Analysis of the double mutant ae du showed a combination of these enzyme changes indicating that the two mutants had acted independently. In the sugary (su) mutant, which contains phytoglycogen and a lower content of starch, there were also three types of branching enzyme. Fraction I had an altered pattern of activity when amylose was the substrate, giving a product

Table IX

Enzyme Activity in Maize Genotypes[a]

Genotype	Amylose content of starch(%)	Starch synthases		Branching enzymes		
		I	II	I	IIa	IIb
Normal	26	+	+	+	+	+
ae	66	−	+	+	+	−
du	42	+	−	+	−	+
ae du	—	−	−	+	−	−

[a]Data from Boyer and Preiss (1981). +, normal level; −, reduced level of activity.

staining with iodine in the range 400–500 nm, i.e., a phytoglycogen. In terms of current enzyme nomenclature, fraction I would be an amylopectin-branching enzyme, while fractions IIa and IIb would be Q-enzymes.

All of this work has established that normal maize and mutants contain two starch synthases and three branching enzymes. The biosynthesis of amylopectin therefore involves not simply a two-enzyme complex of a starch synthase and a branching enzyme, but various pairs of these enzymes, which presumably interact to form different regions of the amylopectin macromolecule.

E. The Biosynthesis of Starch

The previous sections have described current knowledge of various starch-synthesizing enzymes, much of which was reported during the last decade. It is now possible to synthesize *in vitro* both amylose and amylopectin-type polysaccharides, but not a two-component starch granule.

Many problems remain for solution. For example, the amylose content of most starches, including those from relatively uncommon plants, e.g. Japanese wild plants (Fujimoto *et al.*, 1981a–d), is remarkably constant at 20–30%, but the mechanism of the control of the amylose content is unknown. If, as has been suggested, two starch synthases are present in most tissues, one of which acts together with Q-enzyme to form amylopectin, then the synthesis of high-amylose starches in barley, maize, and wrinkled peas represent major variations in the mechanism of this control. The synthesis of both a starch and phytoglycogen in sweet corn continues to be an intriguing problem.

The initiation of starch synthesis has been the subject of much speculation. If true *de novo* synthesis is not possible with starch synthase, then some mechanism for the production of suitable primers by another enzyme system is required. In this respect, the ability of a potato starch enzyme preparation to transfer glucose from UDPG, but not ADPG, to a glycoprotein acceptor (Lavintman and Cardini, 1973) is relevant. The product of this transferase action could then serve as an acceptor for glucose residues transferred from ADPG.

While some reviewers have attempted to produce a unified scheme for the synthesis of plant starches, this may well be impossible in view of the metabolic differences between leaf starches and those from other tissues. In leaves, the starch has a transient existence, being deposited in the chloroplasts during periods of active photosynthesis, and being degraded during periods of darkness to provide metabolic energy. By contrast, starch in storage organs, such as tubers, rhizomes, and cereal endosperms, can be stored over several months, and can be broken down more slowly than the leaf starches. The kinetic properties of the leaf and storage-tissue enzymes and their regulatory mechanisms would therefore be expected to be different. Moreover, in several plant species, the photosynthate partitioning into chloroplast starch is a programmable process that can be manipulated by varying the length of the daily photosynthetic period (Chatterton and Silvius, 1980).

From the point of view of the plant physiologist, starch synthesis continues to pose problems. It is known that the granules grow by apposition, i.e. the successive and gradual addition of new polysaccharide from the outside, and evidence for the location of starch synthase on the granule surface has already been cited (see Section III,C,1). During the development of many plants, there are increases in the average size of the granules, their starch content, their amylose content, and in the molecular size of the two components. Typical results for potato starch are shown in Table X. It follows that the complete starch synthesizing system is in a dynamic state, especially in leaves, and can respond to changes in the physiological state of the plant tissue. The situation is even more complex in the cereals such as wheat and barley, where there are bimodal distributions of granule size and, to some extent, the synthesis of the large and small granules is independent of each other.

The plant physiologist is also interested in the synthesis of starch from sucrose, which is a major carbon assimilation product in chloroplasts. In theory, sucrose synthase could provide ADPG from sucrose and ADP (see Section III,C), but the enzyme in some tissues, e.g. ripening sorghum grains (Sharma and Bhatia, 1980) is more active with UDP than with ADP. From this and other observations, these authors suggest that the sugar nucleotides required for starch synthesis arise mainly from pyrophosphorylase reactions, rather than the reversal of sucrose synthase. Nevertheless, they were able to demonstrate an efficient transfer of $[^{14}C]$glucose from sucrose to starch in the presence of ADP and, to a lesser extent, UDP.

Table X

The Properties of the Starches from Various Sizes of Potatoes[a]

Property	Potato size (cm)						
	1	1–2	3–4	4–5	6–7	8–9
Size of granule (μm)	18	22	34	38	46	54
Iodine affinity[b]	2.95	3.35	3.75	4.30	4.36	4.42
Gelatinization temperature (°C)	52–57	—	—	54–57	—	54–56
Amylose component						
β-Amylolysis limit (%)	92	86	84	83	79	72
Degree of polymerization	750	1000	1800	2200	2700	3400
Phosphorus content (%)	0.000	0.000	0.000	0.000	0.002	0.005
Amylopectin component						
Average chain length	21.0	22.4	—	24.6	25.0	26.0
β-Amylolysis limit (%)	49	52	—	56	58	59
Exterior chain length[c]	12–13	13–14	—	16	16–17	17
Interior chain length[d]	7–8	8	—	7–8	7–8	8
Phosphorus content (%)	0.029	0.039	0.045	0.048	0.049	0.049

[a] Data from Geddes et al. (1965).
[b] Expressed as mg iodine bound/100 mg starch.
[c] Number of D-glucose residues removed by β-amylase + 2.0.
[d] Number of D-glucose residues between interchain linkages.

The pathway from sucrose to starch may therefore be less direct than that shown in Section III,C and could involve the following reactions:

$$\text{Sucrose} + \text{UDP} \xrightarrow{\text{sucrose synthase}} \text{UDPG} + \text{fructose}$$

$$\text{UDPG} + \text{PP}_i \xrightarrow{\text{UDPG} - \text{pyrophophorylase}} \text{UDP} + \text{glucose 1-phosphate}$$

$$\text{Glucose 1-phosphate} + \text{ATP} \xrightarrow{\text{ADPG} - \text{pyrophosphorylase}} \text{ADPG} + \text{PP}_i$$

$$\text{ADPG} + [\text{G}]_x \xrightarrow{\text{starch synthase}} [\text{G}]_{x+1} + \text{ADP}$$

A comprehensive review of the sucrose–starch transformation in photosynthetic tissues of rice, maize, wheat, barley and sorghum and in potato tubers has been provided by Avigad (1982).

The genetic control of the starch synthesizing system must not be overlooked. This aspect can be illustrated by the studies of Ozbun et al. (1973) on developing maize kernels. During development from 8 to 28 days after pollination, the activity of both ADPG-pyrophosphorylase and starch synthase was more than adequate to account for starch synthesis. However, in kernels of the mutant *shrunken-4*, the starch content was only 25–30% of normal and the ADPG pyrophosphorylase activity was only 10–12% of normal. This shows that in normal maize endosperms, the major part, if not all, of the starch is synthesized by the ADPG pathway.

The presence of two types of noncarbohydrate constitutents of starch granules—lipid and phosphate—merits comment at this stage.

Many starch granules contain up to 1% of lipid (Vieweg and de Fekete, 1980), which in cereals is mainly lysolecithin, in maize leaves mainly fatty acids, and in the potato lecithin and serine–cephalin. Most of this lipid is present as a complex with amylose, and the role of lipids on the *in vitro* action of various enzymes was therefore studied. These authors found that starch synthase in maize leaves was activated by various lipids to such an extent that they suggested starch synthase required a lipid–amylose complex for activity. Maize endosperm Q-enzyme was completely inhibited by the phospholipids from egg yolk, and the synthetic action of *Vicia faba* phosphorylase was inhibited by lysolecithin. Since amylopectin does not complex with phospholipids, the authors suggested that amylose may be protected from Q-enzyme by the phospholipids, which could regulate the concomittant rates of synthesis of amylose and amylopectin. Further studies on the effect of physiological concentrations of plant lipids on the starch synthesizing systems are required.

Many samples of amylopectin, but not amylose (Table X), contain small amounts of esterified phosphate groups, mainly as residues of glucose 6-

phosphate but with some glucose 3-phosphate residues. These have a marked effect on the physical properties of the amylopectin and the extent of degradation by α- and β-amylase (Takeda and Hizukuri, 1982). The origin of this phosphate is not known. In general, glucose 6-phosphate and other glycolytic intermediates are not substrates for starch-synthesizing enzymes, and it remains to be determined whether the glucose 6-phosphate residues are incorporated into a growing $(1 \rightarrow 4)$-α-D-glucan chain or arise by the enzymic phosphorylation of certain glucose residues in amylopectin. This latter is not random, since one-third of the phosphate groups are present in the inner regions of B chains and two-thirds in the outer parts of the B chains and in the A chains.

The literature contains several outstanding problems concerning starch synthesis. Three examples are given: (a) What is the *in vivo* role of D-enzyme, the disproportionating enzyme in potato juice, which can synthesize iodine-staining maltosaccharides from maltotriose (Peat *et al.*, 1959)? (b) What is the *in vivo* role of T-enzyme, also from potato juice, which can transform $(1 \rightarrow 4)$-α-D-glucosidic linkages into $(1 \rightarrow 6)$ linkages in low-molecular-weight substrates (Abdullah and Whelan, 1960)? (c) What is the origin of the small proportion of D-fructose residues that have been reliably reported in some samples of waxy maize starch (Whelan and Roberts, 1952)?

IV. THE ENZYMIC DEGRADATION OF STARCH

A. Introduction

Studies on this aspect of starch metabolism have been carried out for more than 50 years and show no sign of ceasing. In particular, the use of amylases for the industrial saccharification of starch for the production of glucose, maltose, and related glucose syrups is becoming of increased importance, although this subject is not described in detail here. This section describes the properties of the β- and α-amylases that act only on the $(1 \rightarrow 4)$-α-D-glucosidic linkages in starch, the debranching enzymes, which hydrolyse the $(1 \rightarrow 6)$-α-D-glucosidic inter-chain linkages, and the various α-D-glucosidases and glucose-producing amylases that degrade the oligosaccharide end products of amylase action to the monosaccharide level.

B. β-Amylase

1. Occurrence and Isolation

The presence of β-amylase in extracts of many higher plants is well documented (for example, French, 1960). The earlier literature contained some reports of microbial β-amylases, and in the last decade these reports have been substantiated by the isolation and purification of true β-amylases from various microorganisms, particularly *Bacillus* spp. (for a review, see Fogarty, 1983). However, this discussion is limited to plant β-amylases.

The classical work on β-amylase was carried out on highly purified preparations from malted barley, sweet potatoes, and soya beans (French, 1960). In recent years, β-amylase has been isolated and purified from other plant tissues (see Table XI). In many cases, this has involved separation from large amounts of α-amylase; the removal of the last traces of α-amylase impurity is not easy and, in the period 1950–1960, led to the erroneous postulation of a new starch-degrading enzyme termed Z-enzyme (Peat *et al.*, 1952a,b; Banks *et al.*, 1960; Cunningham *et al.*, 1960). Many β-amylases are SH-enzymes and are therefore inactivated by oxidation or metal ions; the SH groups may be involved in the *in vivo* regulation of enzymic activity (Spradlin *et al.*, 1969).

The molecular weight of β-amylase varies with the plant source (Table XI), and there is evidence that the enzymes from sweet potato and *Vicia faba* exist as tetramers (Spradlin *et al.*, 1969; Chapman *et al.*, 1972). However, the mode of action of β-amylase appears to be the same, irrespective of the plant source.

In some plant tissues, multiple forms of the enzyme exist (Manners, 1974b). In wheat and barley, the enzyme occurs in both free and bound forms; the latter can be released with thioglycerol. During germination, the amount of free enzyme increased, while the amount of bound enzyme decreased (Pollock and Pool, 1958; Kruger, 1970). In barley the situation is more complex, since β-amylase I and II in extracts of the ungerminated cereal were transformed into β-amylase III and IV in the malted barley (LaBerge and Meredith, 1969). β-Amylase II and III had similar chromatographic properties on CM-cellulose. β-Amylase I and IV were the major components in the ungerminated and germinated cereal, respectively.

Table XI

The Molecular Weights of Some Purified Plant β-Amylase Preparations

Plant source	Molecular weight	Reference
Barley	55,500	Nummi *et al.* (1966)
Japanese radish	58,000	Aibara *et al.* (1978)
Rice	53,000	Okamoto and Akazawa (1978)
Soya bean	61,700	Gertler and Birk (1965)
Sweet potato	198,000	Spradlin *et al.* (1969)
Broad bean	107,000	Chapman *et al.* (1972)
Wheat	64,200	Tkachuk and Tipples (1966)

While a chemical enzymologist may be content to extract enzymes from homogenates of whole plant tissues, the plant physiologist is concerned with the subcellular distribution of the enzyme in these tissues. In the cereals, β-amylase occurs in the endosperm. In broad bean (*Vicia faba*), enzyme activity in the epidermal layer of the leaves was 3–5 times greater than that in other parts of the leaves and was absent from chloroplast preparations (Chapman *et al.*, 1972). Hence the enzyme was located in the cytoplasmic fraction of the epidermal cells.

2. Mode of Action

Enzyme action on a linear or branched $(1{\rightarrow}4)$-α-D-glucan involves a stepwise hydrolysis of alternate linkages, starting from the nonreducing end, with the liberation of β-maltose as the first product. If the substrate is linear, then β-amylolysis will be complete; if branched, hydrolysis will be confined to those parts of a chain between the nonreducing end group and the outermost branch point, i.e. to the outer or exterior chains. As already described, this property enabled the branched nature of some samples of amylose and the external chain lengths of amylopectins to be examined.

Care is required in studying β-amylase action in the vicinity of a branch point. It is generally accepted that in a β-limit dextrin, the A-chain stubs contain two or three glucose residues and the B-chain stubs contain one or two residues. If the amylopectin contains equal numbers of A and B chains, the average size of the stubs is two glucose residues. If there are more A chains than B chains, the precision of this type of analysis is lost. Moreover, if the enzyme concentration is too low, A-chain stubs contain three or four glucose residues, since a stable intermediate dextrin can be isolated (Lee, 1971). This dextrin can be degraded further by higher concentrations of enzyme to give the conventional β-limit dextrin.

Residues of glucose 6-phosphate that are present in some starch fractions also serve as a barrier to β-amylase (Takeda and Hizukuri, 1982). Enzyme action on phosphorylated linear dextrins ceased, leaving a "stub" of either one or no glucose residue.

Unlike α-amylase, β-amylase has little or no action on starch granules (Dunn, 1974). This important observation has a bearing on the physiological role of the enzyme.

While the hydrolytic action of β-amylase has been known for many decades, the ability of the enzyme to catalyse the transfer of maltosyl residues is a more recent finding (Hehre *et al.*, 1979). Earlier work from this laboratory had shown the synthesis of maltotetraose from β-maltose by sweet potato β-amylase. In later work, β-maltosyl fluoride was used as the substrate, and the transfer of maltosyl residues from it to methyl β-maltoside and p-nitrophenyl α-D-glucoside was observed. β-Amylase can also act on α-maltosyl fluoride, an unusual example of a carbohydrase acting on both anomeric forms of the substrate. Hehre *et al.* (1979) consider that β-amylase catalyses the hydrolysis of α-maltosyl substrates to give β-maltose, and the transfer of maltosyl residues from β-maltosyl substrates to give oligosaccharide chains containing new (1→4)-α-D-glucosidic linkages. Hence β-amylase, like α-amylase, is both a glycoside hydrolase and a glycosyl transferase. While maltosyl fluoride is clearly not a physiological substrate, the concept that β-amylase can catalyse maltosyl transfer should be borne in mind when considering enzyme action in situations where the enzyme and substrate concentration is much higher than that in normal laboratory experiments—e.g. within certain plant cells, and with industrial immobilized enzyme systems.

3. Role of β-Amylase

In some higher plants, β-amylase appears to have a role in the rapid mobilization of starch during germination or other periods of increased metabolic activity. This assumes that the enzyme and substrate are present within the same organelle. In other plants, e.g. soya beans, the function of β-amylase is unknown, especially since little starch is present.

In an industrial brewing situation, β-amylase plays a role in increasing the saccharification of starch during mashing (Manners, 1974c), but this is subordinate to that of α-amylase, which appears in this and other instances to be the most important starch-hydrolysing enzyme.

C. α-Amylases

1. Occurrence and Isolation

α-Amylase is very widely distributed in Nature. Its presence can be expected in plant tissues that actively metabolize starch, for example, germinating cereals and the leaves of many species of plants (Gates and Simpson, 1968). It is also present in lesser or trace amounts in other tissues where its biological function is less obvious, for example, in extracts of some ungerminated cereals and almond seeds. These plant tissues are the traditional sources of β-amylase and α-glucosidase, respectively, and the α-amylase impurity can interfere in the laboratory use of these enzyme preparations unless its presence is recognised and steps are then taken to eliminate it (Section IV,B,1).

Although α-amylase has been highly purified from a wide range of animal and microbial sources (Fischer and Stein, 1960; Fogarty, 1983), this discussion will be confined to plant α-amylases. A range of improved methods are now available for the purification and analysis of α-amylases, including specific precipitation as a glycogen complex (Schramm and Loyter, 1966), the use of cycloheptaamylose derivatives (Weselake and Hill, 1982) for affinity chromatography, and isoelectric focusing and immunochemical techniques (MacGregor and Daussant, 1981). These latter methods are particularly important, since many plant α-amylases occur in multiple forms (Manners, 1974b). Some of these, but not all, are true isoenzymes and are synthesized in different parts of a plant.

Many α-amylases are water soluble proteins having a molecular weight of ~ 50,000 and containing at least 1 gram-atom of calcium per mole, which is essential for enzymic activity (Fischer and Stein, 1960). Most α-amylases do not contain cysteine or cystine and therefore are not affected by SH reagents. However, spinach leaf α-amylases are an exception to these generalisations; they do not appear to contain calcium ions, and are inhibited by N-ethyl maleimide (Okita and Preiss, 1980).

The presence of multiple forms of α-amylase in extracts of germinated barley, malted wheat, malted rye, and malted sorghum has been reviewed previously (Manners, 1974b). The most detailed studies are those of MacGregor and his colleagues, who have identified three components in germinated barley. The major components are α-amylase II and III, which are immunochemically identical (MacGregor and Daussant, 1981). During kilning, there was a substantial conversion of α-amylase III into α-amylase II. α-Amylase I is a minor component that represents only about 5% of the total α-amylase activity (MacGregor, 1977) and is antigenically different from α-amylase II. The overall specificity of the three components is generally similar, but there are differences in the detailed catalytic activity, since α-amylase I could digest large barley starch granules more quickly than α-amylase II (MacGregor, 1980).

2. Mode of Action

α-Amylases catalyse an essentially random hydrolysis of amylose and amylopectin during the initial stages of enzyme action. There is a rapid decrease in viscosity and iodine staining power, but only a limited production of reducing sugars. In the later stages, enzyme action becomes less random, since it is unable to hydrolyse readily $(1 \rightarrow 4)$-α-D-glucosidic linkages that are either terminal or in the vicinity of a $(1 \rightarrow 6)$-α-D-glucosidic linkage. Hence, the products from amylose are maltose ($\sim 90\%$) and either maltotriose or glucose, depending on the enzyme concentration. The products from amylopectin are qualitatively similar, but with the addition of a series of branched α-dextrins (see Fig. 7), which contain the original interchain linkages.

The smallest α-dextrin obtained by the action of α-amylase from malted rye or barley on amylopectin was 6^3-α-D-glucosylmaltotriose (Manners and

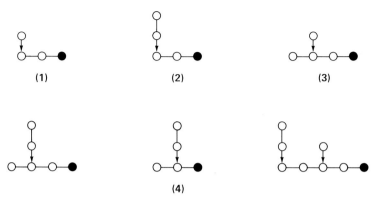

Fig. 7. Typical structures of α-dextrins: **1,** 6^3-α-D-glucosylmaltotriose; **2,** 6^3-α-maltosylmaltotriose; **3,** 6^3-α-D-glucosylmaltotetraose; **4,** 6^2-α-maltosylmaltotriose. Key as in Fig. 5, and ● represents a free reducing group.

Marshall, 1971; Hall and Manners, 1978). Pentasaccharides such as 6^3-α-D-glucosylmaltotetraose and 6^3-α-maltosylmaltotriose were also produced. α-Amylase from *Bacillus subtilis* gave 6^2-α-maltosylmaltotriose as the smallest α-dextrin (French *et al.*, 1972), showing that the detailed specificity of all α-amylases is not identical.

The course of α-amylolysis of amylopectin β-limit dextrins has been examined by gel permeation chromatography (Kruger and Marchylo, 1982). This technique showed an initial and rapid depolymerization stage, followed by the continuing production of low-molecular-weight oligosaccharides. The initial stages could involve the liberation of "clusters" of chains from the amylopectin macromolecule. These authors also found that the patterns of hydrolysis by α-amylases from malted wheat and *Aspergillus oryzae* were not identical—further evidence of the variation in detailed α-amylase specificity with the biological source.

A most important property of α-amylase is its ability to degrade whole starch granules. The current literature contains many elegant electron-microscopic studies that give a clear visual representation of the characteristic pitting of the granules, leading eventually to complete dissolution (see, for example, MacGregor, 1980). Typical results for some cereal starches are shown in Fig. 8. Enzyme action on large and small barley starch granules was significantly different (Bertoft and Henriksnäs, 1982), suggesting that the arrangement of the amylopectin components within the two sizes of granules was different. The results also supported the cluster model for amylopectin (see Section II,C,3 and Fig. 3), since the initial attack by the enzyme involved the preferential hydrolysis of certain linkages with the liberation of α-D-glucan of molecular weight about 3×10^4. Overall, enzyme action on the granules was nonrandom.

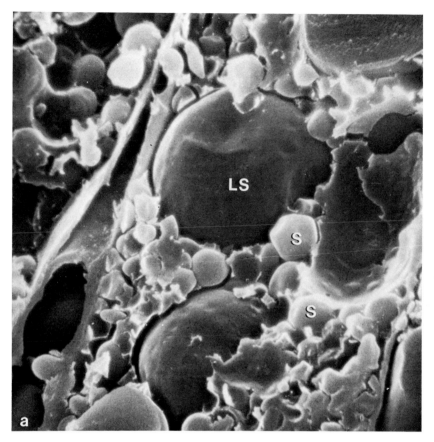

Fig. 8. Scanning electron micrographs of barley starch granules in the endosperm. (a) SEM showing the large (LS) and small (s) granules that are embedded in a storage protein matrix. C represents part of a cell wall (× 1400).

It has been assumed so far that α-amylases act only as hydrolytic enzymes. However, α-amylases from mammalian and microbial sources can transfer glucosyl residues from starch to *p*-nitrophenyl α-D-glucoside with the formation of *p*-nitrophenyl α-maltosaccharide glucosides (Takeshita and Hehre, 1975) and can also act on α-maltosyl fluoride with the formation of maltosaccharides of DP 3–6 (Okada *et al.*, 1979). It seems highly probable that plant α-amylases could also catalyse glycosyl-transfer reactions with certain substrates, and this possibility should not be overlooked in considering their detailed mechanism of action.

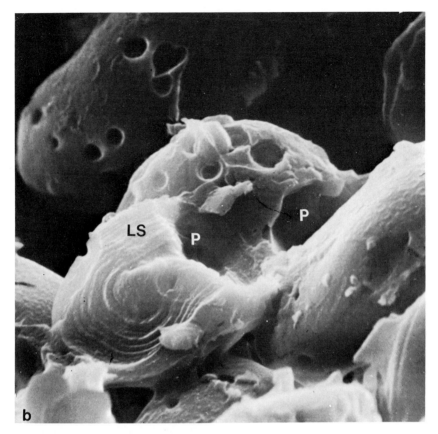

Fig. 8b. SEM of germinated barley showing α-amylolytic pitting (P) of the large starch granules (LS). Most of the small starch granules have been hydrolysed by amylase activity (× 1750).

3. Role of α-Amylase

It is now generally accepted that α-amylase plays the major role in most, but not necessarily all, plant tissues in the *in vivo* breakdown of starch. The literature has been reviewed by Dunn (1974), who emphasized that α-amylase was the only amylolytic enzyme that had any action *in vivo* on starch granules. Other amylolytic enzymes (β-amylase, limit dextrinase, α-glucosidase) act to various extents on the intermediate α-dextrins that are released in a soluble form during the α-amylolysis of the granule.

The major end product of α-amylase action is maltose, together with linear and branched α-dextrins. These oligosaccharides are not metabolised as such

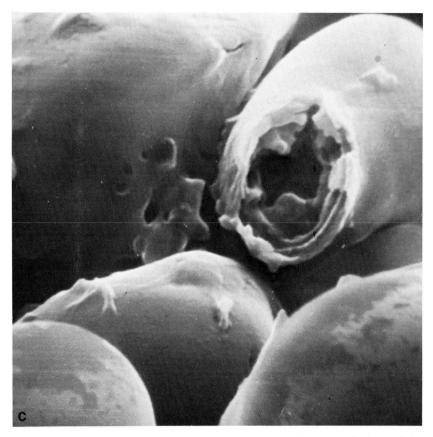

Fig. 8c. SEM of germinated barley showing extensive and relatively localized α-amylolytic attack on a large starch granule. This has exposed a lamella structure within the granule (× 4200).

and require further degradation by α-glucosidases and limit dextrinase to the monosaccharide level. The localization of these enzymes requires consideration, since chloroplast membranes are not readily permeable to glucose and oligosaccharides, which have to be converted into phosphorylated sugars, e.g. 3-phosphoglycerate, for transport out of the chloroplast. Moreover, in some tissues, e.g. pea chloroplasts, there is evidence of starch degradation by phosphorolysis rather than α-amylolysis (Stitt et al., 1978). It is therefore clear that while α-amylase has a key role in the degradation of starch, its complete breakdown requires the participation of several other enzymes.

The development of α-amylase during the germination of cereals and during the growth of other plants has been extensively studied. The biosynthesis of α-amylase and the effect of gibberellic acid on this process has

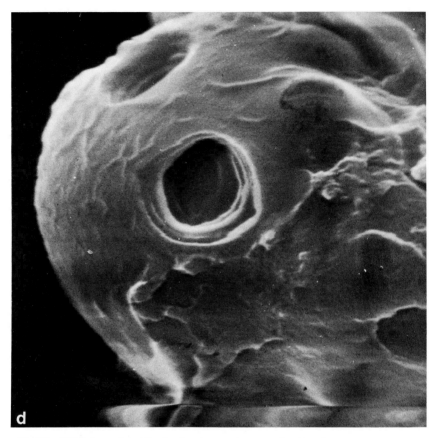

Fig. 8d. SEM of germinated barley showing another example of localized α-amylolytic pitting of a large granule. The progressive amylolytic attack is reflected in the exposure of the lamella rings of the granule (× 2800).

been reviewed by Akazawa and Miyata (1982). However, details of a current controversy (see, for example, Gibbons and Nielsen, 1983; Palmer, 1983) on the actual site of *de novo* synthesis of α-amylase and the mode of dissolution of the starchy endosperm are not given here.

Finally, some *in vitro* applications of α-amylase merit comment. In the laboratory, α-amylase has been widely used for the structural analysis of starch-type polysaccharides, especially from algae and protozoa, and for an indirect method for the determination of the average chain length of branched (1→4)-α-D-glucans (Manners and Wright, 1962). On an industrial scale, α-amylase is the major enzyme for starch degradation during the mashing stage of the brewing process (Manners, 1974c) and is a key enzyme in the initial stages in various enzymic processes for the saccharification of starch (Underkofler *et al.*, 1965).

D. Phosphorylase

1. The Degradative Role of Phosphorylase

In the presence of inorganic phosphate, phosphorylase will degrade chains of (1→4)-linked α-D-glucose residues with the formation of glucose 1-phosphate. Hence an energetically favourable route for the conversion of starch into hexose phosphate is theoretically available. However, the enzyme is unable to attack starch granules, unless the granules can be partly degraded or solubilized by other enzymes. Enzyme action on amylose should be essentially complete, although any interchain linkages would arrest phosphorolysis unless traces of α-amylase were also present in the plant cell. The phosphorolysis of amylopectin would also be incomplete, and confined to the exterior chains unless a debranching enzyme such as limit dextrinase was also present (see Section IV,E,2). In laboratory experiments with purified potato phosphorylase (Liddle *et al.*, 1961), the extent of degradation of various amylopectins (33–45%) was significantly less than by β-amylolysis (56–58%). The size of the exterior chain stubs in a plant phosphorylase limit dextrin has not been determined, but with the corresponding mammalian phosphorylase, these are maltotetraose units. In the presence of limit dextrinase, the barriers to phosphorolysis would be removed, giving virtually complete degradation of amylopectin.

The intracellular location of phosphorylase must be considered. It is present in many chloroplasts, e.g., from spinach leaves (Levi and Gibbs, 1976), and, provided the concentration of inorganic phosphate is high enough, it could account for much of the starch degradation. In fact, both phosphorolytic and amylolytic pathways are involved in spinach chloroplasts (Peavey *et al.*, 1977). However, in other tissues the enzyme is also in the cytoplasm, and since it is not in direct contact with the starch granules, its function remains to be determined (Okita *et al.*, 1979). It is interesting to note that in spinach leaves the chloroplastic and cytoplasmic phosphorylases are different enzymes, showing significant differences in molecular weight and in certain kinetic properties (Preiss *et al.*, 1980).

E. Debranching Enzymes

1. Introduction

Debranching enzymes are a relatively new class of carbohydrase which hydrolyse (1→6)-α-D-glucosidic linkages in amylopectin, glycogen and related α- and β-limit dextrins. Their existence was first suspected during the 1940s when certain plant amylolytic preparations (sorghum malt) showed activities that could not be explained in terms of α- and β-amylase action; this was attributed to a "dextrinase" (Kneen, 1945) and later to "limit dextrinase"

(Kneen and Spoerl, 1948), since it was suggested that enzyme action involved the hydrolysis of the interchain linkages in limit dextrins. The first enzymic debranching of amylopectin by a plant enzyme was reported in 1950–1951; enzyme preparations from the potato and broad bean increased the iodine-staining power and β-amylolysis limit of amylopectin and amylopectin β-dextrin (Hobson et al., 1951). The enzyme was an impurity in certain Q-enzyme preparations and was described as R-enzyme. It had no action on glycogen, but hydrolysed the interchain linkages in α-limit dextrins prepared from glycogen and therefore also showed limit dextrinase activity (Peat et al., 1954). Debranching enzymes have since been isolated from various animal, plant, and microbial sources. This discussion will be confined to hydrolytic enzymes acting on interchain linkages; other enzymes [e.g., amylo-$(1\rightarrow6)$-glucosidase] from animal tissues and yeast that act by a combination of transfer and hydrolysis will not be considered.

The specificities of the plant and microbial enzymes are not always clear-cut, and attempts to classify them, and name them, have not always been entirely successful. Notwithstanding these reservations, there appear to be three main classes of debranching enzyme (Table XII). In an alternative scheme (Lee and Whelan, 1971), only two classes were given, namely, isoamylases and pullulanases; the latter class included both plant and

Table XII

Classification of Debranching Enzymes

Characteristic	Trivial Name		
	Isoamylase	Pullulanase	Limit dextrinase (R-enzyme)
Source	Bacteria, yeast, some plants	Bacteria	Many higher plants
Substrates			
Amylopectin	$+^d$	$+$	$+$
Glycogen[a]	$+$	$\pm$	$-$
Pullulan[b]	$-$	$+$	$+$
α-Dextrins[c]	$+$	$+$	$+$

[a] Glycogen is defined here as the amorphous glucan extracted from animal tissues. It excludes the very high molecular weight particulate glucan isolated by differential centrifugation.

[b] A glucan synthesized by *Pullularia pullulans* (*Aureobasidium pullulans*) consisting essentially of a linear chain of $(1\rightarrow6)$-linked α-maltotriose residues.

[c] Branched α-dextrins with side chains containing more than one glucose residue. Those with maltosyl side chains are not readily attacked by isoamylases, but are good substrates for pullulanases and limit dextrinases.

[d] $+$, readily hydrolysed; $\pm$, slowly hydrolysed; $-$, not hydrolysed.

bacterial enzymes, despite the fact that they differ in their activity towards glycogen (Table XII).

It is recognised that the classification in Table XII does not cover all examples in the literature. For example, *Escherichia coli* produced an isoamylase that had very little action on glycogen, almost no activity on pullulan, but readily acted on glycogen α- and β-limit dextrins (Jeanningros *et al.*, 1976). Germinating rice endosperm contained a limit dextrinase that hydrolysed oyster glycogen and corn phytoglycogen, but at only 4 and 2% of the rate of pullulan, respectively (Iwaki and Fuwa, 1981); these results are therefore compatible with Table XII. Moreover, the actual enzyme and substrate concentrations used in the various laboratories have varied considerably, so that the assessment of rapid or slow hydrolysis cannot always be entirely accurate.

The following discussion will be largely limited to the plant debranching enzymes. The widespread use of the bacterial isoamylases and pullulanases in the laboratory for the structural analysis of $(1\rightarrow4)$-α-D-glucans and their use on an industrial scale for the partial degradation of starch have been described elsewhere (Lee and Whelan, 1971; Norman, 1979, 1983).

2. Occurrence, Isolation, and Properties of Limit Dextrinase (R-Enzyme)

The discovery of limit dextrinase and R-enzyme has already been described. It should be noted that the inability of broad bean R-enzyme to act on glycogen provided a simple test for distinguishing between amylopectins and glycogens (Peat *et al.*, 1954). However, since the R-enzyme could act on α-dextrins derived from glycogen, it could hydrolyse both polysaccharide and oligosaccharide substrates.

Similar enzymic activities were described in extracts of malted barley (MacWilliam and Harris, 1959), and since the activity towards amylopectin could be separated from that towards α-dextrins by alumina adsorption chromatography, it was suggested that two debranching enzymes were present. That acting only on amylopectin was termed "R-enzyme," and that acting on α-dextrins "limit dextrinase." Although first results from this laboratory supported the idea of two distinct debranching enzymes (Manners and Sparra, 1966; Manners and Rowe, 1971), this concept was questioned by Drummond *et al.* (1970), and later work (Manners and Hardie, 1977) led to the conclusion that amylopectin, pullulan, and α-dextrins were hydrolysed by a single enzyme. The view that germinated barley contained two distinct debranching enzymes was no longer justified.

In this and related work, the enzyme has been named limit dextrinase rather than R-enzyme, since it reflects the natural substrate. Although the enzyme acts on the fungal α-glucan pullulan (Table XII), it does not seem logical to change the name to plant pullulanase simply because it will hydrolyze this additional *in vitro* substrate (see Lee and Whelan, 1971). Moreover, the bacterial pullulanases and the plant limit dextrinases differ in their relative ability to degrade glycogen (Abdullah *et al.*, 1966; Lee and Whelan, 1971). For the remainder of this discussion, the name "limit dextrinase" will therefore be used instead of R-enzyme or plant pullulanase.

Limit dextrinases have now been isolated and purified from many plant sources (see Table XIII). In all cases the enzyme could readily hydrolyze α-dextrins, amylopectin β-dextrin, and pullulan, and could slowly hydrolyze amylopectin, but had little or no action on glycogen. The rate of hydrolysis of amylopectin was only 15–23% of that of the corresponding β-limit dextrin (Dunn *et al.*, 1973). It should be noted that amylopectin β-dextrin is not an *in vivo* substrate.

An unusual feature of limit dextrinase action is its ability to *increase* the iodine-staining power of amylopectin and its β-dextrin, in contrast to other amylolytic enzymes that *decrease* this property. The distinction has not been appreciated by some workers (e.g., Swain and Dekker, 1966; Perez *et al.*, 1971; Vlodawsky *et al.*, 1971) who claim to have assayed debranching enzyme activity from the *decrease* in iodine-staining power. Their conclusions on the rate of development of this enzyme in plant tissues such as peas and rice are therefore erroneous.

3. Occurrence, Isolation, and Properties of Plant Isoamylases

Since limit dextrinase has no action on animal glycogens or plant phytoglycogens, it seemed probable that sweet corn, which contains both starch and phytoglycogen, might contain another type of debranching

Table XIII

The Molecular Weights of Some Limit Dextrinase Preparations

Plant Source	Molecular weight	Reference
Barley, malted	80,000[a], 103,000[b]	Maeda *et al.* (1979)
Broad beans	80,000[a]	Gordon *et al.* (1975)
Oats, ungerminated	80,000[a]	Dunn and Manners (1975)
Peas, ungerminated	180,000[a]	Yellowlees (1980)
Potato tubers	87,000–220,000[a]	Ishizaki *et al.* (1983b)
Rice, germinated	58,000[a], 100,000[b]	Iwaki and Fuwa (1981)
Sorghum, malted	90,000[a]	Hardie *et al.* (1976)
Sweet corn	110,000[a]	Lee *et al.* (1971)

[a]Determined by gel filtration.
[b]Determined by SDS–gel electrophoresis.

enzyme. Sweet corn extracts were shown (Manners and Rowe, 1967) to contain a mixture of two debranching enzymes, one acting on amylopectin but not glycogen (i.e., a limit dextrinase) and the other acting on both poly-saccharides (i.e., an isoamylase). The latter increased the β-amylolysis limit of phytoglycogens by 13%.

More recently, a debranching enzyme of the isoamylase type has been isolated from potato tubers (Ishizaki et al., 1978, 1983a). This enzyme occurs together with two forms of limit dextrinase (Ishizaki et al., 1983b); it had no action on pullulan, but readily attacked glycogen. The presence of an isoamylase in potatoes is unexpected, since the anticipated substrate—phytoglycogen—does not occur in this plant. The physiological role of the enzyme is therefore uncertain. An examination of other plant tissues for isoamylase-type debranching enzymes would be worthwhile.

4. Role of Plant Debranching Enzymes

Although in laboratory experiments the plant debranching enzymes act effectively on polysaccharide substrates, it seems unlikely that this is their *in vivo* function. Moreover, limit dextrinase, by itself, has no action on starch granules.

It seems probable that in many plant tissues, the natural substrates are intermediate α-dextrins produced during the α-amylolysis of amylopectin. The hydrolysis of the interchain linkages in these α-dextrins would then permit further degradation by the action of either α-amylase or a mixture of α- and β-amylase, which would also be present in the plant tissues. The overall result will be the complete conversion of amylopectin into a mixture of maltose with smaller amounts of glucose and the lower DP linear maltosaccharides, depending upon the relative *in vivo* concentration of the various enzymes.

A metabolic relationship between α-amylase and limit dextrinase in cereals might be expected, since both enzymes are synthesized during germination in the aleurone layer of barley by a process that is stimulated by gibberellic acid; experiments with cycloheximide and density labelling in deuterium oxide suggest that the observed increases in activity are the result of *de novo* protein synthesis (Hardie, 1975).

In other plant tissues, e.g. pea chloroplasts, limit dextrinase could enhance the rate of phosphorolysis of amylopectin, In this event, the initial substrate would be high-molecular-weight phosphorylase limit dextrins, and the combined action of phosphorylase and limit dextrinase would completely convert the polysaccharide into glucose 1-phosphate.

In an analogous industrial situation, an exo-acting enzyme (β-amylase) together with a debranching enzyme is used for the production of maltose from starch, in an extremely high yield (Sakai, 1981).

F. Plant α-D-Glucosidases

1. Occurrence and Isolation

The final group of enzymes in this section is the α-D-glucosidases, which hydrolyze maltose and other low-molecular-weight maltosaccharides to glucose. These enzymes occur in many plant tissues, and their specificity is not absolute, in that nigerose and isomaltose can be hydrolysed under laboratory conditions, although these disaccharides are not *in vivo* substrates (Hutson and Manners, 1965). In many plant tissues, the optimum pH of the α-D-glucosidase activity is about 5.0, but in sweet corn, it is about 3.0 (Marshall and Taylor, 1971). The precise function of such an acid α-D-glucosidase is not yet known.

In general, the plant α-D-glucosidases have not been extensively purified; examples where significant purifications have been obtained include enzyme preparations from buckwheat (Takahashi and Shimomura, 1968), rice seeds (Takahashi *et al.*, 1971), sweet corn (Marshall and Taylor, 1971), and sugar beet (Chiba *et al.*, 1978). Multiple forms of certain α-D-glucosidases have been reported (Manners, 1974b), and particularly in germinated green gram (Yamasaki and Suzuki, 1979).

2. Role of α-D-Glucosidases

These enzymes catalyse the conversion of the end products of amylase action on starch into glucose. A limited number of plant α-D-glucosidase preparations, e.g. from malted barley (Jorgensen, 1964) and sugar beet (Chiba *et al.*, 1978), also act directly on soluble starch to give glucose. This activity has been described as glucoamylase or amyloglucosidase; whether it occurs *in vivo* on native starch granules remains to be determined.

V. SUMMARY

Present knowledge of the structure of starch and of the major starch-metabolizing enzymes has been reviewed in this chapter, and the relevant metabolic pathways are summarized in Figs. 9 and 10. It should be emphasised that these metabolic maps contain information from a variety of plant tissues; not all the pathways need be operative within a particular plant or, indeed, within a particular organ of that plant.

The current literature contains a wealth of information on many additional aspects of starch metabolism, particularly in certain specialized tissues. It is

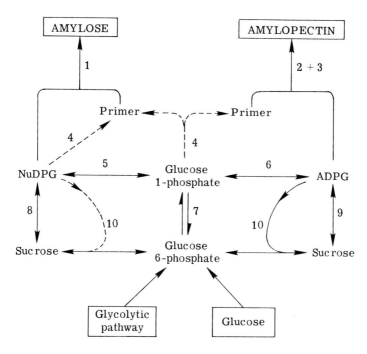

Fig. 9. Metabolic pathways for starch synthesis. NuDPG either UDPG or ADPG. Speculative pathways are shown by a broken line. 1. Starch synthase; 2. Second starch synthase(s) differing from 1 in solubility and in being unable to use UDPG as a donor substrate; 3. Q-Enzyme(s); 4. Phosphorylase or UDPG: protein transglucosylase (?); 5. Pyrophosphorylase with UTP or ATP; 6. Pyrophosphorylase with ATP; 7. Phosphoglucomutase; 8. Sucrose synthase with UDP or ADP; 9. Sucrose synthase with ADP; 10. Sucrose phosphate synthase and sucrose phosphatase, with fructose 6-phosphate. Note that for the sake of clarity, many metabolites, e.g., ATP, ADP, P_i, etc., have been omitted, and that this diagram does not imply that there are two separate metabolic pools of sucrose.

not possible to discuss these here, and attention is drawn to several recent reviews *inter alia*: the metabolism of starch in leaves (Preiss and Levi, 1979), the regulation of starch biosynthesis and degradation (Preiss and Levi, 1980; Preiss, 1982), starch metabolism during cereal grain development (Duffus and Cochrane, 1982), the storage of starch (Jenner, 1982), and factors controlling the mobilization of starch in higher plants (Halmer and Bewley, 1982).

ACKNOWLEDEGMENT

The author is indebted to his colleague, Dr. G. H. Palmer, for the hitherto unpublished electron micrographs shown in Fig. 8.

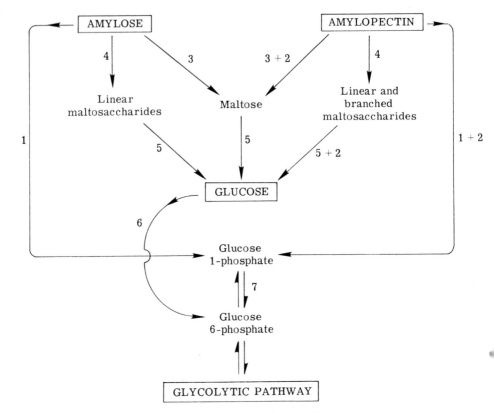

Fig. 10.   Metabolic pathways for starch degradation. 1. Phosphorylase; 2. Limit dextrinase; 3. β-Amylase; 4. α-Amylase; 5. α-Glucosidase; 6. Hexokinase; 7. Phosphoglucomutase.

REFERENCES

Abdullah, M. and Whelan, W. J. (1960). *Biochem. J.* **75**, 12P.

Abdullah, M., Catley, B. J., Lee, E. Y. C., Robyt, J., Wallenfels, K. and Whelan, W. J. (1966). *Cereal Chem.* **43**, 111–118.

Adkins G. K. and Greenwood, C. T. (1966) *Starch/Staerke* **18**, 171–176.

Aibara, S., Yamashita, H. and Morita, Y. (1978). *Agric. Biol. Chem.* **42**, 179–180.

Akatsuka, T. and Nelson, O. E. (1966). *J. Biol. Chem.* **241**, 2280–2286.

Akazawa, T. and Miyata, S. (1982). *Essays Biochem.* **18**, 40–78.

Avigad, G. (1982). *In* "Encyclopedia of Plant Physiology, New Series" (F. A. Loewus and W. Tanner, eds.), Vol. 13A, pp. 217–347. Springer-Verlag, Berlin

Baba, T., Arai, Y., Ono, T., Munakata, A., Yamaguchi, H. and Itoh, T. (1982). *Carbohydr. Res.* **107**, 215–230.

Badenhuizen, N. P. (1969). "The Biogenesis of Starch Granules in Higher Plants." Appleton New York.

Badenhuizen, N. P. (1973). *Ann. N. Y. Acad. Sci.* **210**, 11–16.
Banks, W. and Greenwood, C. T. (1959). *J. Chem. Soc.* pp. 3436–3439.
Banks, W. and Greenwood, C. T. (1966). *Arch. Biochem. Biophys.* **117**, 674–675.
Banks, W. and Greenwood, C. T. (1968). *Carbohydr. Res.* **6**, 241–244.
Banks, W. and Greenwood, C. T. (1975). "Starch and Its Components." Edinburgh Univ. Press, Edinburgh.
Banks, W. and Muir, D. D. (1980) *In* "The Biochemistry of Plants" (J. Preiss, ed.), Vol. 3, pp. 321–369. Academic Press, New York.
Banks, W., Greenwood, C. T. and Jones, I. G. (1960). *J. Chem. Soc.* pp. 150—155.
Barker, S. A., Bourne, E. J. and Peat, S. (1949). *J. Chem. Soc.* pp. 1705—1711.
Barker, S. A., Bourne, E. J., Peat, S. and Wilkinson, I. A. (1950). *J. Chem. Soc.* pp. 3022–3027.
Bathgate, G. N. and Manners, D. J. (1966). *Biochem. J.* **101**, 3C–5C.
Bender, H., Siebert, R. and Stadler-Szoke, A. (1982). *Carbohydr. Res.* **110**, 245–259.
Bertoft, E. and Henriksnäs, H. (1982). *J. Inst. Brew.* **88**, 261–265.
Borovsky, D., Smith, E. E. and Whelan, W. J. (1975). *Eur. J. Biochem.* **59**, 615–625.
Borovsky, D., Smith, E. E. and Whelan, W. J. (1976). *Eur. J. Biochem.* **62**, 307–312.
Borovsky, D., Smith, E. E., Whelan, W. J., French, D. and Kikumoto, S. (1979). *Arch. Biochem. Biophys.* **198**, 627–631.
Bourne, E. J. and Peat, S. (1945). *J. Chem. Soc.*, pp. 877–882.
Boyer, C. D. and Preiss, J. (1978a). *Carbohydr. Res.* **61**, 321–334.
Boyer, C. D. and Preiss, J. (1978b). *Biochem. Biophys. Res. Commun.* **80**, 169–175.
Boyer, C. D. and Preiss, J. (1981). *Plant Physiol.* **67**, 1141–1145.
Brown, D. H. and Brown, B. I. (1966). *Biochim. Biophys. Acta* **130**, 263–266.
Chapman, G. W., Pallas, J. E. and Mendicino, J. (1972). *Biochim. Biophys. Acta* **276**, 491–507.
Chatterton, N. J. and Silvius, J. E. (1980). *Physiol. Plant.* **49**, 141–144.
Chiba, S. Inomata, S., Matsui, H. and Shimomura, T. (1978). *Agric. Biol. Chem.* **42**, 241–245.
Cowie, J. M. G. and Greenwood, C. T. (1957). *J. Chem. Soc.* pp. 2862–2866.
Cowie, J. M. G., Fleming, I. D., Greenwood, C. T. and Manners, D. J. (1957). *J. Chem. Soc.* pp. 4430–4437.
Cunningham, W. L., Manners, D. J., Wright, A. and Fleming, I. D. (1960). *J. Chem. Soc.* pp. 2602–2613.
Dais, P. and Perlin, A. S. (1982). *Carbohydr. Res.* **100**, 103–116.
de Fekete, M. A. R. and Cardini, C. E. (1964). *Arch. Biochem. Biophys.* **104**, 173–184.
de Fekete, M. A. R. and Vieweg, G. H. (1973). *Ann. N. Y. Acad. Sci.* **210**, 170–180.
de Fekete, M. A. R. and Vieweg, G. H. (1974). *In* "Plant Carbohydrate Biochemistry" (J. B. Pridham, ed.), pp. 127–144. Academic Press, New York.
Dickinson, D. B. and Preiss, J. (1969). *Arch. Biochem. Biophys.* **130**, 119–128.
Doi, A. (1969). *Biochim. Biophys. Acta* **184**, 477–485.
Doi, A., Doi, K. and Nikuni, Z. (1966). *Biochim. Biophys. Acta* **113**, 312–320.
Drummond, G. S., Smith, E. E. and Whelan, W. J. (1970). *FEBS Lett.* **9**, 136–140.
Drummond, G. S., Smith, E. E. and Whelan, W. J. (1972). *Eur. J. Biochem.* **26**, 168–176.
Duffus, C. M. and Cochrane, M. P. (1982). *In* "The Physiology and Biochemistry of Seed Development, Dormancy and Germination" (A. A. Khan, ed.), pp. 43–66. Elsevier Biomedical, Amsterdam.
Dunn, G. (1974). *Phytochemistry* **13**, 1341–1346.
Dunn, G. and Manners, D. J. (1975). *Carbohydr. Res.* **39**, 283–293.
Dunn, G., Hardie, D. G. and Manners, D. J. (1973). *Biochem. J.* **133**, 413–416.
Enevoldsen, B. S. (1980). *Abstr. 10th Int. Symp. Carbohydr. Chem.*, 1L4.
Fischer, E. H. and Stein, E. A. (1960). *In* "The Enzymes" (P. D. Boyer, H. Lardy and K. Myrback, eds.), 2nd rev. ed., Vol. 4A, pp. 313–343. Academic Press, New York.

Fogarty, W. M. (1983). *In* "Microbial Enzymes and Biotechnology" (W. M. Fogarty, ed.), pp. 1–92. Applied Science, London.

Fox, J., Kennedy, L. D., Hawker, J. S., Ozbun, J. L., Greenberg, E., Lammel, C. and Preiss J. (1973). *Ann. N. Y. Acad. Sci.* **210**, 90–103.

Fredrick, J. F. (1971). *Physiol. Plant.* **24**, 55–58.

French, D. (1960). *In* "The Enzymes" (P. D. Boyer, H. Lardy and K. Myrbäck, eds.), 2nd rev. ed., Vol. 4A, pp. 345–368. Academic Press, New York.

French, D. (1973). *MTP Int. Rev. Sci.: Biochem., Ser. One* **5**, 267–335.

French, D., Smith, E. E. and Whelan, W. J. (1972). *Carbohydr. Res.* **22**, 123–134.

Frydman, R. B. and Cardini, C. E. (1964a). *Biochem. Biophys. Res. Commun.* **14**, 353–357.

Frydman, R. B. and Cardini, C. E. (1964b). *Biochem. Biophys. Res. Commun.* **17**, 407–411.

Frydman, R. B. and Cardini, C. E. (1965). *Biochim. Biophys. Acta* **96**, 294–303.

Frydman, R. B. and Cardini, C. E. (1966). *Arch. Biochem. Biophys.* **116**, 9–18.

Frydman, R. B. and Cardini, C. E. (1967). *J. Biol. Chem.* **242**, 312–317.

Frydman, R. B. and Slabnik, E. (1973). *Ann. N. Y. Acad. Sci.* **210**, 153–169.

Fujimoto, S., Sugimura, K., Nakashima, S., Suganuma, T. and Nagahama, T. (1981a). *J. Jpn. Soc. Starch Sci.* **28**, 166–173.

Fujimoto, S., Nakashima, S., Kubo, Y., Suganuma, T. and Nagahama T. (1981b). *J. Jpn. Soc. Starch Sci.* **28**, 174–179.

Fujimoto, S., Nakashima, S., Kubo, Y., Suganuma, T. and Nagahama, T. (1981c). *J. Jpn. Soc. Starch Sci.* **28**, 180–187.

Fujimoto, S., Nakashima, S., Kubo, Y., Suganuma, T., and Nagahama, T. (1981d). *J. Jpn. Soc. Starch Sci.* **28**, 188–193.

Furlong, C. E. and Preiss, J. (1969). *Prog. Photosynth. Res.* **3**, 1604–1617.

Gates, J. W. and Simpson, G. M. (1968). *Can. J. Bot.* **46**, 1459–1462.

Geddes, R., Greenwood, C. T. and Mackenzie, S. (1965). *Carbohydr. Res.* **1**, 71–82.

Gertler, A. and Birk, Y. (1965). *Biochem. J.* **95**, 621–627.

Gibbons, G. C. and Nielsen, E. B. (1983). *J. Inst. Brew.* **89**, 8–14.

Gilbert, G. A. and Patrick, A. D. (1952). *Biochem. J.* **51**, 181–186.

Gold, A. M., Johnson, R. M. and Sanchez, G. R. (1971). *J. Biol. Chem.* **246**, 3444–3450.

Gordon, R. W., Manners, D. J. and Stark, J. R. (1975). *Carbohydr. Res.* **42**, 125–134.

Greenwood, C. T. (1956). *Adv. Carbohydr. Chem.* **11**, 335–385.

Greenwood, C. T. and Mackenzie, S. (1966). *Carbohydr. Res.*, **3**, 7–13.

Griffin, H. L. and Wu, Y. V. (1971). *Biochemistry* **10**, 4330–4335.

Gunja-Smith, Z., Marshall, J. J., Mercier, C., Smith, E. E. and Whelan, W. J. (1970). *FEBS Lett.* **12**, 101–104.

Hall, R. S. and Manners, D. J. (1978). *Carbohydr. Res.* **66**, 295–297.

Halmer, P. and Bewley, J. D. (1982). *In* "Encyclopedia of Plant Physiology, New Series" (F. A. Loewus and W. Tanner, eds.), Vol. 13A, pp. 748–793. Springer-Verlag, Berlin and New York.

Hardie, D. G. (1975). *Phytochemistry* **14**, 1719–1722.

Hardie, D. G., Manners, D. J. and Yellowlees, D. (1976). *Carbohydr. Res.* **50**, 75–85.

Hawker, J. S., Ozbun, J. L. and Preiss, J. (1972). *Phytochemistry* **11**, 1287–1293.

Hawker, J. S., Ozbun, J. L., Ozaki, H., Greenberg, E. and Preiss, J. (1974). *Arch. Biochem. Biophys.* **160**, 530–551.

Hehre, E. J., Brewer, C. F. and Genghof, D. S. (1979). *J. Biol. Chem.* **254**, 5942–5950.

Hizukuri, S., Takeda, Y. and Yasuda, M. (1981). *Carbohydr. Res.* **94**, 205–213.

Hobson, P. N., Whelan, W. J. and Peat, S. (1951). *J. Chem. Soc.* pp. 1451–1459.

Hodges, H. F., Creech, R. G. and Loerch, J. D. (1969). *Biochim. Biophys. Acta* **185**, 70–79.

Holló, J., Laszlo, E. and Hoschke, A. (1971). "Plant α-1,4-Glucan Phosphorylase." Akadémiai Kiadó, Budapest.

Hutson, D. H. and Manners, D. J. (1965). *Biochem. J.* **94**, 783–789.

Ishizaki, Y., Taniguchi, H. and Nakamura, M. (1978). *Agric. Biol. Chem.* **42**, 2433–2435.

Ishizaki, Y., Taniguchi, H., Maruyama, Y. and Nakamura, M. (1983a). *Agric. Biol. Chem.* **47**, 771–779.

Ishizaki, Y., Taniguchi, H., Maruyama, Y. and Nakamura, M. (1983b). *J. Jpn. Soc. Starch Sci.* **30**, 19–29.

Iwaki, K. and Fuwa, H. (1981). *Agric. Biol. Chem.* **45**, 2683–2688.

Jeanningros, R., Creuzet-Sigal, N., Frixon, C. and Cattaneo, J. (1976). *Biochim. Biophys. Acta* **438**, 186–199.

Jenner, C. F. (1982). *In* "Encyclopedia of Plant Physiology, New Series" (F. A. Loewus and W. Tanner, eds.), Vol. 13A, pp. 700–746. Springer-Verlag, Berlin and New York.

Jorgensen, O. B. (1964). *Acta Chem. Scand.* **18**, 1975–1978.

Kjolberg, O. and Manners, D. J. (1983). *Biochem. J.* **86**, 258–262.

Kneen, E. (1945). *Cereal Chem.* **22**, 112–134.

Kneen, E. and Spoerl, J. M. (1948). *Proc. Am. Soc. Brew. Chem.* pp. 20–27.

Kruger, J. E. (1970). *Cereal Chem.* **47**, 79–85.

Kruger, J. E. and Marchylo, B. A. (1982). *Cereal Chem.* **59**, 488–492.

Kumar, A. and Sanwal, G. G. (1977). *Phytochemistry* **16**, 327–328.

LaBerge, D. E. and Meredith, W. O. S. (1969). *J. Inst. Brew.* **75**, 19–25.

Lavintman, N. (1966). *Arch. Biochem. Biophys.* **116**, 1–8.

Lavintman, N. and Cardini, C. E. (1973). *FEBS Lett.* **29**, 43–46.

Lavintman, N. and Krisman, C. R. (1964). *Biochim. Biophys. Acta* **89**, 193–196.

Lee, E. Y. C. (1971). *Arch. Biochem. Biophys.* **146**, 488–492.

Lee, E. Y. C. (1972). *FEBS Lett.* **27**, 341–344.

Lee, E. Y. C. and Braun, J. J. (1973a). *Ann. N. Y. Acad. Sci.* **210**, 115–128.

Lee, E. Y. C. and Braun, J. J. (1973b). *Arch. Biochem. Biophys.* **156**, 276–286.

Lee, E. Y. C. and Whelan, W. J. (1971). *In* "The Enzymes" (P. D. Boyer, ed.), Vol. 5, pp. 192–234. Academic Press, New York.

Lee, E. Y. C. Mercier, C., and Whelan, W. J. (1968). *Arch. Biochem. Biophys.* **125**, 1028–1030.

Lee, E. Y. C., Marshall, J. J. and Whelan, W. J. (1971). *Arch. Biochem. Biophys.* **143**, 365–374.

Lee, Y. P. (1960a) *Biochim. Biophys. Acta* **43**, 18–24.

Lee, Y. P. (1960b). *Biochim. Biophys. Acta* **43**, 25–30.

Leloir, L. F. (1964). *Biochem. J.* **91**, 1–8.

Leloir, L. F., de Fekete, M. A. R. and Cardini, C. E. (1961). *J. Biol. Chem.* **235**, 636–641.

Levi, C. and Gibbs, M. (1976). *Plant Physiol.* **57**, 933–935.

Liddle, A. M., Manners, D. J. and Wright, A. (1961). *Biochem. J.* **80**, 304–309.

McCracken, D. A. and Badenhuizen, N. P. (1970). *Starch/Staerke* **22**, 289–291.

MacGregor, A. W. (1977). *J. Inst. Brew.* **83**, 100–103.

MacGregor, A. W. (1980). *MBAA Tech. Q.* **17**, 215–221.

MacGregor, A. W. and Daussant, J. (1981). *J. Inst. Brew.* **87**, 155–157.

MacWilliam I. C. and Harris, G. (1959). *Arch. Biochem. Biophys.* **84**, 442–454.

Maeda, I., Jimi, N., Taniguchi, H. and Nakamura, M. (1979). *J. Jpn. Soc. Starch Sci.* **26**, 117–127.

Mangat, B. S. and Badenhuizen, N. P. (1970). *Starch/Staerke* **22**, 329–333.

Manners, D. J. (1962). *Adv. Carbohydr. Chem.* **17**, 371–430.

Manners, D. J. (1974a). *Essays Biochem.* **10**, 37–71.

Manners, D. J. (1974b). *In* "Plant Carbohydrate Biochemistry" (J. B. Pridham, ed.), pp. 109–125. Academic Press, New York.

Manners, D. J. (1974c). *Brew. Dig.* **49**, 56–62.

Manners, D. J. and Hardie, D. G. (1977). *MBAA Tech. Q.* **14**, 120–125.

Manners, D. J. and Marshall, J. J. (1971). *Carbohydr. Res.* **18**, 203–209.

Manners, D. J. and Matheson, N. K. (1981). *Carbohydr. Res.* **90**, 99–110.

Manners, D. J. and Rowe, J. J. M. (1964). *Chem. Ind.* (*London*) pp. 1834–1835.
Manners, D. J. and Rowe, K. L. (1967). *Arch. Biochem. Biophys.* **119**, 585–586.
Manners, D. J. and Rowe, K. L. (1971). *J. Inst. Brew.* **77**, 358–365.
Manners, D. J. and Sparra, K. L. (1966). *J. Inst. Brew.* **72**, 360–365.
Manners, D. J. and Wright, A. (1962). *J. Chem. Soc.* pp. 1597–1602.
Manners, D. J., Rowe, J. J. M. and Rowe, K. L. (1968). *Carbohydr. Res.* **8**, 72–81.
Marshall, J. J. (1975). *Starch/Staerke* **27**, 377–383.
Marshall, J. J. and Taylor, P. M. (1971). *Biochem. Biophys. Res. Commun.* **42**, 173–179.
Marshall, J. J. and Whelan, W. J. (1974). *Arch. Biochem. Biophys.* **161**, 234–238.
Murata, T. and Akazawa, T. (1968). *Arch. Biochem. Biophys.* **126**, 873–879.
Murata, T. and Akazawa, T. (1969a). *Plant Cell Physiol.* **10**, 457–460.
Murata, T. and Akazawa, T. (1969b). *Arch. Biochem. Biophys.* **130**, 604–609.
Nelson, O. E. and Rines, H. W. (1962). *Biochem. Biophys. Res. Commun.* **9**, 297.
Nelson, O. E. and Tsai, C. Y. (1964). *Science* **145**, 1194–1195.
Norman, B. E. (1979). *In* "Microbial Polysaccharides and Polysaccharases" (R. C. W. Berkeley, G. W. Gooday and D. C. Ellwood, eds.), pp. 339–376. Academic Press, London.
Norman, B. E. (1983). *J. Jpn. Soc. Starch Sci.* **30**, 200–211.
Nummi, H., Vilhunen, R. and Enari, T. M. (1966). *Proc. 10th Congr. Eur. Brew. Conv. 1965* pp. 52–69.
Okada, G., Genghof, D. S. and Hehre, E. J. (1979). *Carbohydr. Res.* **71**, 287–298.
Okamota, K. and Akazawa, T. (1978). *Agric. Biol. Chem.* **42**, 1379–1384.
Okita, T. W. and Preiss, J. (1980). *Plant Physiol.* **66**, 870–876.
Okita, T. W., Greenberg, E., Kuhn, D. N. and Preiss, J. (1979). *Plant Physiol.* **64**, 187–192.
Ozbun, J. L., Hawker, J. S. and Preiss, J. (1971). *Plant Physiol.* **48**, 765–769.
Ozbun, J. L., Hawker, J. S. and Preiss, J. (1972). *Biochem. J.* **126**, 953–963.
Ozbun, J. L., Hawker, J. S., Greenberg, E., Lammel, C. and Preiss, J. (1973). *Plant Physiol.* **51**, 1–5.
Palmer, G. H. (1983). *J. Inst. Brew.* **89**, 158–159.
Palmer, T. N., Macaskie, L. E. and Grewal, K. K. (1983). *Carbohydr. Res.* **115**, 139–150.
Parrish, F. W., Smith, E. E. and Whelan, W. J. (1970). *Arch. Biochem. Biophys.* **137**, 185–189.
Peat, S., Pirt, S. J. and Whelan, W. J. (1952a). *J. Chem. Soc.* pp. 705–713.
Peat, S., Thomas, G. J. and Whelan, W. J. (1952b). *J. Chem. Soc.* pp. 722–733.
Peat, S., Whelan, W. J., Hobson, P. N. and Thomas, G. J. (1954). *J. Chem. Soc.* pp. 4440–4445.
Peat, S., Whelan, W. J. and Thomas, G. J. (1956). *J. Chem. Soc.* pp. 3025–3030.
Peat, S., Turvey, J. R. and Jones, G. (1959). *J. Chem. Soc.* pp. 1540–1544.
Peavey, D. G., Steup, M. and Gibbs, M., (1977). *Plant Physiol.* **60**, 305–308.
Perez, C. M., Palmiano, E. P., Baun, L. C. and Juliano, B. O. (1971). *Plant Physiol.* **47**, 404–408.
Pollock, J. R. A. and Pool, A. A. (1958). *J. Inst. Brew.* **64**, 151–156.
Preiss, J. (1982). *In* "Encyclopedia of Plant Physiology, New Series" (F. A. Loewus and W. Tanner, eds.), Vol. 13A, pp. 397–417. Springer-Verlag, Berlin and New York.
Preiss, J. and Greenberg, E. (1967). *Arch. Biochem. Biophys.* **118**, 702–708.
Preiss, J. and Levi, C. (1979). *In* "Encyclopedia of Plant Physiology" New Series, (M. Gibbs and E. Latzko, eds.), Vol. 6, pp. 282–312. Springer-Verlag, Berlin and New York.
Preiss, J. and Levi, C. (1980). *In* "The Biochemistry of Plants" (J. Preiss, ed.), Vol. 3, pp. 371–423. Academic Press, New York.
Preiss, J., Lammel, C. and Sabraw, A. (1971). *Plant Physiol.* **47**, 104–108.
Preiss, J., Okita, T. W. and Greenberg, E. (1980). *Plant Physiol.* **66**, 864–869.
Robin, J. P., Mercier, C., Charbonniere, R. and Guilbot, A. (1974). *Cereal Chem.* **51**, 389–406.
Sakai, S. (1981). *J. Jpn. Soc. Starch Sci.* **28**, 72–78.
Sanwal, G. G. and Preiss, J. (1967). *Arch. Biochem. Biophys.* **119**, 454–469.
Schiefer, S., Lee, E. Y. C. and Whelan, W. J. (1973). *FEBS Lett.* **30**, 129–132.

Schiefer, S., Lee, E. Y. C. and Whelan, W. J. (1978). *Carbohydr. Res.* **61,** 239–252.

Schoch, T. J. (1945). *Adv. Carbohydr. Chem.* **1,** 247–277.

Schramm, M. and Loyter, A. (1966). *In* "Methods in Enzymology" (E. F. Neufeld and V. Ginsburg, eds.), Vol. 8, pp. 533–537. Academic Press, New York.

Sharma, K. P. and Bhatia, I. S. (1980). *Physiol. Plant.* **48,** 470–476.

Slabnik, E. and Frydman, R. B. (1970). *Biochem. Biophys. Res. Commun.* **38,** 709–714.

Spradlin, J. E., Thoma, J. A. and Filmer, D. (1969). *Arch. Biochem. Biophys.* **134,** 262–264.

Stark, J. R., Aisien, A. O. and Palmer, G. H. (1983). *Starch/Staerke* **35,** 73–76.

Stitt, M., Bulpin, P. V. and ap Rees, T. (1978). *Biochim. Biophys. Acta* **544,** 200–214.

Swain, R. D. and Dekker, E. E. (1966). *Biochim. Biophys. Acta* **122,** 87–100.

Takahashi, M. and Shimomura, T. (1968). *Agric. Biol. Chem.* **32,** 929–939.

Takahashi, M., Shimomura, T. and Chiba, S. (1971). *Agric. Biol. Chem.* **35,** 2015–2024.

Takeda, Y. and Hizukuri, S. (1982). *Carbohydr. Res.* **102,** 321–327.

Takeshita, M. and Hehre, E. J. (1975). *Arch. Biochem. Biophys.* **169,** 627–637.

Tanaka, Y. and Akazawa, T. (1969). *J. Jpn. Soc. Starch Sci.* **17,** 229–236.

Tkachuk, R. and Tipples, K. H. (1966). *Cereal Chem.* **43,** 62–79.

Tsai, C. Y. and Nelson, O. E. (1968). *Plant Physiol.* **43,** 103–112.

Tsai, C. Y. and Nelson, O. E. (1969). *Plant Physiol.* **44,** 159–167.

Underkofler, L. A., Denault, L. J. and Hon, E. F. (1965). *Starch/Staerke* **17,** 179–184.

Verhue, W. and Hers, H. G. (1966). *Biochem. J.* **99,** 222–227.

Vieweg, G. H. and de Fekete, M. A. R. (1980). *In* "Mechanisms of Saccharide Polymerization and Depolymerization" (J. J. Marshall, ed.), pp. 175–185. Academic Press, New York.

Vlodawsky, L., Harel, E. and Mayer, A. M. (1971). *Physiol. Plant.* **25,** 363–368.

Weselake, R. J. and Hill, R. D. (1982). *Carbohydr. Res.* **108,** 153–161.

Whelan, W. J. (1963). *Starch/Staerke* **15,** 247–251.

Whelan, W. J. and Roberts, P. J. P. (1952). *Nature (London)* **170,** 748.

Williams, J. M. (1968). *In* "Starch and its Derivatives" (J. A. Radley, ed.), 4th ed., pp. 91–138. Chapman & Hall, London.

Yamaguchi, M., Kainuma, K. and French, D. (1979). *J. Ultrastruct. Res.* **69,** 249–261.

Yamasaki, Y. and Suzuki, Y. (1979). *Agric. Biol. Chem.* **43,** 481–489.

Yellowlees, D. (1980). *Carbohydr. Res.* **83,** 109–118.

Chapter 5

Fructans

HORACIO G. PONTIS and ELENA DEL CAMPILLO
Instituto de Investigaciones Biologicas
Facultad de Ciencias—Exactas, Naturales y Biologicas
Universidad Nacional de Mar del Plata
and Centro de Investigaciones Biologicas
F.I.B.A.
Casilla de Correo, Argentina

I. INTRODUCTION

A few years ago the senior author was asked to give a lecture on fructans in a symposium. The speaker preceding him happened to give a list of all carbohydrate polymers that exist in plants, but he precisely forgot to mention fructans! When the senior author started his lecture he thanked his colleague and remarked that he was going to talk about a nonexisting entity in the plant world: fructans. The audience laughed and for once fructans got the full attention that they deserve. This anecdote illustrates the place that fructan biochemistry occupies in plant polysaccharide research. With the exception of a few enthusiasts, fructans are usually ignored. The present authors, on the other hand, are convinced that fructans are not only a type of polymer very widely distributed in the plant kingdom but also that the knowledge of their biochemistry will bring information about plants' responses to environmental

BIOCHEMISTRY OF STORAGE
CARBOHYDRATES IN GREEN PLANTS

conditions that will open new insights into mechanisms of metabolic control hitherto unknown.

Several reviews on fructans have been published (Akazawa, 1965; Archbold, 1940; Bhatia, 1955; Carles, 1935; Hirst, 1957; Kandler and Hopf, 1980; McDonald, 1946; Meier and Reid, 1982; Schlubach, 1958; Smith, 1973; Whistler and Smart, 1953), but they have mainly described and listed the different polymers that have been isolated, considering only some aspects of their chemistry and biochemistry. In this review we attempt to present the current status of the biochemistry of fructans and to discuss their physiological role.

There exists a confusing situation in the literature regarding the designation of the polymers of fructose that are found in Nature. The polymers elaborated by microorganisms are called levans, while those present in plants are called either fructans or fructosans. The term fructosan very likely started to be used after Schlubach and Sinh (1940) utilized the word "polyfructosan" to describe the fructose polymers present in monocotyledons, because they thought they were derivatives of cyclic structures of fructose called fructosan. Neither of the two terms polyfructosan or fructosan should be used to describe fructose polymers, as the former means polymerized fructosan and the latter corresponds to a very defined type of fructose dianhydride (Wolfrom and Thompson, 1957) that is only produced on acid hydrolysis of fructose polymers. Consequently, by similarity with the polymers of D-glucose found in Nature, the name fructan should be assigned to describe the polymers of D-fructose and it is this, then, that is used in this chapter.

II. OCCURRENCE

Fructans are widely distributed in the plant kingdom. They are present not only in monocotyledons and dicotyledons but also in green algae. There has been a report on the presence of a glucofructan in blue-green algae (Tsusué and Fujita, 1964).

Inulin, the first of fructans to be discovered and the one most thoroughly studied, was isolated by Rose (1804), who obtained it from the rhizomes of *Inula helenium*, although the name inulin was first employed by Thomson (1818) in 1811. Later, Sachs (1864) localized inulin in tubers of *Dahlia variabilis, Inula helenium,* and *Helianthus tuberosus* using a microscopic test.

By the end of the nineteenth century, the numbers of plants known to contain fructans had increased considerably, and fructans were reported in the rhizomes of grasses (Ekstrand and Johanson, 1887, 1888; Harlay, 1901; Lüding and Müller, 1872-1873; Schulze, 1899), bulbs of *Urginea scella*

(Schmiedeberg, 1879), and in cereals such as barley, rye, and wheat (Colin and Belval, 1937; Tanret, 1891).

All these polymers were similar in that they contained D-fructose, but they had differences in molecular structure and in molecular weight. They may be classified in three main types: the inulin group or those with (2→1')-glycosidic linkages, the phlein group or those with (2→6')-glycosidic linkages, and the branched group or those with both types of glycosidic linkages.

The fructans of the dicotyledons are, as far as we know, all of the inulin group, but it should be remembered that virtually all the fructans that have been studied in some detail are from the Compositae (Bacon and Edelman, 1951; Colin, 1925; Colin and Chollet, 1939; Grafe and Vouk, 1912; Hirst *et al.*, 1950; Omslow, 1929). Fructans have been also shown to be present in species of the following families belonging also to the dicotyledons: Asclepiadaceae, Boraginaceae (Bourdu, 1954, 1957), Campanulaceae (Colin and Chollet, 1939; Hegnauer, 1964), Gentianaceae (including the Menyanthaceae) (Hegnauer, 1966; Keegan, 1916), Goodeniaceae (Fischer, 1902; Weber, 1955), Lobeliaceae (Colin and Chollet, 1939), Malpighiaceae (Hegnauer, 1973), Primulaceae, Stylidiaceae, and Violaceae (Meier and Reid, 1982). According to Srepel and Mijatovic (1975), fruits of Berberidaceae also contain fructans.

The fructans of the monocotyledons are of the phlein type or of the branched type, and, in a few cases, also of the inulin type. According to Hegnauer (1963), they can be found in the families belonging to the subclass Liliidae: Liliaceae (Bacon, 1959; Colin and Chaudun, 1933; Darbyshire and Henry, 1981; Das and Das, 1978; du Mérac, 1949; Madam, 1972; Nitsch *et al.*, 1979; Pantanelli, 1949; Schlubach and Flörsheim, 1929; Schlubach and Lopp, 1936; Shiomi *et al.*, 1976; Tomoda *et al.*, 1973), Agavaceae (Aspinall and Gupta Das, 1959; Bhatia and Srinivasan, 1953; Belval, 1939; Boggs and Smith, 1956; Dimick and Christiensen, 1942; Ekstrand and Johanson, 1887; Schlubach and Flörsheim, 1931; Srinivasan and Bhatia, 1953; Srinivasan *et al.*, 1952), Amaryllidaceae (Belval, 1937; Hammer, 1970; Mizuno and Hayashi, 1955; Schlubach and Lendzian 1937), Iridaceae (Carles, 1935; Flood *et al.*, 1947; Schlubach *et al.*, 1933; von Euler and Erdtman, 1925; Wallach, 1886), and Poaceae. In the last family, fructans occur mainly in the subfamily Festucoideae (e.g., *Bromus, Dactylis, Festuca, Poa, Hordeum, Agropyron, Elymus, Lolium, Secale, Triticum, Agrostis, Alopecurus, Phleum, Arrhenatherym, Avena, Trisetum, Phalaris*) (Archbold, 1940; Arni and Percival, 1951; Bell and Palmer, 1952; Colin and Belval, 1922; Grotelueschen and Smith, 1968; Laidlaw and Reid, 1951; Medcalf and Cheung, 1971; Montgomery and Smith, 1956, 1957; Pollock *et al.*, 1979; Schlubach, 1958; Schlubach and Berndt, 1961; Schlubach and Koehn, 1957, 1958; Schlubach and Müller, 1952; Schlubach *et al.*, 1955; Smith, 1973).

In green algae, fructose polymers have been found in the Dasycladales (Meeuse, 1962). Probably the most striking case is presented by *Acetabularia mediterranea*, from which du Mérac (1953) succeeded in isolating an inulin practically indistinguishable from that of *Dahlia variabilis*. Furthermore, in *Acetabularia crenulata* a series of fructans similar in structure to those of the Compositae have been demonstrated (Bourne *et al.*, 1972). The presence of fructose-containing oligosaccharides is not restricted to the Dasycladales. A survey of members of the Cladophorales has also shown the occurrence of an homologous series of fructose polymers in various marine and freshwater species (Percival and Young, 1971).

Whatever their source, most fructans are distributed in an unbroken series between a fructosylsucrose and the member of the highest molecular weight found in that particular source (see Section III). This fact should be taken into account when considering the many fructans that have been given trivial and highly specialized names on the basis of their origin. It is clear now that these names usually refer to the polymer of highest molecular weight isolated, but it should be borne in mind that the highest member of a series may well be a mixture of neighbouring homologues that are difficult to separate by virtue of their similar physical properties.

The presence of fructans in plants shows no obvious correlation with the presence or absence of starch. Thus, although in *Allium* (onion) (Belval, 1939), *Agave vera cruz* (Srinivasan and Bhatia, 1953), *Iris pseudacorus* (Carles, 1935), and *Furcroea gigantea* (Bhatia and Srinivasan, 1953) fructans are the only polymer, in other plants fructans and starch may occur together in a single tissue, as in the rhizomes of *Iris foetidisima* (Colin and Augen, 1927), or may be found in different parts of the same plant, as in *Helianthus tuberosus* and *Dahlia variabilis* (Edelman and Jefford, 1968), where the leaves contain starch and the tubers fructans, and in cereals like *Triticum*, where the leaves contain fructans and the seeds starch (Edelman and Jefford, 1968). No correlation has been found between the presence of starch or fructan and the occurrence of the C4 and C3 photosynthetic pathways (Bender and Smith, 1973).

In general, fructans are distributed throughout the plants in which they occur, although the amounts in different parts of the same plant vary considerably. Usually the amounts are very small in leaves and especially large in roots, bulbs, tubers, rhizomes, and sometimes in immature fruits (Meier and Reid, 1982). In some grasses the lower sections of the stem are quite rich in fructans (Smith, 1973). Important variations occur during the life cycle of the plants (Whistler and Smart, 1953) and, in some cases (e.g., the Graminae) there seems also to exist a difference based on the geographical origin of the plants (Fig. 1). Grasses of tropical and subtropical origin accumulate starch, but grasses of temperate origin accumulate fructans. This separation of species was first proposed by De Cugnac (1931) and has been supported by the work of Ojima and Isawa (1968), Smith (1968a), and others (Meier and Reid, 1982).

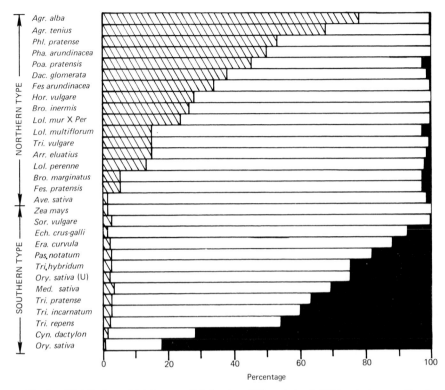

Fig. 1. Distribution of total sugars (□), fructans (▨), and starch (■) in plants according to their geographical origin. From Ojima and Isawa (1968).

III. CHEMICAL NATURE

Fructans are polymers of D-fructose carrying a D-glucosyl residue at the end of the chain attached via a (2→1) linkage as in sucrose. They constitute a series of homologous oligosaccharides which can be considered as derivatives of sucrose (Hirst, 1957).

They are usually represented by the simple formula G-F-(F)$_n$ where G-F denotes a sucrosyl group and n the number of fructose units present in the molecule.

Fructans have a unique structural feature within the family of poly-saccharides in that no bond of the fructose furanose ring is part of the macromolecular backbone (Marchessault *et al.*, 1980). Besides, they are one of the few natural polymers where the carbohydrate exists in the furanose form (Fig. 2). These two structural features play an important role in the final conformation of the molecules in solution. Moreover, the enhanced flexibility

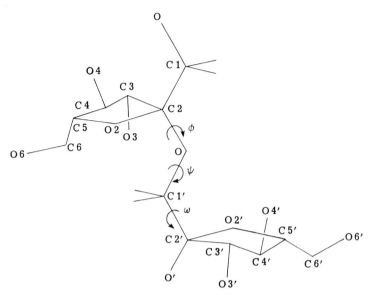

Fig. 2. Inulin-type difructan molecule (inulobiose) in all trans conformation ($\phi = \psi = \omega = 180°$). Ring hydrogen atoms are not shown. From Marchessault *et al.* (1980).

of the furanose ring in comparison with the relatively rigid pyranose ring of the majority of reserve polysaccharides brings additional flexibility to the whole fructan molecule (Marchessault *et al.*, 1980).

As indicated before, there are three main groups of fructans: (a) the inulin type or those with β-(2→1)-D-fructofuranosyl units, (b) the phlein type or those with β-(6→2)-D-fructofuranosyl units, and (c) the branched type or those with both kinds of glycosidic linkages. A minor group is formed by those fructans with β-(2→1)-D-fructofuranosyl linkages, but that are derivatives of neo-kestose or 6^G-fructofuranosyl sucrose. By prolongation of the chain at either of the two terminal fructoses, fructans with a nonterminal glucose are obtained (Tomoda *et al.*, 1973).

The general properties of fructans can be summarized as follows: they are laevorotatory, amorphous or microcrystalline, of varying solubility in cold water, very soluble in hot water, insoluble in absolute alcohol, not precipitated by lead acetate except in alkaline solution, but precipitated by baryta, either directly or on addition of alcohol (Colin and Belval, 1940; Whelan, 1955). They are also nonreducing, not hydrolyzed by yeast invertase (this may not be true of low-molecular-weight fructans), resistant to amylase action, but very susceptible to hydrolysis by acid (Archbold, 1940). They do not colour iodine, but hydrochloric acid vapours impart a purple colour that distinguishes them from polysaccharides not containing fructose (Colin and De Cugnac, 1926).

The chemical structures of fructans have been elucidated mainly by the work of Schlubach (1953), Hirst (1957), Montgomery and Smith (1956, 1957), and others (Aspinall and Telfer, 1953; Bell and Palmer, 1952; Tomasic *et al.*, 1978). The structure of the fructans belonging to the inulin group is shown in **1**, while those of the phlein group correspond to the structure indicated in **2**.

F2—1F2—1F2—1F2 —————1F1—1F1—1G

1

F2—6F2—6F2—6F2————6F2—6F2—1G

2

The branched group is characterized by having a backbone of either the inulin (**3**) or the phlein (**4**) type, to which single fructose residues are linked as indicated schematically.

F2—6F2—6F2—6F2—6F2—1G   1F2—1F2—1F2—1F2—1F2—1G
1   1   6   6
2F   2F   2F   2F

    3   **4**

X-Ray and electron diffraction studies suggest that the conformation of inulin is a helix of either four or five fructofuranose units per turn (Marchessault *et al.*, 1980; Middleton, 1977; Phelds, 1965; Streefkerk and Glaudemans, 1977).

The degree of polymerization (DP) varies with plant species and with their life cycle, but all fructans have a low degree of polymerization compared with starch. The number of fructose units range from ~ 10 in the fructans of onion (Bacon, 1957) to ~ 260 in *Phleum pratense* (Grotelueschen and Smith, 1968).

IV. ISOLATION, SEPARATION, AND DETERMINATION

A. Isolation

It is necessary to prepare the plant material in order to prevent enzyme action that may alter the composition and the total amount of fructans present. This is accomplished by freezing the plant material in dry ice, covering the specimen with alcohol, or by freeze drying (Laidlaw and Wylam, 1952). Enzymes are then inactivated by boiling with alcohol, or alternatively the material may be extracted directly in boiling water (see below). Drying at a low oven temperature does not inactivate the enzymes (Archbold, 1938), while drying at 105°C causes a change in the fructan content (Laidlaw and Wylam, 1952; Raguse and Smith, 1965).

The subsequent procedure depends upon whether low-molecular-weight fructans are present. If they are absent, then extraction with boiling 80% ethanol serves both to inactivate enzymes and to extract sugars. When a series of low-molecular-weight fructans are present the extraction with boiling alcohol effects an arbitrary fractionation of the fructans. In any case, in their absence, boiling with 80% alcohol is an essential step. However, in this case the alcoholic extract must be combined with the subsequent water extract (Archbold, 1938). Alternatively, sugars and fructans are both extracted with boiling water (Bacon and Loxley, 1952). In each case the problem becomes one of detecting fructans in the presence of sugars. This can be accomplished by analytical procedures or by separation of the accompanying sugar and the fructans.

B. Separation

Extracts are conveniently prepared for analysis or for separation by eliminating proteins by precipitation with neutral lead acetate, followed by removal of the excess with potassium oxalate (Archbold, 1938). The deproteinized extract is deionized by passage through columns of resins (mixed bed or Amberlite IR 100 and IR4B) and concentrated by evaporation *in vacuo* or by freeze drying.

The difficulty in the resolution of fructans lies in their extremely close structural characteristics. The only difference between two consecutive members of any of the homologous fructan series is a single fructosyl residue. Separations of fructans are based on their differences either in size or in solubility.

The separation of sugars and fructans and the resolution of fructan mixtures was initially carried out with alcohol solutions, where they have different solubilities. The procedure was long and required many precipitations without obtaining clear-cut separations (Schlubach, 1958).

Exclusion chromatography exploits the small difference in size that exists among fructans. It was initially developed by Pontis (1968), who was able to separate the fructans from *Dahlia* tubers by gel filtration on a Bio-Gel P-2 (200–400 mesh) column 210 cm long. This method allowed a clear-cut resolution of low molecular weight members of the series up to DP 6. The procedure was improved by Darbyshire and Henry (1978), who used Bio-Gel P-2 (minus 400 mesh) obtaining a separation of onion fructans up to DP 10. However, even though this technique is very clean and elegant, it will be difficult to improve the resolution already obtained as the small difference that represents a fructosyl residue is only significant among the first 10 or 12 members of the fructan series.

On the other hand, descending paper chromatography when run for 5 days using propanol-ethyl acetate–water (6:1:3) allows the separation of fructans up to DP 14. Increasing the chromatographic time to 3 weeks and modifying the solvent to propanol–ethyl acetate–water (11:2:7) will result in the separation of higher polymers up to inulin (Edelman and Dickerson, 1966).

Thin-layer chromatography (TLC) permits faster separations. Collins and Chandorkar (1971) described a TLC procedure that requires approximately 90 min to separate homologues of DP 2–20. Similarly, Karlsson (1969) has reported unidimensional multiple chromatography of fructans on phosphate-buffered cellulose thin-layer plates, requiring 8 hr to separate members up to DP 8.

Adsorption chromatography has been successfully used by Labourel and Péaud-Leonël (1969) using a silica gel column. Separation was achieved with a gradient solvent system of *n*-butanol:ethanol:water exponentially enriched in water and ethanol and gradually impoverished in butanol. This procedure separated fructans up to DP 20. On the other hand, gas chromatography allows only the separation of the first members of the series (DP 3–5). It requires the preparation either of fructan trimethylsilyl derivatives (de Miniac, 1970) or of fructan methylated alditol acetates (Pollock *et al.*, 1979).

C. Determination

The determination of fructans in an extract may be accomplished by measuring the change in reducing power brought about by mild acid hydrolysis. This gives the amount of free sugars plus that of combined sugars, which would include sucrose, raffinose, and stachyose when present. The initial reducing power gives free sugar content, and hypoiodite oxidation may be used to distinguish free aldoses from free ketoses. Alternatively, glucose present may be estimated with glucose oxidase or by coupling hexokinase and glucose-6-phosphate dehydrogenase, while fructose may be measured in a similar way using the same enzymes with the addition of phosphoglucoisomerase.

Total fructans, expressed by their fructose content, may be determined by using any of the colorimetric methods specific for ketoses such as the thiobarbituric acid method (Percheron, 1962). Sucrose and free sugars (glucose and fructose) may be eliminated prior to measuring total fructose content by hydrolyzing sucrose with yeast invertase and destroying the resulting glucose and fructose by boiling with NaOH (Pontis, 1966). Invertase may act on the lower fructan members of the inulin series, but Pontis (1966) has determined that the attack is rather slow, and by selecting a proper time it is possible to hydrolyze sucrose completely while hardly affecting the fructans.

Procedures have also been developed for the automated analysis of nonreducing sugars and fructans present in plant tissues (Wolf and Ellmore, 1975).

A novel approach to the determination of fructans may lie in the antifructan activity possessed by some myeloma inmunoglobulins (Glaudemans, 1975). They are specific for the inulin type of fructans [i.e. those containing β-D-(2→1) linked D-fructofuranosyl residues] (Allen and Kabat, 1957; Streefkerk and Glaudemans, 1977), while others only react with the phlein type [i.e. β-D-(2→6) linked D-fructofuranosyl residues] (Cisar et al., 1974; Grey et al., 1971; Lundblad et al., 1972). No attempt has been made yet to use the reaction for the quantification of fructans but it certainly opens the possibility of detection and measurement of higher molecular weight fructans in a highly specific way.

The position of fructans on paper chromatograms and TLC plates may be ascertained by developing or spraying with reagents specific for fructose. The more used ones are: resorcinol-HCl (Forsyth, 1950), naphthoresorcinol-HCl (Forsyth, 1948), urea-HCl (Dedonder, 1952a) and ureametaphosphoric acid (Dedonder, 1952b). These reagents do not distinguish between free and combined fructose. So far there is no test that can accomplish this result.

V. BIOSYNTHESIS AND DEGRADATION

Our knowledge of the biosynthesis and degradation of fructans comes mainly from the work of Edelman and his associates (Dickerson and Edelman, 1966; Edelman, 1960; Edelman and Dickerson, 1966; Edelman and Jefford, 1964; Edelman and Popov, 1962; Edelman et al., 1963; Jefford and

Edelman, 1961, 1963; Recaldin, 1961; Scott, 1968; Scott *et al.*, 1966), carried out with the inulin type of polymers present in Jerusalem artichoke tubers (*Helianthus tuberosus*). On the other hand, our knowledge of the enzymes involved in the synthesis and degradation of the phlein type of fructans is very scanty, and the metabolism of the phlein fructans has only recently begun to be studied. There is no information about fructan metabolism in algae or the enzymes involved.

In Jerusalem artichoke tubers, Edelman and his group found three enzymes associated with fructan metabolism:

1. Sucrose-sucrose transfructosylase (SST), which forms 1^F-fructosyl sucrose (GF_2), the initial homologue of the $(2 \rightarrow 1)$-fructan series, from sucrose (Scott, *et al.*, 1966), according to

$$\text{Sucrose + sucrose} \overset{\text{SST}}{\rightleftharpoons} \text{fructosyl sucrose + D-glucose} \tag{1}$$

2. β-$(2 \rightarrow 1)$-Fructan: β-$(2 \rightarrow 1)$-fructan-1-fructofuranosyltransferase (Edelman, 1960), which transfers single terminal β-D-fructofuranosyl residues to the same position of another molecule according to

$$\text{Sucrose Fru}_n + \text{sucrose Fru}_m \overset{\text{FFT}}{\rightleftharpoons} \text{sucrose Fru}_{n-1} + \text{sucrose Fru}_{m+1} \tag{2}$$

where n may be any number from 1 (trisaccharide) to ~ 35, and m for the acceptor molecule is any number from 0 (sucrose) to ~ 35.

3. β-$(2 \rightarrow 1)$-Fructan 1-fructanohydrolase (Edelman and Jefford, 1968). Two enzymes with very similar properties have been described. Their activity is expressed by the equation

$$\text{Sucrose F}_n + H_2O \overset{\text{FH}}{\longrightarrow} \text{sucrose F}_{n-1} + F \tag{3}$$

These enzymes break only the β-$(2 \rightarrow 1)$ linkage between a terminal fructosyl group and its adjacent fructose residue. The rate of the reaction depends on the degree of polymerization and on the nature of the group at the nonreactive end of the molecule. The enzymes showed no transfructosylase activity.

The most studied enzyme is sucrose-sucrose transfructosylase (SST), which has been isolated from tubers of Jerusalem artichoke (Scott *et al.*, 1966) and of *Polianthes tuberosa* (Bhatia, 1955), roots of chicory (Singh and Bhatia, 1971), of dandelion (Srepel and Mijatovic, 1975) and of asparagus (Shiomi and Izawa, 1980), bulbs of onion (Scott, 1968), leaf midvein–petiole tissues of lettuce (Chandorkar and Collins, 1974), and stems of *Agave* (Satyanarayana, 1976).

The only enzyme extensively purified to the stage of giving one single protein band on disc electrophoresis is the one isolated from asparagus roots

(Shiomi and Izawa, 1980). The molecular weight of the *asparagus* enzyme was estimated to be about 65,000. This value is similar to the one reported for the *Agave* enzyme (62,000) but smaller than the 100,000 found for the partially purified SST of lettuce leaves. This difference could be ascribed to difficulty in ascertaining the position of SST by gel filtration owing to its contamination with invertase, or it may be an indication of a difference in the enzyme origin. Thus, the lower molecular weight enzymes belong to families of the monocotyledons while the higher molecular weight enzymes belong to the Compositae, which are dicotyledons. Furthermore, the specificity seems to be different. The *asparagus* enzyme can catalyze fructosyl transfer from sucrose to neokestose homologues, while Jerusalem artichoke SST cannot (Scott *et al.*, 1966; Shiomi and Izawa, 1980).

SST seems to occupy a key role in the initiation of fructan synthesis. Thus, Scott (1968) found a definite correlation between the occurrence of SST and the net synthesis of fructans in Jerusalem artichoke tubers and onion bulbs. Pontis (1970) ascribed to sucrose a regulatory role in the induction of SST activity. The production of 1^F-fructosyl sucrose in explants of Jerusalem artichoke tubers was clearly dependent on the sucrose level (Pontis *et al.*, 1972). Similar results have been reported for lettuce leaf discs incubated on a sucrose medium (Chandorkar and Collins, 1974).

The equivalent enzyme catalyzing the formation of the trisaccharide in the Gramineae has recently been measured for the first time in *Festuca rubra* by del Campillo and Pontis (1983). Preliminary experiments seemed to indicate a correlation between the level of the enzyme and the environmental conditions under which plants had been growing. Thus, *Festuca* plants grown at 25°C, when moved to 4°C, showed an increased activity of the enzyme in just 24 hr of exposure to the lower temperature. It is too clearly to know if this very interesting result is due to a *de novo* synthesis of the enzyme owing to the plant's cold exposure or to an activation.

The enzyme that promotes polymerization above the trisaccharide level, β-$(2\rightarrow1)$-fructan:β-$(2\rightarrow1)$-fructan fructosyltransferase (FFT) has been much less studied. Most of the information about its properties comes from the work of Edelman and Dickerson (1966), with the enzyme isolated and partially purified from Jerusalem artichoke tubers. A similar enzyme has been isolated from dandelion roots (Rutheford and Deacon, 1974). According to the general reaction mechanism already indicated for the *Helianthus tuberosus* enzyme, the synthesis of higher fructan homologs is initiated by

$$\text{Fructosyl-sucrose} + \text{fructosyl-sucrose} \overset{\text{FFT}}{\rightleftharpoons} \text{sucrose} + \text{fructosyl}_2\text{-sucrose} \qquad (4)$$

However, it has been found that sucrose, which can act as an acceptor of fructosyl residues, competes with the trisaccharide so successfully that continued transfer to form material of higher degree of polymerization is first partially and then totally prevented. Thus, the following exchange reaction is dominant in the presence of equal concentrations of sucrose and fructosylsucrose:

$$\text{Fructosyl-sucrose} + [^{14}C]\text{sucrose} \xrightleftharpoons{\text{FFT}} \text{sucrose} + [^{14}C]\text{fructosylsucrose} \tag{5}$$

Similarly, polymers of high degree of polymerization also show great affinity, with respect to FFT, as acceptors. Therefore, in a mixture of fructosyl sucrose and such primers, higher polymers are formed, instead of oligosaccharides of a low degree of polymerization that would arise in the absence of such primers by self transfer.

In order to avoid the interference of sucrose in the reaction catalyzed by FFT, Edelman and Jefford (1968) have speculated that the reaction catalyzed by SST may take place in the cytoplasm whereas FFT might be localized at the tonoplast. Thus, the trisaccharide formed in the cytoplasm may be converted to the tetrasaccharide by FFT at the tonoplast and then released into the vacuole, while the sucrose also produced in the reaction [see Eq. (2)] may be released back into the cytoplasm. Polymers in the vacuole may undergo further transformations by the action of the membrane-bound FFT, whereby fructans of higher DP are again released into the vacuole and the terminally resulting sucrose into the cytoplasm. However, the hypothesis does not take into account the pH optima determined for the two enzymes involved, sucrose–sucrose fructosyltransferase and fructan fructanfructosyltransferase. The values presented in Table I indicate that all SST preparations so far isolated exhibit an optimum pH about 5.5, whereas FFT exhibits an optimum pH in the neutral or alkaline region. It seems, then, rather unlikely that SST would be localized in the cytoplasm. It is also difficult to visualize the localization of FFT at the tonoplast, as in order to interact with fructan molecules impinging on it—as Edelman and Jefford (1968) proposed—the enzyme should be in the inner face of the tonoplast in contact with the very acidic vacuolar contents. In any case, no studies of the exact localization of the enzymes of fructan metabolism have been carried out, and the Edelman and Jefford hypothesis should be reviewed when this information becomes available.

There is a property of FFT which must be mentioned that is more clearly observed when the enzyme catalyzes self transfer with polymers of a degree of polymerization 8 or 9. In these cases, the reaction approaches an equilibrium based on the weight of fructose, not the molarity of the various fractions. This means that the fructose ratio between different polymers will always be unity (Edelman and Dickerson, 1966; Jefford and Edelman, 1961). The practical result is that when a chromatogram of the polymers is observed, all spots

Table I

Optimum pH Determined for Fructosyltransferases

Enzyme	Source	Optimum pH	Reference
Sucrose–sucrose fructosyltransferase (SST)	Jerusalem artichoke tubers	5.2	Scott *et al.* (1966)
	Chicory roots	5.6	Singh and Bhatia (1971)
	Dandelion roots	5.0	Rutheford and Deacon (1974)
	Asparagus roots	5.0	Shiomi and Izawa (1980)
	Onion bulbs	5.0–5.2	Scott *et al.* (1966)
	Lettuce leaves	5.2	Chandorkar and Collins (1974)
	Agave stems	5.5	Satyanarayana (1976)
β-(2→1)-Fructan: β-(2→1)-fructan fructosyltransferase (FFT)	Jerusalem artichoke tubers	6.1	Edelman and Dickerson (1966)
	Dandelion roots	8.5	Rutheford and Deacon (1974)

appear, when developed with a fructose reagent, to have the same intensity, i.e., the same amount of fructose.

A rather different mechanism of transfructosylation is claimed to operate in stems of *Agave americana*. Bhatia and Nandra (1979) studied fructan-forming enzymes present in stem extracts that seemed to use only sucrose as the fructosyl donor and that apparently could synthesize fructans up to DP ~ 11. The proposed mechanism is similar to the one existing in bacteria for the formation of levans (Akazawa, 1965), and further investigations are required to ascertain its existence in plants.

Except for the results of del Campillo and Pontis (1983), discussed already, there has not been any determination of enzymes catalyzing transfructosylation reactions in members of the Gramineae.

Analysis of the pattern of turnover of fructans in leaf bases of *Dactylis glomerata* after a pulse of $^{14}CO_2$ (Pollock, 1982) seems to indicate that there is a fructosyl transferase that acts in a way similar to the one isolated from members of the Compositae (FFT): it produces an equilibration between chains of different molecular weight. Moreover, the same data is consistent with the existence of a hydrolase that acts on the terminal fructose residues of fructans (to be discussed further). Nevertheless, in earlier experiments, Pollock (1979) found evidence, also in *Dactylis glomerata*, that suggests that fructan synthesis occurs in leaf bases by direct transfer of fructosyl residues from sucrose to a fructan polymer, without the appearance of the intermediate series of oligosaccharides. On the other hand, Pollock (1982) has also reported recently that, in *Lolium temulentum*, feeding experiments indicated the presence of fructosyl-sucrose as well as tetra and pentasaccharides that behaved chromatographically like the fructose polymers from *Helianthus tuberosus*. He suggested that this evidence indicates that fructosyl-sucrose occupies a role in fructan synthesis in *Lolium temulentum*, similar to its role in the Compositae. Nevertheless, the picture is complicated because in *L. temulentum* two different series of fructans appear to coexist; one appears to be the same as the $(2\rightarrow1)$-linked fructans in *Helianthus tuberosus*, but the identity of the other is unknown.

More detailed structural analysis will be required in order to ascertain the existence of β-$(2\rightarrow1)$-linked fructans in *Lolium temulentum*, as it is generally accepted that the major linkage in Gramineae fructans is β-$(2\rightarrow6)$ (Hirst, 1957; Smith, 1973).

The depolymerization of fructans occurs in the Compositae by the action of specific β-$(2\rightarrow1)$-fructan 1-fructanohydrolases as already indicated. Edelman and Jefford (1964) isolated two enzymes that could be distinguished by their elution from DEAE–cellulose columns and by their relative activities on inulin and fructans of smaller molecular weight. The enzymes were not invertases, as they were both inactive against sucrose, but their action against their substrates was inhibited by this sugar. Rutheford and Deacon (1972a,b) isolated two hydrolases from *Taraxacum officinale* that showed similar characteristics to those of *Helianthus tuberosus* but that were not inhibited by sucrose.

We have little information regarding the depolymerization of fructans present in monocotyledons. The only direct evidence that fructan degradation proceeds in a way similar to that outlined for the Compositae has been given by Smith (1976) and by Mino and Maeda (1974). Smith showed that a β-fructofuranosidase is present in the culm base of tall fescue (*Festuca arundinacea*). This enzyme degrades phlein starting at the fructose end and removes only one fructose residue at a time until a molecule of sucrose remains. The enzyme is specific for the cleavage of $(2\rightarrow6)$ linkages and does not attack inulin. The evidence of Mino and Maeda (1974) indicates also that, in the haplocorn of timothy, a phlein hydrolase exists that does not act on inulin. However, their preparation was a very crude one and no other information about its properties was given.

The action of plant hormones on the metabolism of fructans has been studied by Rutheford and co-workers (Deacon and Rutheford, 1972; Rutheford and Deacon, 1974; Rutheford et al., 1969). They found that 2,4-D enhances the activity of fructan hydrolases present in dandelion roots and in Jerusalem artichoke tubers. Working with explants of Jerusalem artichoke, Pontis (1966) has shown that the *de novo* synthesis of fructans is very likely to be regulated through the interplay of auxins, kinetin, and gibberellic acid. The action of auxin is not limited to enhancing the activity of fructan hydrolyases, as Pontis (1970) has also shown that auxin seems to increase the level of sucrose synthetase, which, in turn, affects the concentration of sucrose and consequently modifies all fructan metabolism.

When Gonzalez and Pontis (1963) and Umemura *et al.* (1967) demonstrated the presence of uridine diphosphate fructose in Dahlia and Jerusalem artichoke tubers, respectively, it was thought that this sugar nucleotide may be involved as glycosyl donor in the biosynthesis of fructans. Furthermore, the isolation of fructose-containing nucleotides led Pontis and Fischer (1963) to synthesize fructofuranose 2-phosphate, which could be an intermediate in the degradation and biosynthesis of fructans by analogy with the role of glucose 1-phosphate in starch and glycogen metabolism. However, investigations carried out with extracts of different parts of Dahlia, chicory, and lettuce plants did not provide conclusive results in favor of this hypothesis (Pontis, 1966). Nevertheless, the participation of phosphorylated fructose derivatives in fructan metabolism should not be completely ruled out, especially if we take into account the fact that the discovery of fructose 2,6-diphosphate has opened up a new field in the area of glycolytic regulation (Hers and Van Schaftingen, 1982). It should be recalled that, earlier β-D-fructofuranosyl phosphates were not recognized as naturally occurring intermediates (Feingold and Avigad, 1980).

VI. PHYSIOLOGY

Fructans are considered another type of reserve polysaccharide, but they may also play the role of "osmotic buffers" (Pontis, 1971), as plants that contain them are, in general, species that endure a cold or a dry period during their life cycle when the osmotic activity of fructans may temper these conditions by increasing resistance to freezing or dessication (Edelman and Jefford, 1968).

The reserve role is clearly fulfilled in many fructan-containing plants. Thus, fructans are accumulated in tubers of *Helianthus tuberosus* and *Dahlia variabilis,* and in roots of *Chicorium intybus, Taraxacum officinale, Aster tripolium,* and *Symphytum officinale* (Whelan, 1955; Whistler and Smart, 1953). The polymers are usually deposited in the reserve organs during the summer and early autumn. They can reach very high levels exceeding 50% of the dry weight of the particular tissue (Kandler and Hopf, 1980; Whistler and Smart, 1953).

In the monocotyledons a similar picture is found. Fructans are distributed in all parts of the plants, but the amount that occurs in a given tissue will depend on the stage of development (Smith, 1967, 1973). The accumulation phase seems to be as in the dicotyledons, in late summer, but the picture is not too clear, as Pollock and Jones (1979) found that in several forage grasses the major period of fructan synthesis was late autumn and early winter.

In every case, whether monocotyledons or dicotyledons, important variation in the fructan content occurs during the life cycle, as can be expected for reserve polymers, but simultaneously these changes are also associated with temperature modifications or with seasonal temperature variations (Colby *et al.*, 1964; McKell *et al.*, 1969; Rutheford and Weston, 1968; Vartha and Bailey, 1980).

In tubers of *Helianthus tuberosus*, seasonal changes of fructans were studied by Bacon and Loxley (1952), who measured variations in the optical rotation of water extracts made at different times of the year. Extracts from late summer and autumn had the most negative rotation (about $-22°$), indicating the presence of large amounts of high-molecular-weight fructans, whereas extracts from winter and early spring had the most positive optical rotation (about $+5°$), indicating the presence of large amounts of fructan oligosaccharides of low molecular weight. This has been explained by Edelman and Jefford (1968), who have pointed out that changes in the average chain length can occur during tuber dormancy without a net change in carbohydrate content. This implies that the fructose residues released by the action of the fructan hydrolases must be converted to glucose, from which more sucrose may be produced, thus allowing the fructan–fructan fructosyltransferase (FFT) to catalyze a redistribution of fructose residues and increase the amount of fructans of smaller molecular size at the expense of larger ones. This mechanism would permit a very rapid change in the size and number of molecules, and hence a rapid change in osmotic pressure. The modification in osmotic pressure could, in turn, increase the resistance to frost injury.

The property mentioned earlier, namely that the reaction catalyzed by the enzyme fructan–fructan fructosyltransferase approaches equilibrium based on the fructose weight of each polymer, may be connected to this facility for modifying the osmotic pressure. Pontis (1983) has proposed a model that explains this curious property as a necessary condition to allow the modification of the molecular population in an orderly way and, at the same time, to produce an amplifying effect on the osmotic pressure, conteracting the changes that may be brought about by a decrease in temperature. He also found that during winter there is a linear correlation between soil temperature and the weight of fructose that can be found in each polymer in dormant *Helianthus tuberosus* tubers (Pontis, 1983).

The correlation between temperature and the level of fructans, as well as their degree of polymerization, has been studied in the Gramineae by Smith (1968b). He showed that in timothy (*Phleum pratense* L.) temperature influenced the levels of fructans accumulated but had little effect on the levels of total sugars and starch. He grew plants to early anthesis under two different temperature regimes: 18.5/10°C and 29.5/21°C (day/night). The percentages of total nonstructural carbohydrates (mainly fructans) within each of the two temperatures regimes were highest in the stem bases and were then greatest in the descending order of stems and sheaths, heads or green leaf blades, stems, and roots (see Table II). It is of particular interest that virtually no fructans were accumulated in the green leaf blades under warm temperatures but a very high content (21.5%) accumulated under the cool temperature. Moreover, not only did fructans accumulate but modifications of fructan chain length also occurred. In leaf blades, stems, and roots, there was a clear increase in chain length towards higher molecular weights.

The studies of Eagles (1967) with two climatic races of *Dactylis glomerata* (cocksfoot) and those of Pollock and Ruggles (1976) also pointed to the relationship between fructan content and temperature. Eagles studied the variations of fructan content in Mediterranean (Portuguese) and North European (Norwegian) population of cocksfoot at four different temperatures. He reached the conclusion that the ecological significance of the accumulation of fructans in the Norwegian population at low temperatures (5°C) may be associated with the high degree of cold resistance of this population.

Table II

Grams of Dry Matter and Percentages of Carbohydrates in Various Plant Parts of Timothy (*Phleum pratense*) Grown to Early Anthesis under 18.5/10°C (Cool) and 29.5/21°C (Warm) Day Night Temperatures[a,b]

Plant parts	Temperature regime	Dry matter (g)	Total sugars (%)	Fructans (%)	Starch (%)	Total nonstructural (%)
Heads	Cool	1.1	8.6	1.3	4.1	14.0
	Warm	0.2	8.6	0.8	3.3	12.7
Leaf blades, green[c]	Cool	7.5	4.7	21.5	2.3	28.5·
	Warm	3.5	4.8	0.7	1.4	6.9
Stems and sheaths	Cool	19.5	6.0	30.2	2.6	38.8
	Warm	3.6	4.2	9.7	4.4	18.3
Stem bases	Cool	6.0	4.2	38.6	2.7	45.5
	Warm	1.8	4.1	28.1	3.3	35.5
Roots	Cool	3.9	2.3	6.4	0.4	9.1
	Warm	0.6	1.4	0.6	—	—

[a] From Smith (1968b).

[b] Anthesis reached in 54 days under cool and 30 days under warm temperatures.

[c] Yellow leaf blade dry matter amounted to 2.3 g under cool and 0.5 g under warm temperatures.

In onions (*Allium cep L.*), Darbyshire and Henry (1978) found that fructans accumulated in the inner younger base leaves. If their role was only that of storage carbohydrates, one would have expected to find a higher concentration in older tissue. They suggested that in onions, in addition to being storage carbohydrates, fructans may have a further function in providing osmotic adjustment during bulbing.

All these facts strongly support the case that the presence of fructans in plants is advantageous for cold hardiness. Nevertheless, it is still valid to ask why fructans are more useful than starch in providing a defense for the plant to environmental hard conditions, as both types of polymers are intracellular. The difference may lie more in the physicochemical properties of the two types of polymers. Fructans are present in the cells as solutes; hence they are osmotically active, and by polymerization and depolymerization their osmotic potential and therefore their contribution to cold hardiness could be easily altered.

VII. CONCLUDING REMARKS

The experiments and data reviewed show that we are only beginning to understand the mechanism of synthesis and degradation of fructans. Even less is known about the regulation of their metabolism. We know that hormones play a part in it, but we do not know their exact role; for example, we do not know if hormones are connected with the mobilization of fructans when plants are exposed to adverse environmental conditions. At the enzymological level the information available about the properties of the enzymes that participate in their metabolism is not complete. For example, we do not know what makes fructan–fructan fructosyltransferase terminate chain elongation when DP reaches 30 in Jerusalem artichoke tubers while it acts at DP 11 in onion bulbs. It should be remembered that if a similar enzyme acts in the Gramineae, polymers there can reach up to DP 260. The biochemistry and physiology of fructans is then a rich, though difficult, field waiting to be plowed.

ACKNOWLEDGEMENTS

The authors are indebted to their colleagues at the Instituto de Investigaciones Biológicas for their criticism. Horacio G. Pontis is a Career Investigator of the Consejo Nacional de Investigaciones Científicas y Técnicas, Argentina, and Elena del Campillo a Fellow of the Comisión de Investigaciones Científicas de la Provincia de Buenos Aires, Argentina.

REFERENCES

Akazawa, T. (1965). *In* "Plant Biochemistry" (J. Bonner and J. E. Varner, eds.), 2nd ed., pp. 258–289. Academic Press, New York.

Allen, R. Z. and Kabat, E. A. (1957), *J. Exp. Med.* **105**, 383.

Archbold, H. K. (1938). *Ann. Bot. (London)* **2**, 183–187.

Archbold, H. K. (1940). *New Phytol.* **39**, 185–219.

Arni, P. C. and Percival, E. G. V. (1951). *J. Chem. Soc.* pp. 1822–1830.

Aspinall, G. O. and Gupta Das, P. C. (1959). *J. Chem. Soc.* pp. 718–722.

Aspinall, G. O. and Telfer, R. G. J. (1953). *Chem. Ind. (London)* p. 490.

Bacon, J. S. D. (1957). *Biochem. J.* **67**, 5p–6p.

Bacon, J. S. D. (1959). *Biochem. J.* **73**, 507–513.

Bacon, J. S. D. and Edelman, J. (1951). *Biochem. J.* **48**, 114–125.

Bacon, J. S. D. and Loxley, R. (1952). *Biochem. J.* **51**, 208–213.

Bell, D. J. and Palmer, A. (1952). *J. Chem. Soc.* pp. 3763–3765.

Belval, H. (1937). *Bull. Soc. Chim. Biol.* **19**, 1158–1162.

Belval, H. (1939). *Bull. Soc. Chim. Biol.* **21**, 294–296.

Bender, M. M. and Smith, D. (1973). *J. Br. Grassl. Soc.* **28**, 97–100.

Bhatia, I. S. (1955). *J. Sci. Ind. Res. Sect. A.* **14**, 552–530.

Bhatia, I. S. and Nandra, K. S. (1979). *Phytochemistry* **18**, 923–927.

Bhatia, I. S. and Srinivasan, M. (1953). *Curr. Sci.* **22**, 236–237.

Boggs, L. A. and Smith, F. (1956). *J. Am. Chem. Soc.* **78**, 1880–1885.

Bourdu, R. (1954). *Co. R. Hebd. Seances Acad. Sci.* **239**, 1524–1526.

Bourdu, R. (1957). *Rev. Gen. Bot.* **64**, 153–192, 197–260.

Bourne, E. J., Percival, E. and Smestad, B. (1972). *Carbohydr. Res.* **22**, 75–77.

Carles, J. (1935). *Rev. Gen. Bot.* **47**, 5–22, 87–95, 144–159, 215–229, 294–304, 363–477.

Chandorkar, K. R. and Collins, F. N. (1974). *Can. J. Bot.* **52**, 1369–1377.

Cisar, J., Kabat, E. A., Liad, J. and Potter, M. (1974). *J. Exp. Med.* **139**, 159–179.

Colby, W. G., Drake, M., Field, D. L. and Kreowski, G. (1964). *Agron. J.* **56**, 169–173.

Colin, H. (1925). *Bull. Soc. Chim. Biol.* **7**, 173–175.

Colin, H. and Augen, A. (1927). *C. R. Hebd. Seances Acad. Sci.* **185**, 473–474.

Colin, H. and Belval, H. (1922). *C. R. Hebd. Seances Acad. Sci.* **175**, 1441–1442.

Colin, H. and Belval, H. (1937). *Bull. Soc. Chim. Biol.* **19**, 65–68.

Colin, H. and Belval, H. (1940). *Bull. Soc. Bot. Fr.* **87**, 341–345.

Colin, H. and Chaudun, A. (1933). *Bull. Soc. Chim. Biol.* **15**, 1520–1523.

Colin, H. and Chollet, M. M. (1939). *C. R. Hebd. Seances Acad. Sci.* **208**, 549–550.

Colin, H. and De Cugnac, A. (1926). *Bull. Soc. Chim. Biol.* **8**, 621–627.

Collins, F. W. and Chandorkar, K. R. (1971). *J. Chromatogr.* **56**, 163–167.

Darbyshire, B. and Henry, R. J. (1978). *New Phytol.* **81**, 29–35.

Darbyshire, B. and Henry, R. J. (1981). *New Phytol.* **87**, 249–256.

Das, N. N. and Das, A. (1978). *Carbohydr. Res.* **64**, 155–167.

Deacon, A. G. and Rutheford, P. P. (1972). *Phytochemistry* **11**, 3143–3148.

De Cugnac, A. (1931). *Bull. Soc. Chim. Biol.* **13**, 125.

Dedonder, R. (1952a). *Bull. Soc. Chim. Biol.* **34**, 144–182.

Dedonder, R. (1952b). *Bull. Soc. Chim. Fr.* **19**, 874–879.

del Campillo, E. and Pontis, H. G. (1983). *Abstr. S.A.I.B. 19th Meet.* p. m 111.

de Miniác, M. (1970). *C. R. Hebd. Seances Acad. Sci.* **270**, 1583–1587.

Dickerson, A. G. and Edelman, J. (1966). *J. Exp. Bot.* **17**, 612–619.

Dimick, K. P. and Christiensen, B. E. (1942). *J. Am. Chem. Soc.* **64**, 2501.

du Mérac, M. L. R. (1949). Thesis. Fac. Sci. University of Paris.

du Mérac, M. L. R. (1953). *Rev. Gen. Bot.* **60**, 689–696.

Eagles, C. F. (1967). *Ann. Bot. (London)* [N.S.] **31**, 645–651.

Edelman, J. (1960). *Bull. Soc. Chim. Biol.* **52**, 1737–1744.

Edelman, J. and Dickerson, A.G. (1966). *Biochem. J.* **98**, 787–794.

Edelman, J. and Jefford, T. G. (1964). *Biochem. J.* **93**, 148–161.

Edelman, J. and Jefford, T. G. (1968). *New Phytol.* **67**, 517–531.

Edelman, J. and Popov, K. (1962). *C. R. Acad. Bulg. Sci.* **15**, 627–630.

Edelman, J., Recaldin, D. A. C. L. and Dickerson, A. G. (1963). *Bull. Res. Counc. Isr. Sect. A* **11**, 275–278.

Ekstrand, A. G. and Johanson, C. J. (1887). *Ber. Dtsch. Chem. Ges.* **20**, 3310–3317.

Ekstrand, A. G. and Johanson, C. J. (1888). *Ber. Dtsch. Chem. Ges.* **21**, 594–597.

Feingold, D. S. and Avigad, G. (1980). *In* "The Biochemistry of Plants" (J. Preiss, ed.), Vol. 3, pp. 101–170. Academic Press, New York.

Fischer, H. (1902). *Beitr. Biol. Pflanz.* **8**, 53–110.

Flood, A. E., Hirst, E. L. and Jones, J. K. N. (1947). *Nature (London)* **160**, 86–87.

Forsyth, W. G. C. (1948). *Nature (London)* **161**, 239–241.

Forsyth, W. G. C. (1950). *Biochem. J.* **46**, 141–145.

Glaudemans, C. P. J. (1975). *Adv. Carbohydr. Chem. Biochem.* **31**, 313–346.

Gonzalez, N. S. and Pontis, H. G. (1963). *Biochim. Biophys. Acta* **69**, 179–181.

Grafe, V. and Vouk, V. (1912). *Biochem. Z.* **43**, 424–433.

Grey, H. M., Hirst, J. W. and Cohn, M. (1971). *J. Exp. Med.* **133**, 289–304.

Grotelueschen, R. D. and Smith, D. (1968). *Crop Sci.* **8**, 210–212.

Hammer, H. (1970). *Acta Chem. Scand.* **24**, 1294–1300.

Harlay, V. (1901). *C. R. Hebd. Seances Acad. Sci.* **132**, 423–424.

Hegnauer, R. (1963). "Chemotaxonomie der Pflanzen," Vol. 2. Birkhaeuser, Basel.

Hegnauer, R. (1964). "Chemotaxonomie der Pflanzen," Vol. 3. Birkhaeuser, Basel.

Hegnauer, R. (1966). "Chemotaxonomie der Pflanzen," Vol. 4. Birkhaeuser, Basel.

Hegnauer, R. (1973). "Chemotaxonomie der Pflanzen," Vol. 6. Birkhaeuser, Basel.

Hers, H. G. and Van Schaftingen, E. (1982). *Biochem. J.* **206**, 1–12.

Hirst, E. L. (1957). *Proc. Chem. Soc., (London)* pp. 193–204.

Hirst, E. L., McGilvray, D. I. and Percival, E. G. V. (1950). *J. Chem. Soc.* pp. 1297–1302.

Jefford, T. G. and Edelman, J. (1961). *J. Exp. Bot.* **12**, 177.

Jefford, T. G. and Edelman, J. (1963). *J. Exp. Bot.* **14**, 56–62.

Kandler, O. and Hopf, H. (1980). *In* "The Biochemistry of Plants" (J. Preiss, ed.), Vol. 3, pp. 246–252. Academic Press, New York.

Karlsson, G. (1969). *J. Chromatogr.* **44**, 413–414.

Keegan, P. Q. (1916). *Chem. News* **113**, 85–87.

Labourel, G. and Péaud-Lenoël, C. (1969). *Chem. Zvesti* **23**, 765–769.

Laidlaw, R. A. and Reid, S. G. (1951). *J. Chem. Soc.* pp. 1830–1833.

Laidlaw, R. A. and Wylam, C. B. (1952). *J. Sci. Food Agric.* **3**, 494–497.

Lüding, H. and Müller, H. (1872–1873). *Jahrl. Chem.* **803**, 832–836.

Lundblad, A., Stellar, R., Kabat, E. A., Hirst, J. W. and Cohn, M. (1972). *Immunochemistry* **9**, 535–538.

McDonald, E. J. (1946). *Adv. Carbohydr. Chem.* **2**, 253–277.

McKell, C. M., Younger, V. B., Nudge, F. J. and Chatterton, N. J. (1969). *Crop Sci.* **9**, 534–537.

Madam, V. K. (1972). *Z. Pflanzenphysiol.* **68**, 272–280.

Marchessault, R. H., Bleha, T., Deslandes, Y. and Revol, J. F. (1980). *Can. J. Chem.* **58**, 2415–2417.

Medcalf, D. G. and Cheung, P. W. (1971). *Cereal Chem.* **48**, 1–8.

Meeuse, B. J. D. (1962). *In* "Physiology and Biochemistry of Algae" (R. A. Levin, ed.), pp. 289–311. Academic Press, New York.

Meier, H. and Reid, J. S. G. (1982). *In* "Encyclopedia of Plant Physiology, New Series" (A. Pirson and M. H. Zimmermann, eds.), Vol. 13A, pp. 435–450. Springer-Verlag, Berlin and New York.

Middleton, E. (1977). *J. Membr. Biol.* **34**, 93–101.

Mino, Y. and Maeda, K. (1974). *J. Jpn. Soc. Grassl. Sci.* **20**, 6–10.

Mizuno, T. and Hayashi, K. (1955). *Nippon Nogei Kagaku Kaishi* **29**, 533; *Chem. Abstr.* **52**, 18,682.

Montgomery, R. and Smith, F. (1956). *Agric. Food Chem.* **4**, 716–720.

Montgomery, R. and Smith, F. (1957). *J. Am. Chem. Soc.* **79**, 446–450.

Nitsch, E., Iwanov, W. and Lederer, K. (1979). *Carbohydr. Res.* **72**, 1–12.

Ojima, K. and Isawa, T. (1968). *Can. J. Bot.* **46**, 1507–1511.

Omslow, M. W. (1929). "Practical Plant Biochemistry. "Cambridge Univ. Press, London and New York.

Pantanelli, E. (1949). *Atti Relaz.—Accad. Pugliese Sci., Part 2* **4**, 43–45.

Percheron, F. (1962). *C. R. Hebd. Seances Acad. Sci.* **255**, 2521–2522.

Percival, E. and Young, M. (1971). *Phytochemistry* **10**, 807–812.

Phelds, C. F. (1965). *Biochem. J.* **95**, 41–47.

Pollock, C. J. (1979). *Phytochemistry* **18**, 777–780.

Pollock, C. J. (1982). *Phytochemistry* **21**, 2461–2466.

Pollock, C. J. and Jones, T. (1979). *New Phytol.* **83**, 9–15.

Pollock, C. J. and Ruggles, P. A. (1976). *Phytochemistry* **15**, 1643–1646.

Pollock, C. J., Hall, M. A. and Roberts, D. P. (1979). *J. Chromatogr.* **171**, 411–415.

Pontis, H. G. (1966). *Arch. Biochem. Biophys.* **116**, 416–423.

Pontis, H. G. (1968). *Anal. Biochem.* **23**, 331–333.

Pontis, H. G. (1970). *Physiol. Plant.* **23**, 1089–1100.

Pontis, H. G. (1971). *An. Soc. Cient. Argent.* sspl. 59–63.

Pontis, H. G. (1983). *Abstrs, S.A.F.V. 15th Meet.* p. 8.

Pontis, H. G. and Fischer, C. (1963). *Biochem. J.* **89**, 452–459.

Pontis, H. G., Wolosiuk, R. A. and Fernandez, L. M. (1972). *In* "Biochemistry of the Glycosidic Linkage" (R. Piras and H. G. Pontis, eds.), pp. 239–265. Academic Press, New York.

Raguse, C. A. and Smith, D. (1965). *J. Agric. Food Chem.* **13**, 306–312.

Recaldin, D. A. C. L. (1961). Ph.D. Thesis, University of London.

Rose, V. (1804). *Neues Allg. J. Chem. (Gehlens)* **3**, 217–219.

Rutheford, P. P. and Deacon, A. C. (1972a). *Biochem. J.* **126**, 569–573.

Rutheford, P. P. and Deacon, A. C. (1972b). *Biochem. J.* **129**, 511–512.

Rutheford, P. P. and Deacon, A. C. (1974). *Ann. Bot. (London)* [N.S.] **38**, 251–260.

Rutheford, P. P. and Weston, E. W. (1968). *Phytochemistry* **7**, 175–180.

Rutheford, P. P., Weston, E. W. and Flood, A. E. (1969). *Phytochemistry* **8**, 1859–1866.

Sachs, J. (1864). *Bot. Ztg.* **22**, 77, 85–89.

Satyanarayana, M. N. (1976). *Indian J. Biochem. Biophys.* **13**, 261–265.

Schlubach, H. H. (1953). *Experientia* **9**, 230–232.

Schlubach, H. H. (1958). *Prog. Chem. Org. Nat. Prod.* **15**, 1–27.

Schlubach, H. H. and Berndt, J. (1961). *Justus Liebigs Ann. Chem.* **647**, 41–50.

Schlubach, H. H. and Flörsheim, W. (1929). *Ber. Dtsch. Chem. Ges.* **62**, 1491–1494.

Schlubach, H. H. and Flörsheim, W. (1931). *Hoppe-Seyler's Z. Physiol. Chem.* **198**, 153–156.

Schlubach, H. H. and Koehn, H. O. A. (1957). *Justus Liebigs Ann. Chem.* **606**, 130–137.

Schlubach, H. H. and Koehn, H. O. A. (1958). *Justus Liebigs Ann. Chem.* **614**, 126–136.

Schlubach, H. H. and Lendzian, H. (1937). *Justus Liebigs Ann. Chem.* **532**, 200–207.

Schlubach, H. H. and Lopp, W. (1936). *Justus Liebigs Ann. Chem.* **532**, 130–136.

Schlubach, H. H. and Müller, H. (1952). *Justus Liebigs Ann. Chem.* **578**, 194–198.

Schlubach, H. H. and Sinh, O. K. (1940). *Justus Liebigs Ann. Chem.* **544**, 111–116.

Schlubach, H. H., Knoop, H. and Liu, M. Y. (1933). *Justus Liebigs Ann. Chem.* **504**, 30–35.
Schlubach, H. H., Lubbers, H. and Borowski, H. (1955). *Justus Liebigs Ann. Chem.* **595**, 229–236.
Schmiedeberg, J. E. O. (1879). *Hoppe. Seyler's Z. Physiol. Chem.* **3**, 112–114.
Schulze, E. (1899). *Hoppe. Seyler's Z. Physiol. Chem.* **27**, 267–271.
Scott, R. W. (1968). Ph.D. Thesis, University of London.
Scott, R. W., Jefford, T. G. and Edelman, J. (1966). *Biochem. J.* **100**, 23P–24P.
Shiomi, N. and Izawa, M. (1980). *Agric. Biol. Chem.* **44**, 603–614.
Shiomi, N., Yamada, J. and Izawa, M. (1976). *Agric. Biol. Chem.* **40**, 567–575.
Shiomi, N., Yamada, J. and Izawa, M. (1979). *Agric. Biol. Chem.* **43**, 2233–2244.
Singh, R. and Bhatia, I. S. (1971). *Phytochemistry* **10**, 495–502.
Smith, A. E. (1976). *J. Agric. Food Chem.* **24**, 476–478.
Smith, D. (1967). *Crop Sci.* **7**, 62–67.
Smith, D. (1968a). *J. Br. Grassl. Soc.* **23**, 306–309.
Smith, D. (1968b). *Crop Sci.* **8**, 331–334.
Smith, D. (1973). *In* "Chemistry and Biochemistry of Herbaces" (G. W. Butler and R. W. Bailey, eds.), Vol. 1, pp. 105–155. Academic Press, London.
Srepel, B. and Mijatovic, D. (1975). *Acta Pharm. Jugosl.* **25**, 189–191.
Srinivasan, M. and Bhatia, I. S. (1953). *Biochem. J.* **55**, 286–289.
Srinivasan, M., Bhalerao, V. R. and Subramanian, N. (1952). *Curr. Sci.* **21**, 159–160.
Streefkerk, D. G. and Glaudemans, C. P. J. (1977). *Biochemistry* **16**, 3760–3765.
Tanret, C. (1891). *C. R. Hebd. Seances Acad. Sci.* **112**, 293.
Thomson, T. (1818). "A System of Chemistry," pp. 4–65. Abraham Small, Philadelphia, Pennsylvania.
Tomasic, J., Jennings, H. J. and Glaudemans, C. P. J. (1978). *Carbohydr. Res.* **62**, 127–133.
Tomoda, M., Satoh, N. and Sugiyama, A. (1973). *Chem. Pharm. Bull.* **21**, 1806–1810.
Tsusué, Y. and Fujita, Y. (1964). *J. Gen. Appl. Microbiol.* **10**, 283–285.
Umemura, Y., Nakamura, M. and Funahashi, S. (1967). *Arch Biochem. Biophys.* **119**, 240–252.
Vartha, E. W. and Bailey, R. W. (1980). *N. Z. J. Agric. Res.* **23**, 93–96.
von Euler, H. and Erdtman, H. (1925). *Hoppe. Seyler's Z. Physiol. Chem.* **145**, 261–264.
Wallach, O. (1886). *Justus Liebigs Ann. Chem.* **234**, 364–367.
Weber, H. (1955). *Ber. Dtsch. Bot. Ges.* **68**, 408–412.
Whelan, W. J. (1955). *In* "Handbuch der Pflanzen physiologie" (W. Ruhland, ed.), Vol. 6, pp. 184–196. Springer-Verlag, Berlin and New York.
Whistler, R. L. and Smart, C. L. (1953). "Polysaccharide Chemistry," pp. 276–291. Academic Press, New York.
Wolf, D. D. and Ellmore, T. L. (1975). *Crop Sci.* **15**, 775–777.
Wolfrom, M. L. and Thompson, A. (1957). *In* "The Carbohydrates" (W. Pigman, ed.), pp. 188–224. Academic Press, New York.

β-(1→3)-Linked Glucans from Higher Plants

R. A. DIXON

Department of Biochemistry
Royal Holloway College (University of London)
Egham Hill, Egham
Surrey, England

I. INTRODUCTION

Starch represents the classical example of a higher plant storage poly-saccharide. It accumulates in a separate organelle, the plastid, during periods of intense photosynthetic activity, and is later degraded to carbohydrate

BIOCHEMISTRY OF STORAGE
CARBOHYDRATES IN GREEN PLANTS

monomers, which then enter the plant's intermediary metabolism and serve as energy sources. Other polysaccharides with clearly defined storage functions occur in the vacuoles of vegetative tissues; these are usually fructans and mannans. In addition, a number of polysaccharides associated with the plant cell wall may serve a storage function; these include mannans, glucomannans and galactomannans, xyloglucans, and galactans (Meier and Reid, 1982). Detailed discussion of the biochemistry of these reserve materials is found in the present volume in the chapters by Bewley and Grant-Reid, Grant-Reid and Brinson and Dey. Of more limited occurrence, and perhaps less clearly defined storage function, are the wall-associated β-glucan polymers, of which barley β-D-glucan is the most studied example. In a recent review, the cereal grain glucans were classified as "minor cell wall storage polysaccharides" (Meier and Reid, 1982). However, there is sufficient evidence for a significant, albeit limited, contribution of such molecules to the energy reserves of certain plants, and this, in addition to the interesting enzymology of β-(1→3)-glucan synthesis and degradation, appears sufficient justification for the "elevation" of these, and related molecules, to a chapter of their own.

Although starch accounts for between 50 and 75% by weight of dry cereal grains, most cereals also contain endospermic hemicellulose fractions, rich in glucan, which may serve storage functions (Preece, 1957). Barley and oats are probably the best sources of such β-glucans, with only small amounts being found in the gums from rye, wheat, and maize. Furthermore, during malting, barley water-soluble β-glucan levels increase, to be followed by extensive degradation; such degradation is not characteristic of the pentosan component of the water-soluble hemicellulose (Preece, 1957). More recently, it has been shown that cell wall and gum polysaccharides of the barley aleurone layer, testa, and endosperm contribute 18.5% of the endospermic polysaccharide degraded during the first 6 days of germination, the remainder being due to starch degradation (Morrall and Briggs, 1978). Over the 6-day period there was a 40-fold increase in embryo wall polysaccharide (initially mainly composed of glucose, followed later by xylose and arabinose), associated with a major decline in the glucose content of endosperm walls, presumably as a result of the loss of β-glucan (Morrall and Briggs, 1978).

Early analytical studies of wheat endosperm cell walls indicated that approximately 75% of the wall was polysaccharide, of which about 10% was β-glucan (Mares and Stone, 1973b). It was suggested that water-soluble components (β-glucan and glucomannan) may be layered on to the surface of the wall, and that this was probably a common feature of the walls of storage tissues. A dual function of (a) storage and (b) physical reinforcement of the wall against cracking during dehydration and dormancy was ascribed to these components (Mares and Stone, 1973b).

A study of the cell walls of *Lolium multiflorum* endosperm tissue cultures has likewise indicated the presence of wall-bound *β*-glucans (Anderson and Stone, 1978). It has been argued that the *Lolium* glucans may have a storage function in view of the lack of such components in the cell walls of tissue cultures from vegetative tissues of other monocots (Burke *et al.*, 1974). However, the extent of their involvement is probably not as great as in some grass seeds, e.g. *Bromus sterilis* (MacLeod and Sandie, 1961), where the endosperm walls are much thicker than those of *Lolium* (Smith and Stone, 1973c).

A histological study of French bean cotyledons demonstrated swelling, followed by a decrease in the thickness of the wall, as material, believed to be noncellulosic polysaccharide, was removed (Smith, 1974). However, the exact nature of the mobilised material was not elucidated. Similar loss of noncellulosic polysaccharides has been reported in other legumes (Meyer and Polyakoff-Mayber, 1963; Flinn, 1969).

The following sections on structure and physiology will mainly be concerned with the mixed-linkage *β*-(1→3)-glucans from cereals, although *β*-(1→3)-glucan synthesizing and degrading enzymes will also be described from other sources. It must be emphasised that other important *β*-(1→3)-glucans exist in plants, examples being the totally *β*-(1→3)-linked polymer callose found, for example, in the sieve tubes and pollen grains of many higher plants, and the algal *β*-(1→3)-glucans such as paramylon, laminarin, and chrysolaminarin (Percival and McDowell, Chapter 9, this volume). Mention might also be made of fungal *β*-(1→3)-glucans such as the phytotoxic mycolaminarins from *Phytophthora* species (Keen *et al.*, 1975). Introduction to the early literature on higher plant *β*-(1→3)-glucans can be found in several review articles (Preece, 1957; Clarke and Stone, 1963; Bull and Chesters, 1966).

II. SOURCES OF *β*-(1→3)-LINKED GLUCANS

A. The Barley Endosperm and Aleurone Layer

Barley endosperm cell walls contain approximately 75% *β*-glucan, the remainder being pentosan (containing small amounts of hexuronic acid), starch, and protein (Forrest and Wainwright, 1977; Ballance and Manners, 1978a). Information on the exact chemical nature and size of the endosperm *β*-glucans is complicated by the different extraction procedures used by different workers (see Section III,A). However, it is clear that at least a fraction of the total *β*-glucan is bound covalently to the cell wall, possibly by linkages to protein via serine, threonine, aspartic acid, or glutamic acid, or by ferulic acid cross-linking (Ballance and Manners, 1978a).

Detailed analyses of the composition of barley endosperm cell walls have been published (Ballance and Manners, 1978a; Thompson and La Berge,

1981). One of the major impediments to modification of the endosperm tissue (e.g., during malting) is the endosperm cell wall, and it has been suggested that the rate of disruption of cell wall structure may in part control the rate of endosperm modification (Thompson and La Berge, 1981; Eastwell and Spencer, 1982). In fact, it was shown as early as the end of the last century that the cell contents in germinating barley remained undergraded as long as the cell walls were intact (Brown and Morris, 1890).

The aleurone layer is a specialised secretory tissue situated at the periphery of the starchy endosperm in seeds of the Graminae. Aleurone cell walls are extremely thick (3–5 μm) and, in barley, undergo extensive degradation during early germination. It has been suggested that intact aleurone walls present a barrier to enzyme mobilization (Varner and Mense, 1972). Early work on the structure of barley aleurone layers reported a composition of 85% arabinoxylan, 8% cellulose, and 6% non-hydroxyproline-containing protein (McNeil et al., 1975). However, the presence of some β-(1→3)-glucan was suggested by the staining of the walls with aniline blue. This apparent contradiction was resolved when it was realised that the amylase used in the purification of the walls also contained a highly active β-glucanase (Huber and Nevins, 1977). Fluorescence microscopy of the aleurone–subaleurone junction in ungerminated seeds of barley indicated that the subaleurone walls, immediately adjacent to the aleurone layer, were thicker than the other endosperm walls and also contained aniline-blue-positive material (Fulcher et al., 1977). This material was relatively unaffected by periodate, but could be removed by β-(1→3)-glucanase.

A more recent analysis of the barley aleurone wall has indicated that it is a bilayer and that there is only extensive degradation of the outer wall layer during germination (Bacic and Stone, 1981). This work confirmed the presence of β-(1→3)-glucan at the subaleurone interface, and indicated that the composition of the barley aleurone walls was similar to that of walls from other monocots. The walls contained mixed-linkage glucans [98% β-(1→3) (1→4) and 1% β-(1→3)], heteroxylans, and a small amount of glucomannan and cellulose. Glucose accounted for approximately 29% of the wall monosaccharides.

B. The *Lolium* Endosperm

Seeds of perennial ryegrass (*Lolium multiflorum*) contain a water-soluble β-glucan (MacLeod and McCorquodale, 1958). Endosperm tissue cultures were isolated with a view to studying β-glucan synthesis in *Lolium* (Smith and Stone, 1973a). The cultured endosperm walls were thicker than seed endosperm walls at maximum growth stage, but cultured cells contained less starch and protein than the mature endosperm cells (Mares and Stone, 1973a). In particular, starch grains and protein bodies did not persist in the later stages of growth in the cultures. The cell walls had a very low protein content, and consisted of approximately 16% (w/w) of β-1,3:1,4-glucan (Smith and Stone, 1973c; Anderson and Stone, 1978). The monosaccharide composition of the walls was 50% glucose, 19% arabinose, 20% xylose, and 5% galactose (Smith and Stone, 1973c). The β-glucan component could be effectively extracted with 7–8 *M* urea (Smith and Stone, 1973c; Anderson and Stone, 1978), thus suggesting a lack of covalent association with other wall components.

C. Cereal Coleoptiles

The endosperm and aleurone glucans of *Hordeum, Lolium,* and *Avena* have been studied in view of their possible roles as (a) storage and (b) structural components. Many monocots are also known to contain noncellulosic β-glucans in their coleoptile walls. To date, the main interest in these molecules has been associated with their possible involvement in cell elongation, as plant growth-regulator treatments have often produced positive correlations between increased wall glucanase activity and extension growth (Masuda and Yamamoto, 1970; Huber and Nevins, 1979; Sakurai and Masuda, 1979). However, other functions (e.g. storage) may exist for these molecules, as recent taxonomic studies have indicated that β-(1→3)(1→4)-D-glucans are confined to grasses (Stinard and Nevins, 1980). Studies using specific endoglucanases to release characteristic oligosaccharides from plant cell walls have failed to show any release from walls of dicots, although a wide range of monocot subfamilies contained noncellulosic β-glucans; furthermore, in the Graminae, the structure of these glucans appeared to be highly conserved (Stinard and Nevins, 1980).

Oat (*Avena sativa*) coleoptile walls have been most studied from the point of view of the cell-elongation hypothesis. Treatment of coleoptile sections with an exo(1→3)-β-D-glucanase from *Sclerotinia libertiana* induced cell elongation, and this effect was inhibited by the glucanase inhibitor nojirimycin (5-amino-5-deoxy-D-glucopyranose) (Yamamoto and Nevins, 1978). In addition, incubation of isolated corn (*Zea mays*) coleoptile walls in buffer for 36 hr resulted in approximately 10% autolysis of the walls, 90% of which could be accounted for by β-glucan release (Kivilaan *et al.,* 1971; Huber and Nevins, 1979). Active glucanases are therefore present in the wall, and these cannot be removed by detergents or 4 *M* LiCl (Huber and Nevins, 1979).

Avena coleoptile cell walls contain between 7 and 9% glucan on a dry weight basis (Nevins *et al.,* 1977). Most of the β-glucan cannot be extracted from the walls by hot water or protease treatment (Nevins *et al.,* 1977). The main matrix wall components, in addition to glucan, are glucoarabinoxylans (Labavitch and Ray, 1978). After separation from the glucoarabinoxylan, much of the β-glucan becomes insoluble. It has been suggested that the glucan may be a structural component for binding together the microfibrillar phase of the wall (Labavitch and Ray, 1978). In addition to insoluble β-glucan, two soluble, virtually unbranched β-glucans have been isolated from *Avena* coleoptile walls (Wada and Ray, 1978).

Other monocotyledonous species where coleoptile cell wall β-glucans have been investigated include, in addition to *Zea mays* (Kivilaan *et al.*, 1971; Nevins *et al.*, 1978; Huber and Nevins, 1979), *Hordeum vulgare* (4.1% glucan on a dry weight of wall basis), *Secale cereale* (5.7% glucan), *Triticum vulgare* (6.5% glucan), and *Sorghum bicolor* (9.7% glucan in mesocotyl walls) (Nevins *et al.*, 1978).

D. Other Sources

Other sources of β-(1→3)-linked glucans to which a storage function has been ascribed include wheat endosperm (Mares and Stone, 1973b) and aleurone (Bacic and Stone, 1981) cell walls, sorghum endosperm walls (Woolard *et al.*, 1976), and French bean cotyledons (Smith, 1974). In the latter case, glucan was only demonstrated as aniline-blue-fluorescent material. Wheat endosperm walls contain 75% polysaccharide, of which 85% is arabinoxylan, 7.5% is glucomannan, and 7.5% is β-glucan (Mares and Stone, 1973b). In wheat aleurone walls, glucose accounts for 31% of the mono-saccharide composition, with 98% of the total wall glucose being present in a mixed-linkage β-(1→3)(1→4)-glucan (Bacic and Stone, 1981). In sorghum endosperm, a water-soluble β-glucan accounts for 4.8% of the total endo-sperm hemicellulose and 0.015% of the whole grain (Woolard *et al.*, 1976).

Developing cotton fibres have been extensively studied in the context of β-(1→4)-glucan (cellulose) synthesis. However, cotton fibres also contain both soluble and insoluble wall-associated β-(1→3)-glucans, the proportion of the soluble polymer varying as a function of fibre age (Maltby *et al.*, 1979). This system is further discussed in Section IV in relation to β-(1→3)-glucan synthesis.

III. STRUCTURES OF MIXED-LINKAGE β-(1→3)-GLUCANS

A. Extraction Procedures

Cell wall preparations for subsequent extraction of β-glucans are usually prepared by repeated sedimentation from 70% ethanol (Forrest and Wainwright, 1977; Ballance and Manners, 1978a; Prentice *et al.*, 1980;

Thompson and La Berge, 1981). This prevents loss of water-soluble glucans and also prevents dehydration effects which may alter the hydrogen-bonding properties of the wall. In material where little or no water-soluble glucan appears to be present, water at 4°C may be used for wall preparation (Anderson and Stone, 1978). Release of β-glucans from cell wall preparations is usually by water extraction at 40–65°C to provide the water-soluble fraction, followed by alkali extraction (0.5–1 N KOH, 1% NaBH$_4$) (Anderson and Stone, 1978; Ballance and Manners, 1978a; Thompson and La Berge, 1981). The alkali-insoluble residue generally contains cellulose and glucomannans. Urea (7–8 M) has been used to extract β-glucan from its strong, but noncovalent, association with *Lolium* endosperm walls (Anderson and Stone, 1978).

More complex extraction methods have been used in the separation of hemicellulosic coleoptile glucans. These include repeated ammonium oxalate: oxalic acid extractions at 75–98°C followed by alkali extraction (Labavitch and Ray, 1978; Kato *et al.*, 1981). Attention has been drawn to the inadequacies of selective fractionation procedures when seeking to isolate polysaccharides for comparative studies (Fraser and Wilkie, 1971).

Further purification of water- and alkali-soluble preparations containing β-glucans has been achieved by (a) further selective alkali treatments (Anderson and Stone, 1978; Labavitch and Ray, 1978; Kato *et al.*, 1981), (b) ammonium sulphate or ethanol fractionation (Fraser and Wilkie, 1971; Anderson and Stone, 1978; Ballance and Manners, 1978a; Labavitch and Ray, 1978; Kato *et al.*, 1981), gel filtration on Sepharose or Bio-Gel columns (Woolard *et al.*, 1976; Forrest, 1977; Anderson and Stone, 1978; Kato *et al.*, 1981) or Bio-Glas porous glass beads (Forrest and Wainwright, 1977), and borate electrophoresis (Labavitch and Ray, 1978). Prior to separation, starch may be removed by incubation with α-amylase, although it is important to ensure that amylase preparations are free from contaminating β-glucanase activity (Huber and Nevins, 1977).

B. Chemical and Enzymatic Analysis

Most of the standard methods of carbohydrate analysis have been applied to the structural elucidation of higher plant β-glucans. These methods include (a) total and partial acid hydrolysis followed by paper, thin-layer, gas-liquid, or high-performance liquid chromatography (Valent *et al.*, 1980) or gel filtration on Biogel P-2 (Nevins *et al.*, 1978; Yamamoto and Nevins, 1978, Stinard and Nevins, 1980), (b) periodate oxidation (Fleming and Manners, 1966; Smith and Stone, 1973c), and (c) methylation analysis (e.g., Kato *et al.*, 1981).

Table I

Glucanases Used in the Structural Analysis of β-(1→3) and Mixed-Linkage Glucans

Source	Enzyme	EC number	Action pattern	Typical substrates attacked
Aspergillus niger	β-(1→4)-Glucan endohydrolase	3.2.1.4	Hydrolyses β-(1→4) linkages, but has preference for mixed-linkage glucans	Barley β-glucan, Carboxymethylcellulose
Basidiomycetes QM-806	β-(1→3)-Glucan exohydrolase	3.2.1.58	—	Callose
Bacillus circulans	β-(1→3)-Glucan endohydrolase	3.2.1.6	Attacks next joining point from β-(1→3) linkage of β-glucan	Mixed linkage β-(1→3)-glucans
Bacillus subtilis	β-(1→3)(1→4)- Glucan endohydrolase	3.2.1.73	Hydrolyses (1→4) linkages in β-glucans containing both (1→3) and (1→4) linkages	RS III,[a] Lichenan, barley glucan, oat glucan
Euglena gracilis	β-(1→3)-Glucan exohydrolase (N.B.: two forms exist)	3.2.1.58	Highly specific for β-(1→3)glucosidic linkage; only terminal (1→3)residues of mixed-linked glucans hydrolysed	Laminarin, $L_5{}^b$
Streptomyces QM B814	β-(1→4)-Glucan endohydrolase	3.2.1.4	Attacks β-(1→4)linkages next to another β-(1→4) linkage	Lichenan, oat glucan carboxymethylcellulose
Trichoderma viride	β-Glucanase	—	—	Barley glucan
Rhizopus arrhizus QM 6789	β-(1→3)-Glucan endohydrolase	3.2.1.6	↓ ↓ 3G4G3G4G3G4G3-	RS III,[a] oat glucan, laminarin, carboxymethylpachyman

[a] Reduced pneumococcal sIII polysaccharide.
[b] laminaripentaose. This abbreviation is used throughout, e.g., L_3 is laminaritriose.

In addition, much valuable structural information has come from the use of specific glucanases, followed by analysis of digestion products by the above methods. A variety of glucanases with different substrate specificities and action patterns, mainly from microbial sources, is now available for β-glucan analysis (Table I). Specific examples of the use of such enzymes for structure determination are given in Section III,D. The crude β-glucanase preparation

K_m	pH Optimum	Notes	References
1.8 mg/ml 11 mg/ml	4.8	Purified from commercial glucanase GV-L	Smith and Stone (1973b); Ballance and Manners (1978a); Svensson (1978)
—	5.8	—	Miyamoto and Tamari (1973); Meier et al. (1981)
—	5.8	—	Miyamoto and Tamari (1973)
3.4 mg/ml	6.5	Occurs as impurity in "Sigma type IIa" α-amylase	Smith and Stone (1973b,c); Anderson and Stone (1975, 1978); Huber and Nevins (1977); Bacic and Stone (1981); Henry and Stone (1982)
0.008% (w/v) $11.1 \times 10^{-5}\,M$	4.7–5.2	—	Barras and Stone (1969a,b); Bacic and Stone (1981); Henry and Stone (1982)
—	5.0	—	Henry and Stone (1982); Anderson and Stone (1975); Wada and Ray (1978)
—		Preparation contains cellulase and β-(1→3)-glucanase	Bamforth et al. (1979); Prentice et al. (1980)
—	5.0		Smith and Stone (1973b,c); Anderson and Stone (1975); Wada and Ray (1978); Maltby et al. (1979); Yamamoto and Nevins (1978)

from *Trichoderma viride* has been used for the total digestion, to glucose, of solubilized glucans from cereal grains, serving as an assay for β-glucan levels (Prentice *et al.*, 1980).

C. Physical Methods

X-Ray diffraction and ^{13}C-NMR spectroscopy have provided valuable information concerning the molecular and crystal structures of a variety of β-

(1→3)-glucans. Thus, X-ray diffraction analysis and stereochemical model refinement techniques have shown that crystalline curdlan, a β-(1→3)-glucan of bacterial origin, exists in a right-handed triple helical structure that is internally stabilized by hydrogen bonds between O-2 hydroxyl groups (Deslandes et al., 1980). Higher plant callose may be assumed to exhibit similar properties. The presence of β-(1→4)-linkages in the β-(1→3) chain, as in the case of the lichenan from Iceland moss and the barley and oat glucans, leads to greatly increased water solubility. The chain conformation of crystalline lichenan corresponds to a right-handed threefold helix with an advance per monomer of 1.40 nm, a helix of much lower amplitude than encountered with purely β-(1→3)-glucans (Marchessault and Deslandes, 1981).

The 100-MHz ^{13}C-NMR spectra of the mixed-linkage β-(1→3)(1→4)-glucans from barley, oats, and Iceland moss are virtually superimposable, the spectra indicating the presence of three nonequivalent 4-O-substituted residues with the ratio of these to the 3-O-substituted residues being 2.4–2.5. The patterns of repeating sequences in the three polymers are therefore basically the same (Dais and Perlin, 1982).

D. Structures

1. Ungerminated Barley and Malted Barley Glucans

The use of different extraction methods has resulted in the isolation from barley of several different glucan fractions. These differ mainly in degree of polymerisation, although differences in linkage proportions have also been reported (Fleming and Kawakami, 1977).

Aqueous extracts from ungerminated barley contain a viscous, water-soluble gum containing β-glucan (Preece, 1957), although not all the glucan is water-extractable (Ballance and Manners, 1978a). Early studies on the soluble glucan preparation involving periodate oxidation, borohydride reduction, and acid hydrolysis resulted in the conclusion that the glucan, although a mixed-linkage β-(1→3)(1→4)-glucan, contained blocks of adjacent β-(1→3) linkages (Fleming and Manners; 1966). In a comparison between the glucans from ungerminated barley and 2-day-germinated barley malt, it was shown that the water-extractable barley glucan was of low specific viscosity, contained approximately 74% (1→4) linkages, had a molecular weight of around 0.4×10^6, and underwent only limited hydrolysis by bacterial endoglucanases or by homogenous endo-β-(1→3)-glucanase from malted barley (Bathgate et al., 1974). In contrast, the glucan from wort prepared at 65°C had a high specific viscosity, had a molecular weight of 2×10^6, and was readily hydrolysed by endoglucanases. Its ratio of (1→4) to (1→3) linkages was identical to that of the barley glucan. In a comparison of barley β-glucans extracted at different temperatures, the glucan extracted at 100°C had more sequences of β-(1→3) linkages than the possibly smaller molecules extracted at 40°C (Fleming and Kawakami, 1977). It was suggested that the blocks of β-(1→3) linkages may be minor structural features. The models proposed for the structures of barley and malt glucans are given in Table II. ^{13}C-NMR studies have confirmed a ratio of approximately 2.7:1 for 1,4:1,3 linkages in barley glucan (Dais and Perlin, 1982).

Recently, attention has been given to the structure of the barley β-glucans as they occur in the endosperm cell wall. The molecular weight of the glucan isolated directly from cell walls was very high (approximately 4×10^7); however, rupture of peptide bonds with hydrazine or thermolysin resulted in the release of molecules of molecular weight about 1×10^6 (Forrest, 1977). On CsCl gradients, where the high salt concentration would split hydrogen or ionic bonding, a significant fraction of the wall protein sedimented with β-glucan (Forrest, 1977). After removal of low molecular weight peptides or amino acids on agarose in CsCl, the β-glucan fraction was shown to contain 2.7 nmol of amino nitrogen per 100 nmol glucose. Acid hydrolysis indicated that 70% of the associated protein comprised glutamic acid, serine, aspartic acid and leucine. There was no hydroxyproline present (Forrest, 1977). It now appears that proteolysis may be the first stage in glucan degradation in barley endosperm walls, as the molecular weight of barley glucan after digestion with thermolysin is very similar to the value reported for the glucan *in vivo* after 48-hr germination (Forrest and Wainwright, 1977).

Fractionation of barley endosperm cell walls has yielded two major β-glucan fractions, glucan I (water extractable) with a degree of polymerisation (DP) of approximately 400, and glucan II (extractable with 1 *M* KOH) with a DP of 258 (Ballance and Manners, 1978a). The ratio of (1→3) to (1→4) linkages in both glucans was 3:7. The glucans were extensively digested from intact cell walls by the *Trichoderma viride* endo-β-1,4-glucanase, and it was suggested that the alkali-extractable glucan II might be bound covalently to the wall by linkages to serine, threonine, aspartic acid, or glutamic acid, or by ferulic acid cross-linking (Ballance and Manners, 1978a).

Ammonium sulphate precipitation followed by chromatography on Sepharose CL-6B has yielded a β-(1→3)(1→4)-linked glucan from barley coleoptiles and primary leaves (Kato *et al.*, 1981). The molecular weight was shown to be 180,000, and analysis of oligosaccharides produced by partial acid hydrolysis and *Bacillus subtilis* mixed-linkage glucanase digestion suggested a structure consisting of repeating oligosaccharide units of 3-O-β-cellobiosyl-D-glucose and 3-O-β-cellotriosyl-D-glucose in a molar ratio of 3:4:1 [i.e. 31% β-(1→3) linkages] (Kato *et al.*, 1981) (Table II).

Table II

Proposed Structures for Barley β-Glucans[a]

Source	Structure[a]	Notes	Reference
Ungerminated barley	$[\circ\!-\!\bullet\!-\!\circ\!-\!\circ\!-\!\circ\!-\!\circ\!-\!\circ\!-\!\circ\!]\underbrace{(\!-\!\bullet\bullet\!)_n}\!-\!\circ\,[x]\!-\!(\!\bullet\bullet\!)_n\!-\!\circ\,[x]\!-\!x$	$n \geq 2$	Fleming and Kawakami (1977)
Ungerminated barley and malted barley	$[\circ\!-\!\bullet\!-\!\circ\!-\!\circ\!-\!\circ\!-\!\bullet\!]_9\!\!\underbrace{\left\{\dfrac{(\!-\!\bullet\bullet\!)_n}{(\!-\!\circ\!-\!\circ\!-\!)_{2n}}\right\}}_{x}\!\!-\![x]_9\!\!\left\{\dfrac{(\!-\!\bullet\bullet\!)_n}{(\!-\!\circ\!-\!\circ\!-\!)_{2n}}\right\}\![x]_9$	$n \geq 2$ Amorphous region short (barley) or long (malt)	Bathgate $et\ al.$ (1974)
Barley coleoptiles	$[\ \bullet\bullet\!-\!\circ\ ]_{n_1}\![\!-\!\bullet\bullet\bullet\bullet\!-\!\circ\,]_{n_2}$	$n_1 : n_2 = 3.4 : 1$	Kato $et\ al.$ (1981)

[a] $\circ$ = 4-substituted glucose, $\bullet$ = 3-substituted glucose

2. Lolium and Sorghum Endosperm Glucans

Hydrolysis of the β-(1→4)(1→3)-glucan from *Lolium* endosperm with *B. subtilis* endoglucanase resulted in the formation of two oligosaccharides similar to those produced from barley β-D-glucan (Anderson and Stone, 1978). Methylation analysis gave a ratio of 2.6:1 for the proportion of (1→4) to (1→3) linkages. A much smaller glucan (DP ~ 26) was isolated from sorghum endosperm. This was shown to be a linear β-(1→3)(1→4)-glucan with a (1→3) to (1→4) linkage ratio of 3:2 (Woolard *et al.*, 1976).

3. Avena Coleoptile Glucans

A mixed-linkage β-glucan has been liberated from *Avena* coleoptile walls by treatment with *B. subtilis* β-glucanase (Nevins *et al.*, 1977). The glucan could not be extracted by hot water or protease treatments. After heat-inactivation of wall-bound glucosidases, the *B. subtilis* enzyme released di- and tri-saccharides, suggesting a structure for the glucan of regular repeating units of 3-O-β-cellobiosyl-D-glucose and 3-O-β-cellotriosyl-D-glucose (Nevins *et al.*, 1977). In a separate study, these oligosaccharide products were found to be produced in a ratio of 1.7:1 (Huber and Nevins, 1977). A more detailed analysis of wall glucans from *Avena* coleoptiles indicated the presence of one insoluble and two water-soluble glucans (Wada and Ray, 1978). The soluble glucans were degraded by *Streptomyces* and *Rhizopus* β-glucanases to yield cellobiose and laminaribiose, respectively. The *Rhizopus* digestion also yielded 3-O-β-cellobiosyl-D-glucose, whereas 4-O-β-laminaribiosyl-D-glucose was tentatively identified as a product of the *Streptomyces* glucanase digestion (Wada and Ray, 1978). Taken together, the analyses of the *Avena* glucans indicate the occurrence of both β-(1→3) and β-(1→4) linkages interspersed together within one glucan chain, as in the mixed-linkage β-glucans of cereal seeds.

4. Other Cereal Coleoptile Glucans

As observed for the digestion of barley endosperm and *Avena* coleoptile glucans, treatment of coleoptile cell walls of *Zea mays*, *Triticum vulgare*, *Secale cereale*, and *Sorghum bicolor* (mesocotyl) with *B. subtilis* glucanase resulted in the formation of 3-O-β-cellobiosyl-D-glucose and 3-O-β-cellotriosyl-D-glucose (Nevins *et al.*, 1978). This comparative study indicated that the structures of the β-glucans in the Graminae were highly conserved, as very similar proportions of oligosaccharides (tri:tetra = 2.5–3.3:1) were produced from the walls of each species with both *B. subtilis* and *Rhizopus* glucanases, and the ratio of (1→3) to (1→4) linkages was exactly 3:7 for each species (Nevins *et al.*, 1978).

IV. BIOSYNTHESIS OF β-(1→3)- AND MIXED-LINKAGE GLUCANS

A. Relationship between β-(1→3) and β-(1→4)-Glucan Synthesis

Little attention has so far been given to the biosynthesis of β-glucans in plants where it is believed that they serve a storage function (e.g. barley or *Lolium* endosperm). There is, however, considerable data in the literature concerning both *in vivo* and *in vitro* β-glucan synthesizing systems. These studies are mainly in the contexts of (a) cellulose synthesis (Delmer *et al.*, 1977; Heininger and Delmer, 1977; Helsper *et al.*, 1977; Raymond *et al.*, 1978; Maltby *et al.*, 1979; Pillonel *et al.*, 1980; Meier *et al.*, 1981; Bacic and Delmer, 1981), (b) hemicellulosic structural wall glucans and their relationship to wall loosening and cell extension growth (Chambers and Elbein, 1970; Miyamoto and Tamari, 1973; Montague and Ikuma, 1978; Klein *et al.*, 1981), and (c) callose synthesis (Southworth and Dickinson, 1975; Beltran and Carbonell, 1978; Meier *et al.*, 1981).

It was apparent, by the late 1960s, that the enzymatic synthesis of β-glucans was of greater complexity than had at first seemed likely. Thus, although synthesis of cellulose-like material by particulate enzyme preparations from *Phaseolus aureus* had been reported (Elbein *et al.*, 1964; Barber *et al.*, 1964), some workers were claiming to obtain cellulose formation from both uridine diphosphate-glucose (UPDG) and guanosine diphosphate-glucose (GDPG), while others reported β-(1→4)-glucan formation from GDPG and β-(1→3)-glucan formation from UDPG (Chambers and Elbein, 1970). There are differences in substrate specificity, and it is also now clear that several β-(1→3)-glucan synthetase preparations catalyse the formation of polymers with different proportions of (1→3) to (1→4) linkages at different UDPG concentrations (e.g. Péaud-Lenöel and Axelos, 1970; Tsai and Hassid, 1971, 1973; Miyamoto and Tamari, 1973; Smith and Stone, 1973b; Henry and Stone, 1982). Furthermore, studies on cellulose biosynthesis in cotton fibres have suggested a causal relationship between callose synthesis and cellulose synthesis (Maltby *et al.*, 1979; Pillonel *et al.*, 1980, Meier *et al.*, 1981). The often complex relationship between β-(1→3), β-(1→4), and mixed-linkage glucan synthesis will be apparent from the following descriptions of higher plant β-glucan synthetases.

B. Synthesis and Accumulation of β-(1→3)-Glucans in Intact Systems

The pattern of β-glucan accumulation has been studied in developing barley endosperm (Coles, 1979). Although experimental results varied according to moisture content during development and barley variety used, it was observed that initiation of synthesis occurred in the order hemicellulose

→ starch → mixed-linkage β-glucan at 10, 11, and 17 days after fertilization, respectively (Fig. 1).

Detached cotton fibres incorporated UDP-[^{14}C]glucose into a linear β-(1→3)-glucan in a similar manner to particulate preparations (Heiniger and Delmer, 1977), although it appeared that some damage to fibres was necessary in order for significant activity to be detected (Delmer et al., 1977). The kinetics of glucan synthetase activity in detached fibres were complex and indicated activation by substrate. The pH optimum for glucan formation was broad, between pH 6.8 and 8.4, and the activity was stimulated by Ca^{2+} and Mg^{2+} at high UDPG concentrations. The enzyme activity was also strongly stimulated by β-linked glucosides (e.g. cellobiose) alone at high UDPG concentrations, or in the presence of uridine triphosphate (UTP) at low UPDG concentrations. The apparent K_m for UPDG in the intact fibre system was 5 mM, decreasing to 1.7 mM in the presence of 10 mM cellobiose (Delmer et al., 1977).

The deposition of β-(1→3)-glucan during cotton fibre development coincided closely with the onset of secondary wall cellulose synthesis (Maltby et al., 1979) (Fig. 2). The kinetics of incorporation of [^{14}C]glucose into glucans in cultured ovules or fibres indicated a linear rate of formation of β-(1→3)-glucan, with a lag of around 1 hr in the synthesis of cellulose (Maltby et al.,

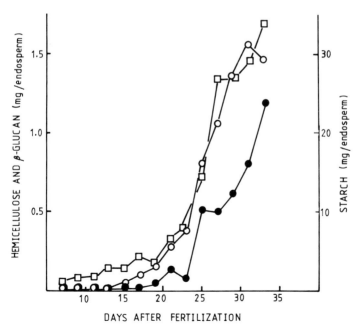

Fig. 1. Appearance of starch (○), hemicellulose (□), and mixed-linkage β-glucan (●) in developing embryos of barley cv. Bomi. Data redrawn with permission from Coles (1979).

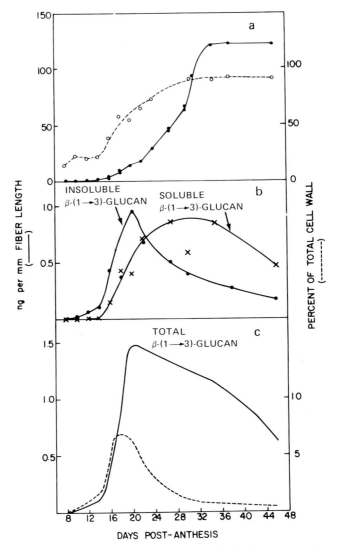

Fig. 2. Variation in the content of cellulose (a) and β-(1→3)-glucan in cotton fibres as a function of fibre age. The total β-(1→3)-glucan (c) was calculated as the sum of the areas under the curves in (b). Reproduced with permission from Maltby *et al.* (1979).

1979). It was suggested that β-(1→3)-glucan could be a precursor for cellulose synthesis, although pulse–chase experiments indicated that neither β-(1→3)-glucan nor cellulose exhibited significant turnover after synthesis (Maltby *et al.*, 1979). Later pulse–chase experiments with $^{14}CO_2$, in which precautions were taken to prevent wounding of fibres, indicated that callose did have a high rate of turnover and may therefore be a prerequisite for cellulose synthesis (Meier et al., 1981). When seed clusters of individual locules with adhering fibres were fed UDPG, GDPG, glucose or sucrose, it was shown that 1 m*M* UPDG was the best precursor for the synthesis of primary wall fibres, whereas sucrose was a better precursor for secondary wall fibres. In both cases, exoglucanase digestion of the fibres indicated the formation, within 1 hr, of approximately 30% cellulose and 70% β-(1→3)-glucan (Pillonel et al., 1980). The idea of a high turnover rate for β-(1→3)-glucan was supported by the fact that β-(1→3)-glucan synthetic rate was high at all stages of secondary wall formation, although this polymer accounted for no more than 2% of the total wall (Pillonel et al., 1980).

The results of experiments with intact cotton fibres clearly depend on whether or not the fibres are damaged. Details of the synthesis of β-glucans at wounded surfaces have also been reported for pea epicotyl tissue slices (Raymond et al., 1978) and for soybean cell suspension cultures damaged by vigorous stirring (Brett, 1978). In the case of pea epicotyls, the behaviour of the tissue slice system, where UDPG was reacting at accessible surfaces of cut cells, was similar to that of the particulate system (see next section) with respect to pH optimum, optimum Mg^{2+} concentration, and apparent K_m (Raymond et al., 1978). Interesting systems for future study of glucan synthesis in intact tissues may be the formation of β-(1→3)- and β-(1→4)-glucans during wall regeneration in soybean (Klein et al., 1981) and *Vinca rosea* (Takeuchi and Komamine, 1981) protoplasts.

C. Cell-Free Synthesis of β-(1→3)-Glucans

In all cases, attempts at purification of higher plant β-glucan synthetases have resulted in the activity being associated with a particulate fraction, although in some cases it has been possible to solubilise the activity by the use of detergents. A summary of the properties of a selection of such preparations is given in Table III. The following descriptions of the synthetases from five sources summarise the nature of these preparations.

1. Cotton Fibres

The properties of the β-glucan synthesizing system from detached cotton fibres has already been described. A particulate preparation sedimenting between 1,000 and 20,000 *g* maintained between 7 and 15% of the activity measurable in the detached fibre assay (Delmer et al., 1977). Mg^{2+} ions and digitonin were necessary for maximum activity, and the properties (stimulation by β-glucosides, activity with glucose 1-phosphate in the presence of

Table III

Properties of Enzymic Systems Forming β-(1→3)-Linked Glucans in Higher Plants

Source	Particulate (P) or solubilised (S)	Substrate	Main linkages in the product(s)	[S] giving predominantly	
				β-(1→3) (mM)	β-(1→3)(1→4) or β-(1→4) (μM)
Avena coleoptile	P and S	UDPG	β-(1→3); β-(1→4) β-(1→3)(1→4)	1.0	10.0
		CDPG	β-(1→3)		
		GDPG	β-(1→4)		
Citrus phloem	P	UDPG	β-(1→3) [some β-(1→4)?]	1.0	
Gossypium fibres	P	UDPG	β-(1→3)	All [S] give β-(1→3)	
Lilium longiflorum pollen tubes	P	UDPG ADPG G-1-P (+cellobiose)	β-(1→3)	All [S] give β-(1→3)	
Lupinus albus hypocotyls	P and S	UDPG	β-(1→4) β-(1→3)	— —	— —
Lolium multiflorum endosperm	P	UDPG GDPG	β-(1→3)(1→4); β-(1→3); β-(1→4) β-(1→4) (at 10 μM GDPG)	More β-1,3 produced at high [UDPG]; mixed-linkage main product	
Petunia hybrida pollen tubes	P	UDPG	β-(1→3); β-(1→4)	—	—
Phaseolus aureus hypocotyls	P	UDPG GDPG	β-(1→3); β-(1→3)(1→4)	0.4–20	8.5
Pisum sativum epicotyl	P	UDPG	β-(1→3)(1→4); β-(1→3) (high [S])	5.0	"Low"

K_m (mM)	pH Optimum for the synthesis of		Stimulation by Mg^{2+}		Substrate activation of β-(1→3)-linkage formation	References
	β-(1→3)	β-(1→3)(1→4)	High [S]	Low [S]		
0.011 [β-(1→4)]	8.5	8.5	no	yes	yes	Péaud-Lenöel and Axelos (1970); Tsai and Hassid (1971, 1973); Montague and Ikuma (1978)
Non-Michaelis–Menten kinetics; Hill coefficient = 1.2	7.5	—	no	yes	—	Beltran and Carbonell (1978)
5.0	6.8–8.4	—	—	—	yes	Delmer et al. (1977); Heiniger and Delmer (1977) Pillonel et al. (1980)
0.74			Stimulated by Mg^{2+}			Southworth and Dickinson (1975)
—	6.5	8.0	β-(1→4) Activity stimulated, β-(1→3) inhibited by Mg^{2+}			Larsen and Brummond (1974)
—						
—	Depends on [UDPG] 7.0 8.0 (values at high [UDPG])	—	—	—	Henry and Stone (1982) Smith and Stone (1973b)
—	—	8.0	yes	no		Helsper et al. (1977)
1.74	6.5	—	no	yes	no	Miyamoto and Tamari (1973); Chambers and Elbein (1970)
2.0	6.5–7.0	7.0–7.5	—	—	yes	Raymond et al. (1978)

cellobiose and UPT) were similar to those observed for the detached fibres. Sucrose density-gradient centrifugation yielded a very broad activity distribution, suggesting that the activity may be associated with a variety of membrane or organelle fractions (Delmer et al., 1977). The particulate activity reached maximum levels between 16 and 18 days postanthesis, corresponding to the time of onset of secondary wall synthesis. It was concluded that more than one enzyme could be involved in the formation of noncellulosic glucan, in view of the complexity of the kinetics and requirements for different stimulatory factors at different substrate concentrations. In addition, the particulate system could also synthesise surose, acetylated and nonacetylated steryl glycosides, and traces of glucosylpolyprenolphosphate from UDPG (Delmer et al., 1977). The main reaction product of the cotton synthetase was a linear β-(1→3)-glucan, irrespective of UDPG concentration (Heiniger and Delmer, 1977). A later report indicated that treatment of cotton membrane preparations with 50 mM K^+ in the presence of the potassium ionophore valinomycin resulted in a four- to twelvefold stimulation of β-glucan synthesis, the product being predominantly β-(1→3)-linked but with some β-(1→4) linkages also being formed (Basic and Delmer, 1981). This work suggests the possibility of control of glucan synthesis by transmembrane potentials, which may possibly affect the structure or fluidity of the membrane environment of the synthetase enzyme(s) (Basic and Delmer, 1981).

2. Avena Coleoptiles

The particulate (1,000–36,000 g pellet) glucan synthetase preparation from Avena coleoptiles was highly unstable at 4°C, some stabilisation of activity being possible by the addition of 0.5% serum albumin. It produced polymers containing both β-(1→4) and β-(1→3) linkages (Péaud-Lenöel and Axelos, 1970; Tsai and Hassid, 1971). A similar activity could be solubilised by the use of 8% digitonin, with 40–50% of the activity then being retained in the 100,000 g supernatant. With both preparations, high UDPG concentrations (10^{-3} M) favoured β-(1→3)-linkage formation, whereas 40 mM MgCl$_2$ gave an approximately tenfold stimulation of β-(1→4) formation (Tsai and Hassid, 1971). It was possible to separate the β-(1→3) from the β-(1→4) synthesizing activity on hydroxyapatite if column fractions were assayed with 10^{-3} M UDPG. If 10^{-5} M UDPG was used, all fractions synthesised mainly β-(1→4)-glucan (Tsai and Hassid, 1971). It therefore appeared that two separate synthetases existed, the β-(1→3) form being activated by high UDPG concentration (Tsai and Hassid, 1971, 1973). The activation by high substrate concentration was so great that Mg^{2+} became ineffective in reducing the proportion of β-(1→3) linkage formation (Tsai and Hassid, 1973). Synthesis of β-(1→3)-glucan was stimulated up to threefold by increasing concentrations of cellobiose, and cytosine diphosphate-glucose (CDPG) could also serve as substrate in the presence of the β-(1→3)-synthetase (Tsai and Hassid, 1973). Treatment of *Avena* stem segments with gibberellic acid results in elongation and increased wall synthesis. This was associated with a two- to fourfold stimulation of a GDPG-dependent β-(1→4)-glucan synthetase, with no stimulation of the formation β-(1→3)-glucan from UDPG (Montague and Ikuma, 1978).

3. Phaseolous Aureus Hyptocotyls

A crude particulate preparation (1,000–30,000 g pellet) from *Phaseolus aureus* hypocotyls incorporated ^{14}C from UDP-[^{14}C]glucose and GDP-[^{14}C]glucose into alkali-soluble and alkali-insoluble products (Chambers and Elbein, 1970). The reactions were linear for up to 2 min, with UDPG resulting in the incorporation of ^{14}C into 10% alkali-insoluble and 90% alkali-soluble material, and with GDPG giving 60–90% alkali-insoluble material. It was shown that GDPG as substrate led to the formation of a β-(1→4)-linked glucan, whereas both alkali-soluble and alkali-insoluble fractions formed from UDPG contained β-(1→3) linkages (Chambers and Elbein, 1970). Evidence for the existence of separate β-(1→3) and β-(1→4) synthetases included (a) the solubilisation of β-(1→3)-glucan synthetase activity by digitonin (this preparation also producing laminaribiose and laminaritriose); (b) the restoration to the particulate fraction of 70% of the GDPG activity, but none of the UDPG activity, by the addition of 7 mM Mg^{2+} following digitonin treatment; and (c) the removal of 50% of the UDPG activity from the particles by treatment with albumin, thus indicating a very loose association for the β-(1→3)-synthetase activity (Chambers and Elbein, 1970). It was later shown that the nature of the glucan synthesised by a *P. aureus* particulate preparation (0–20,000 g pellet, reprecipitated through 0.4 M sucrose) depended on the substrate concentration, with 0.4–20 mM UDPG resulting in formation of a water-insoluble β-(1→3)-glucan whereas 8.5 μM UDPG resulted in formation of an alkali-insoluble β-(1→3)(1→4) mixed-linkage glucan (Miyamoto and Tamari, 1973). The two activities exhibited differences in pH optimum, temperature optimum, stimulation by Mg^{2+}, and storage stability (Miyamoto and Tamari, 1973).

4. Lolium Multiflorum Endosperm Suspension Cultures

The *Lolium* endosperm glucan synthetase is of particular importance in the context of this chapter, as it appears that the glucan product can serve as a storage carbohydrate. Its properties are, however, similar to those of the synthetases from *Avena*, *Phaseolus aureus*, and *Pisum sativum*. Initial studies on the *Lolium* preparation (10,000–250,000 g pellet) indicated the formation of both β-(1→3)- and β-(1→4)-linked glucans from UDPG, with 1 μM UDPG resulting in formation of 90% of the residues with β-(1→4) linkages, whereas 1 mM UDPG resulted in only 49% β-(1→4)-linkages (Smith and Stone, 1973b). The ratio of (1→3) to (1→4) linkages formed *in vitro* from 100 μM UDPG was similar to that found in the isolated *Lolium* endosperm glucan, the concentration of UDPG in the mid-log phase cultured cells being calculated to be around 30 μM (Smith and Stone, 1973b). More recently, it has been shown that a mixed-linkage glucan is the major product at all substrate concentrations, although the formation of a predominantly β-(1→3)-linked glucan increases with increasing UDPG concentration, and 10 μM GDPG leads to the formation of a β-(1→4)-linked glucan (Henry and Stone, 1982). At low UDPG concentrations, more sequences of three consecutive β-(1→4) linkages were formed. The effects of pH, temperature, and metal ion concentrations were dependent upon UDPG concentration (Henry and Stone, 1982), and these complex relationships led to the proposal of the existence of separate enzymes for the formation of β-(1→3) and β-(1→4) linkages, although it was pointed out that the exact mechanism of formation of β-glucans containing different linkages in the same polymer is still not known.

5. Pisum Sativum

Various membrane fractions isolated from pea roots were shown to be active in *in vitro* β-glucan synthesis (Brett and Northcote, 1975). A fraction rich in Golgi dictyosomes formed, from UDPG, (a) water-insoluble glycolipids containing one glucose residue or oligoglucans, (b) oligoglucosides linked to protein, and (c) alkali-soluble and insoluble β-glucan. Both the glucans and the oligosaccharides contained β-(1→3) and β-(1→4) linkages, and it was proposed that the lipid- and/or protein-linked oligosaccharides were inter-mediates in the synthesis of β-(1→4), β-(1→3), and mixed-linkage glucans (Brett and Northcote, 1975). Golgi vesicle preparations from onion stems showed similar ability to convert UDPG into β-glucan, with formation of β-(1→3)-linkages predominating at high UDPG concentrations (Van der Woude *et al.*, 1974).

The ability of pea epicotyl tissue slices to synthesise β-glucan from UDPG has already been discussed. Particulate preparations were less active in forming alkali-insoluble glucan than the tissue slices, although the properties of the synthetase preparations were similar, with both β-(1→3) and β-(1→4) linkages being formed, (1→4) linkages predominating at low substrate concentrations and the capacity for β-(1→3) synthesis being substrate-activated (Raymond *et al.*, 1978). A heat-stable, dialyzable component from the soluble fraction of pea epicotyls enhanced approximately threefold the synthesis of alkali-soluble and alkali-insoluble products from 0.6 mM UDPG, whereas a heat-labile, nondialyzable component suppressed synthesis, its effects dominating at low (6 μM) UDPG concentration (Chao and Maclachlan, 1978). Both components primarily affected the proportion of β-(1→4) linkages formed, although the data indicated that the absolute levels of β-(1→3) linkages must also have changed in response to the components. The suppressor had a molecular weight of 65,000 and co-purified with a protease during ammonium sulphate and Sephadex G-200 fractionation, and it was suggested that this could explain the apparent lability of the β-(1→4)-glucan synthetase activity after homogenization. This work suggests that at least part of the total β-glucan synthetase activity may be subject to rapid turnover *in vivo*; inactivation *in vitro* occurred rapidly at first, followed by a slower second phase, suggesting the possibility of two synthetases being destroyed at different rates (Chao and Maclachlan, 1978).

6. Others

In addition to the synthetases just described, Table III also summarises the properties of several other higher plant β-glucan synthetases. Points of note include (a) the formation of only β-(1→3)-glucan (callose) by the preparations from *Citrus* phloem (Beltran and Carbonell, 1978) and *Lilium longiflorum* pollen (Southworth and Dickinson, 1975), (b) the ability of the synthetase from *Lilium* to use adenosine diphosphate-glucose (ADPG), and (c) the inability of the *Lilium* and *Petunia* pollen tube synthetases to utilise GDPG for the synthesis of β-glucans.

V. β-(1→3)-GLUCANASES AND THEIR PHYSIOLOGICAL FUNCTIONS

A. General Properties of Higher Plant β-(1→3)-Glucanases

1. Barley and Malted Barley

Four distinct enzyme types are now believed to be involved in the degradation of barley mixed-linkage glucan to glucose (Manners and Marshall, 1969; Meier and Reid, 1982; Woodward and Fincher, 1982a,b). Some of the properties of these enzymes are summarised in Table IV, and two short reviews have also appeared (Thompson and La Berge, 1977; Ghaemi *et al.*, 1978). All the enzymes in Table IV except the β-(1→4)-glucanase increase in activity during germination (Manners and Marshall, 1969; Meier and Reid, 1982); it has recently been suggested that a large part of the β-(1→4)-glucanase activity may be of fungal origin, as barley husks are often contaminated with

Table IV

Enzymes from Barley Capable of Degrading β-Glucosidic Linkages

Enzyme[a]	EC number	Source	Molecular weight	pH Optimum	pI
β-(1→3)(1→4)-D glucan 4-glucanohydrolase (endo) [barley β-(1→3)(1→4)-D glucanase; endo barley β-glucanase; lichenase]	3.2.1.73	Malted barley; Germinating barley	15,150; I = 28,000; II = 33,000	4.8 4.7 4.7	— 8.5 >10.0
β-(1→3)-D-Glucan glucanohydrolase (endo) [endo-β-(1→3)-D-glucanase; laminarinase]	3.2.1.39	Malted barley Malted barley	8,920 12,800	5.5 5.0	— —
β-(1→4)-D-glucan 4-glucanohydrolase (endo) [endo-β-(1→4)-D-glucanase]	3.2.1.4	Malted barley	20,400		—
β-Glucosidase	3.2.1.21	Malted barley	I = 760,000 II = 35,000		

[a] In each horizontal column, some data are given for the same enzyme TYPE from different publications. Alternative names for the enzymes are given in square brackets.

Aspergillus fumigatus and other microorganisms (Manners *et al.*, 1982). It has been shown that the barley β-glucosidases (cellobiases) are true glycosidases and not exo-β-glucanases (Manners and Marshall, 1969).

Work in the early 1960s showed the presence of several forms of endo-β-glucanase activity separable from extracts of green barley malt by differential ammonium sulphate fractionation; pH optima from viscometric assays against barley mixed-linkage glucan were in the range 4.5–5.0, although no further details of the type of endoactivity were given (Luschinger, 1962). This was followed up by the demonstration of the presence of a laminarinase (endo-β-(1→3)-D-glucanase) in extracts from germinated barley. This enzyme had no activity against barley β-glucan or carboxymethylcellulose (Chen and Luschinger, 1964). The endo-β-(1→3)-D-glucanase from malted barley was later purified approximately 60-fold by differential ammonium sulphate fractionation and gel filtration on Biogel P-60. Its molecular weight was reported to be 8920; it exhibited a pH optimum of 5.5; and it was activated by Ca^{2+} ions and deactivated by EDTA (Manners and Marshall, 1969). Later work resulted in a more purified enzyme (92-fold), and the molecular weight was shown to be 12,800 (Manners and Wilson, 1974). The enzyme appeared highly specific for β-(1→3)-glucans, converting insoluble laminarin and pachyman into G2–G7 (di–hepta) oligoglucosides (with a trace of glucose). It was also active against laminaripentaose (products were predominantly laminaribiose and triose with traces of glucose) and laminaritriose (giving

Temperature optimum	K_m for preferred substrate	Notes	References
39°C	—	—	Manners and Marshall (1969) Manners and Wilson (1976)
—	3.0 mg/ml ⎱ Barley	—	Woodward and Fincher (1982a,b)
—	3.4 mg/ml ⎰ β-D-glucan		
—	—	—	Manners and Marshall (1969)
37°C	0.92 mg/ml (laminarin)	—	Manners and Wilson (1974)
—	—	May be of fungal origin	Manners and Marshall (1969); Manners et al. (1982)
—	—	True glucosidases, not exo-β-glucanases	Preece (1957); Manners and Marshall (1969)

laminaribiose plus traces of glucose). It was inactivate against $(1→3)(1→4)$ mixed-linkage glucans, including barley glucan, and against $β$-$(1→3)$-linked xylans (Manners and Wilson, 1974). Using laminarin with a DP of 21 as substrate, the K_m was calculated to be $2.7 × 10^{-4}$ M. Studies on the inhibition of the enzyme by Hg^{2+} and mercury phenyl nitrate produced no conclusive evidence for the participation of an SH group in the catalysis (Manners and Wilson, 1974). The activity of the endo-$β$-$(1→3)$-glucanase increased strikingly during germination (Manners and Marshall, 1969). A similar enzyme, described as laminarinase, was detected in barley cell suspension cultures (Gamborg and Eveleigh, 1968).

Barley $β$-$(1→3)(1→4)$-D-glucanase (EC 3.2.1.73, also commonly known as endo barley $β$-glucanase) is possibly the most important enzyme involved in barley glucan degradation, as judged by its abundance in germinating endosperm (Manners and Marshall, 1969). The enzyme from malted barley was separated from the laminarinase activity by chromatography on carboxymethylcellulose and was further purified on Biogel P-30 to yield a 105-fold purified product (Manners and Wilson, 1976). The enzyme had no activity towards laminarin, carboxymethylpachyman or carboxymethylcellulose, and produced mainly 3-O-$β$-cellobiosyl-D-glucose from attack on lichenan or barley mixed-linkage glucan (Manners and Wilson, 1976). Later workers separated two forms of the enzyme, on carboxymethylcellulose, from extracts of germinating barley, achieving a 470-fold purification of enzyme I and a 280-fold purification of enzyme II by further chromatography on carboxymethyl Sepharose and Sephadex G-75 (Woodward and Fincher, 1982b). The properties of the two forms, which were very similar, are shown in Table IV. Form II was more stable than form I and contained approximately 3% carbohydrate. Immunological cross-reactivity was demonstrated between the two forms (Woodward and Fincher, 1982a,b). Both enzymes were only active against barley mixed-linkage β-D-glucan or pneumococcal polysaccharide RSIII, which contains alternating $(1\rightarrow 3)$ and $(1\rightarrow 4)$ linkages. A trace of activity was observed against carboxymethylpachyman and wheat arabinoxylan. Barley β-D-glucan was converted to 3-O-β-cellobiosyl-D-glucose plus 3-O-β-cellotriosyl-D-glucose plus 3-O-β-cellotriosyl-D-glucose, thus suggesting the following action pattern (Woodward and Fincher, 1982a):

$$\begin{array}{cc} \downarrow & \downarrow \\ \ldots G3G4G4G3G4G \text{ (reducing terminal)} \\ \downarrow & \downarrow \\ \ldots G3G4G4G4G3G4G \text{ (reducing terminal)} \end{array}$$

where arrows indicate site of attack and 3 and 4 refer to β-$(1\rightarrow 3)$- and β-$(1\rightarrow 4)$-linked glucosyl residues, respectively. The N-terminal amino acid sequences of the two forms of the enzyme have been determined (Woodward et al., 1982). The only differences in the first 40 amino acids from the N-terminal end were at positions 20, 28 and 36, and these were the result of single base substitution. It was concluded that two separate genes had probably arisen by duplication, and it was pointed out that possession of two genetic loci may be of use in breeding programmes aimed at improving brewing qualities (Woodward et al., 1982). It is still not clear why there is a need in barley for two separate enzymes with identical substrate specificity.

A study of glucanase distribution in barley during development suggests complexity further to that outlined above (Ballance et al., 1976). The β-$(1\rightarrow 3)$ $(1\rightarrow 4)$-D-glucanase was reported to exist in two forms in the endosperm and hull, whereas extracts from whole seeds exhibited four forms on carboxymethylcellulose chromatography. Similarly, four peaks were observed for the endo-β-$(1\rightarrow 3)$-glucanase from whole seed, three forms in the hulls and one form in the endosperm. The relationships between the various forms were not further investigated. The distribution of the enzymes and changes in their activities during germination are shown in Fig. 3.

2. Pisum Sativum Seedlings

The properties of the β-glucanases isolated from pea seedlings and a variety of other plant sources are summarised in Table V. The work on pea seedling glucanases was stimulated by reports of auxin-induced enhancement of β-$(1\rightarrow 3)$-glucanase and extension growth and their inhibition by protein synthesis inhibitors (e.g. Datko and Maclachlan, 1968). Growing regions of pea were shown to contain β-glucosidase, two forms of endo-β-$(1\rightarrow 4)$-D-glucanase (EC 3.2.1.4) and two forms of endo-β-$(1\rightarrow 3)$-D-glucanase. The β-$(1\rightarrow 3)$-

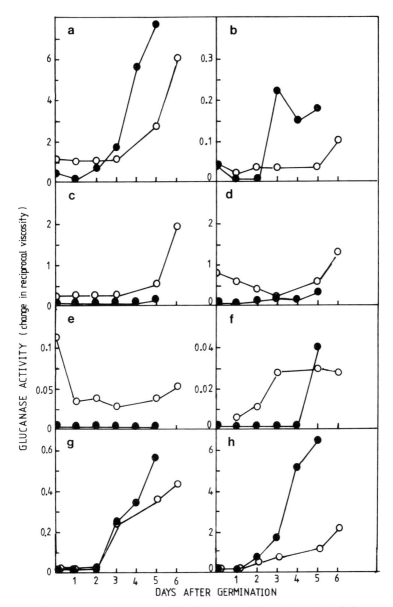

Fig. 3. Changes in activities of endo-β-(1→3)-glucanase (○) and endo barley β-glucanase (●) in barley seed tissues during germination. Enzyme activity is expressed as change in reciprocal viscosity per unit time per unit enzyme aliquot using CM-pachyman (○) or barley β-glucan (●) as substrate. Enzyme units for β-(1→3)-glucanase are × 10^{-2}, for barley β-glucanase × 10^{-1}. Seed tissues were (a) whole seed, (b) hull, (c) embryo, (d) scutellum, (e) pericarp, (f) green layer, (g) aleurone, and (h) endosperm. Data redrawn with permission from Ballance *et al.* (1976).

Table V

Properties of Some β-(1→3)-Glucanases from Higher Plants

Source	Nature	Molecular weight		pH optimum	pI	Laminarin	Pachyman/ CM-pachyman
Lycopersicon	Endo-β-(1→3)-	I	37,500	5.0	5.70	+	+
esculentum	glucanase		63,000?				
(stems)		II	27,000	5.0	5.35	+	+
			to				
		III	74,000?	5.0	5.15	+	+
Nicotiana	Endo-β-(1→3)-		45,000	5.0	4.87	+	+
glutinosa	glucanase						
(leaves)	(EC 3.2.1.39)						
Phaseolus	Endo-β-(1→3)-		11,500–12,500		11.0	+	+
vulgaris	glucanase						
(leaves)	(EC 3.2.1.6)						
Pisum	Endo-β-(1→3)-	I	22,000	6.0 (Laminarin),	5.4	+	+
sativum	glucanase			5.5 (CM-pachyman)			
(stems)	(EC 3.2.1.6)						
		II	37,000	6.0 (Laminarin),	6.8	+	+
				5.5 (CM-pachyman)			
Secale cereale	Endo-β-(1→3)-						
	glucanase						
Ungerminated	(EC 3.2.1.39)			5.2		+	+
seeds							
Germinated	(EC 3.2.1.39)		24,300	5.0 (Laminarin)	9.2	+	+
seeds				5.7 (CM-pachyman)	10.3		
Vitis vinifera	β-(1→3)-					+	+
(phloem and	glucanase						
xylem)	(crude?)						

^{a} ND, not determined.

glucanases were separable on DEAE cellulose (Wong and Maclachlan, 1979b). Both enzymes hydrolysed laminarin (DP 20), laminaridextrans (DP 3–7), carboxymethylpachyman, curdulan, and mixed-linkage glucans, but had no activity against β-glucosides or cellulose. Enzyme I generated reducing groups more readily than enzyme II, hydrolysed lower laminaridextrins at the nonreducing terminal, and converted laminarin to dextrins faster than II while taking longer to achieve complete hydrolysis. Enzyme II hydrolysed lower laminaridextrins at internal linkages (Wong and Maclachlan, 1979b). Further details of substrate specificity and action pattern are given in Table V.

				Substrate specificity					
L5	L4	L3	L2	Barley β-D-glucan	Lichenan	Cellulose/ cellodextran, CM-cellulose	K_m for preferred substrate	Action pattern	References
−	−	−	−	ND[a]	ND	ND	10 mg/ml (Laminarin)		Young and Pegg (1981)
−	−	−	−	ND	ND	ND	2.86 mg/ml (Laminarin)		
−	−	−	−	ND	ND	ND	0.80 mg/ml (Laminarin)	↓	
ND	ND	ND	ND	−	±	−	0.6 mg/100 ml (CM-pachyman)	G-3-G-3-G-	Moore and Stone (1972a,b)
ND	ND	ND	ND	ND	ND	−			Abeles et al. (1970)
+	+	+	±	+	+	−	1.5 mg/ml (Laminarin)	See Wong and Maclachlan (1979b)	Wong and Maclachlan (1979a,b; 1980)
±	±	±	−	+	+	−	0.6 mg/ml (Laminarin)		
ND	+	+	ND	−	−	−		↓ G-3-G-3-G-	Manners and Marshall (1973)
ND	+	+	−	−	−	−	0.25 mg/ml (Laminarin)		Ballance and Manners (1978b)
ND	ND	ND	+	ND	ND	ND			Clarke and Stone (1962)

A study of the distribution of glucanase forms in pea stems indicated that enzyme I was found in the subapical end of the stem, whereas enzyme II was found in the basal regions; the activity of enzyme II increased near the apex following auxin application, and this was inhibited by ethylene (Wong and Maclachlan, 1980). The preferred substrates for enzymes I and II occurred naturally in the subapical and basal regions, respectively. It was observed that the aniline-blue-fluorescent material in pea cell walls could be removed by the glucanases, although it was pointed out that less than 5% of total pea polysaccharide was susceptible to glucanase hydrolysis, thus ruling out any major role of the glucanases in the mobilisation of reserve polysaccharide. However, it was also thought unlikely that either enzyme could be involved in cell expansion, as the increase in enzyme II occurred after the major auxin-mediated effects, and did not occur in the exact area of cell expansion. It was suggested that enzyme I may be involved in maintaining the vertical flow of translocates by preventing blockage of intracellular canals, whereas enzyme II may possibly maintain a route for lateral transport by degrading callose-like material (Wong and Maclachlan, 1980).

3. Others

β-(1→3)-Glucanases have been purified or characterised from a number of other higher plant sources, including *Phaseolus vulgaris* leaves (Abeles *et al.*, 1970), *Nicotiana glutinosa* leaves (Moore and Stone, 1972a,b) and cell cultures (Kato *et al.*, 1973), *Lycopersicon esculentum* stems (Young and Pegg, 1981), germinated and ungerminated *Secale cereale* (Manners and Marshall, 1973; Ballance and Manners, 1978b), *Vitis vinifera* xylem and phloem (Clarke and Stone, 1962), *Avena* coleoptiles (Heyn, 1969), and *Zea mays* seedlings (Huber and Nevins, 1980, 1981). These enzymes hydrolyse either mixed-linkage and homopolymer β-(1→3)-glucans (EC 3.2.1.6) or only β-(1→3)-linked substrates (EC 3.2.1.39). The properties of the best studied examples are summarised in Table V. Functions ascribed to these various glucanases include (a) roles in disease resistance/defence mechanisms (Abeles *et al.*, 1970; Young and Pegg, 1981; Keen and Yoshikawa, 1983) and possibly symptom expression (Moore and Stone, 1972c), (b) roles in auxin-mediated wall loosening and autohydro-lysis (Heyn, 1969; Masuda and Yamamoto, 1970; Goldberg, 1977; Huber and Nevins, 1979; Sakurai and Masuda, 1979; Huber and Nevins, 1981), (c) seasonal removal of sieve tube callose (Clarke and Stone, 1962), and (d) the facilitation of endosperm wall degradation during germination (Ballance and Manners, 1978b).

B. Carbohydrate and Enzyme Release from the Barley Endosperm and Aleurone Layer

As already discussed, the aleurone layer is a specialised tissue whose function is the secretion of enzymes which act upon the adjacent starchy reserves in the endosperm. Aleurone cell walls are thick and undergo extensive degradation during the phase of growth regulator-mediated enzyme release from the aleurone layer. It has been suggested that intact aleurone walls may present a barrier to enzyme mobility (Varner and Mense, 1972).

Light and electron microscopy of aleurone walls has demonstrated their degradation, predominantly in the region of the wall closest to the starchy endosperm (Taiz and Jones, 1970; Eastwell and Spencer, 1982), and it was shown that acid phosphatase could diffuse through channels made in the walls by dissolution of carbohydrate (Eastwell and Spencer, 1982). Furthermore, α-amylase passage is believed to be diffusion-limited (Varner and Mense, 1972). It was therefore postulated that barley β-glucanases may be involved in wall dissolution to aid release of enzymes, which would then act upon the endosperm (Eastwell and Spencer, 1982). This is supported by studies which suggest a large decline in endosperm wall glucan content during germination (Morrall and Briggs, 1978).

The synthesis and release of hydrolases from the aleurone layer is under control by plant growth regulators, including gibberellins, ethylene, and auxins. Details of the recent elegant work on the molecular biology of the α-amylase induction process are outside the scope of the present chapter (see Higgins et al., 1976; Bernal-Lugo et al., 1981; Baulcombe and Buffard, 1983); the effects of growth regulators on the barley glucanases have been investigated in far less detail. Release of barley β-(1→3)-glucanase from the aleurone layer is stimulated by gibberellic acid (Jones, 1971; Fulcher et al., 1977). Density-labelling experiments indicated that there was de novo synthesis of glucanase during imbibition (Bennett and Chrispeels, 1972); the effect of gibberellic acid, however, appeared to be associated with stimulation of release, not synthesis, in spite of an overall requirement for protein and RNA synthesis (Jones, 1971). The release response was initiated at 5×10^{-10} M GA and saturated at 5×10^{-7} M, and was completed before total release of α-amylase had occurred (Taiz and Jones, 1970). GA also induces the formation of endo-β-(1→4)-xylanase, β-xylopyranosidase, and β-arabinofructofuranosidase activities in barley aleurone layers, although there is no stimulation of cellulase (Taiz and Honigman, 1976). Indoleacetic acid (10^{-7} M) has also been reported to stimulate endo-β-glucanase activity in germinating barley (Polyakov et al., 1977).

As already mentioned, there is now evidence that release of glucan from barley endosperm walls requires initial proteolytic attack (Bamforth et al., 1979). An enzyme activity, which increased during malting and which appeared to be a carboxypeptidase with pH optimum of 6.35, was shown to catalyze the release of soluble β-glucan from insoluble endosperm walls; after purification of the carboxypeptidase by isoelectric focussing, it was shown to have no β-glucanase activity (Bamforth et al., 1979).

The role of β-glucanases in the mobilization of storage material in the barley endosperm is clearly complex. At present it may be suggested that glucanase activity has a twofold function in (a) digesting wall glucans, which may serve as a minor, but significant, energy source to the developing embryo, and (b) aiding release of other hydrolytic enzymes from the aleurone layer and their passage to the starchy endosperm.

VI. β-(1→3)-GLUCANS AND GLUCANASES IN BREWING

During the process of malting, whereby barley is germinated under controlled conditions in order to bring about conversion of storage poly-saccharides to more fermentable products, the main sugars formed are sucrose, glucose, and maltose, with smaller amounts of fructose and tri- and oligosaccharides (Preece, 1957). Water-soluble β-glucan levels increase during malting, reaching a maximum level after approximately 7 days on the malting floor, but decline considerably during kilning (Preece, 1957). It is believed that the changes in grain hemicellulose represent the major factor contributing to the physical modification of the grain during malting.

The advantages and disadvantages of the presence of hemicellulosic β-glucans in wort and beer have been reviewed (Preece, 1957). The presence of highly viscous glucan can cause problems associated with loss of extract, problems of separation and filtration, and poor beer clarity (Bamforth et al., 1979). In this respect, several groups of workers have analysed various barley varieties for β-glucan content, the results generally showing a lack of correlation between β-glucan and malting potential (Allison et al., 1979; Bathgate, 1979; Martin and Bamforth, 1980; Prentice et al., 1980; Smith et al., 1980; Prentice and Faber, 1981; Thompson and La Berge, 1981). Although viscosity measurements on aqueous extracts from barley have been regarded as important tests to be made during barley breeding programmes, such measurements may be somewhat artefactual in view of variations in cytolytic potential (Bathgate, 1979). The breakdown of barley endosperm walls may be divided into two processes, the cytoclastic reactions whereby wall components are solubilized, and the cytolytic reactions whereby β-glucan and pentosan are degraded (Martin and Bamforth, 1980). However, high β-glucanase activities may not themselves be a good indicator of β-glucan modification during malting (Martin and Bamforth, 1980; Prentice and Faber, 1981). More important may be the carboxypeptidase, or "β-glucan solubilase," already described in Sections III and V. This activity reaches a maximum level at approximately 3 days after germination (Bamforth et al., 1979). In addition to releasing β-glucan from the cell walls, the enzyme can also decrease the viscosity of barley β-glucan (Martin and Bamforth, 1980).

"Solubilases" and hydrolases are clearly important in the control of soluble β-glucan levels; it is also clear that cell wall degradation is potentially one of the most important sites for regulation of carbohydrate modification during malting and extraction during mashing; in particular, partially degraded or undegraded walls retard run-off after mashing poorly modified malts (Forrest and Wainwright, 1977). In this respect, attention has been given to the development of enzyme adjuncts for use during the mashing process, in particular when using unmalted cereals in brewing. Addition of *Bacillus amyliliquifaciens* β-glucanase to barley extracts has been reported to result in striking effects on wort viscosity and filtration rate, such additions also being an effective filtration aid in mashes from corn and rice (Denault *et al.*, 1981). The *Bacillus* enzyme was probably effective in barley in view of its greater heat stability and more favourable pH optimum during the mashing process than the apparently higher endogenous barley glucanase activity (Denault *et al.*, 1981).

VII. CONCLUDING REMARKS

Although the main theme of this chapter has been the storage β-(1→3)-glucans of higher plants, it is clear that our knowledge of the biochemistry of these compounds is in many areas at an earlier stage of development than our knowledge relating to starch or certain structural polysaccharides. Some potential areas for future research are (a) the biosynthesis of the barley endosperm and aleurone glucans, (b) the exact molecular mechanisms underlying β-glucanase formation and release from the barley aleurone layer and the precise natures of the glucanases concerned, (c) biotechnological applications of glucan enzymology to the various stages of the brewing process, and (d) molecular genetic approaches, based on the emerging details of the multiplicities and structures of glucan-metabolizing enzymes, to barley breeding programmes. The elucidation of β-glucan synthetic and degradative pathways and their physiological functions will also continue to provide fascinating problems for workers in other areas such as cell wall growth and development and molecular plant pathology.

REFERENCES

Abeles, F. B., Bosshart, R. P., Forrence, L. E. and Habig, W. H. (1970). *Plant Physiol.* **47,** 129–134.
Allison, M. J., Borzuchi, R., Cowe, I. A. and McHale, R. (1979). *J. Inst. Brew.* **85,** 86–88.
Anderson, M. A. and Stone, B. A. (1975). *FEBS Lett.* **52,** 202–207.
Anderson, R. L. and Stone, B. A. (1978). *Aust. J. Biol. Sci.* **31,** 573–586.
Bacic, A. and Delmer, D. P. (1981). *Planta* **152,** 346–351.
Bacic, A. and Stone, B. A. (1981). *Aust. J. Plant Physiol.* **8,** 475–495.
Ballance, G. M. and Manners, D. J. (1978a). *Carbohydr. Res.* **61,** 107–118.
Ballance, G. M. and Manners, D. J. (1978b). *Phytochemistry* **17,** 1539–1543.
Ballance, G. M., Meredith, W. O. S. and La Berge, D. E. (1976). *Can. J. Plant Sci.* **56,** 459–466.
Bamforth, C. W., Martin, H. L. and Wainwright, T. (1979). *J. Inst. Brew.* **85,** 334–338.
Barber, G. A., Elbein, A. D. and Hassid, W. Z. (1964). *J. Biol. Chem.* **239,** 4056–4061.
Barras, D. R. and Stone, B. A. (1969a). *Biochim. Biophys. Acta* **191,** 329–341.
Barras, D. R. and Stone, B. A. (1969b). *Biochim. Biophys. Acta* **191,** 342–353.

Bathgate, G. N. (1979). *J. Inst. Brew.* **85**, 326–328.

Bathgate, G. N., Palmer, G. H. and Wilson, G. (1974). *J. Inst. Brew.* **80**, 278–285.

Baulcombe, D. C. and Buffard, D. (1983). *Planta* **157**, 493–501.

Beltran, J. P. and Carbonell, J. (1978). *Phytochemistry* **17**, 1531–1532.

Bennett, P. A. and Chrispeels, M. J. (1972). *Plant Physiol.* **49**, 445–447.

Bernal-Lugo, I., Beachy, R. N. and Varner, J. E. (1981). *Biochem. Biophys. Res. Commun.* **102**, 617–623.

Brett, C. T. (1978). *Plant Physiol.* **62**, 377–382.

Brett, C. T. and Northcote, D. H. (1975). *Biochem. J.* **148**, 107–117.

Briggs, D. E. (1963). *J. Inst. Brew.* **69**, 13–19.

Brown, H. T. and Morris, G. H. (1890). *J. Chem. Soc.* **57**, 458–528.

Bull, A. T. and Chesters, C. G. C. (1966). *Adv. Enzymol.* **28**, 325–364.

Burke, D., Kaufman, P., McNeil, M. and Albersheim, P. (1974). *Plant Physiol.* **34**, 109–115.

Chambers, J. and Elbein, A. D. (1970). *Arch. Biochem. Biophys.* **138**, 620–631.

Chao, H.-Y. and Maclachlan, G. A. (1978). *Plant Physiol.* **61**, 943–948.

Chen, S. and Luchsinger, W. W. (1964). *Arch. Biochem. Biophys.* **106**, 71–77.

Clarke, A. E. and Stone, B. A. (1962). *Phytochemistry* **1**, 175–188.

Clarke, A. E. and Stone, B. A. (1963). *Rev. Pure Appl. Chem.* **13**, 134–156.

Coles, G. (1979). *Carlsberg Res. Commun.* **44**, 439–453.

Dais, P. and Perlin, A. S. (1982). *Carbohydr. Res.* **100**, 103–116.

Datko, A. H. and Maclachlan, G. A. (1968). *Plant Physiol.* **43**, 735–742.

Delmer, D. P., Heininger, H. and Kulow, C. (1977). *Plant Physiol.* **59**, 713–718.

Denault, L. J., Glenister, P. R. and Chau, S. (1981). *J. Am. Soc. Brew. Chem.* **39**, 46–52.

Deslandes, Y., Marchessault, R. H. and Sarko, A. (1980). *Macromolecules* **13**, 1466–1471.

Eastwell, K. C. and Spencer, M. S. (1982). *Plant Physiol.* **69**, 563–567.

Elbein, A. D., Barber, G. A. and Hassid, W. Z. (1964). *J. Am. Chem. Soc.* **86**, 309–310.

Fleming, M. and Kawakami, K. (1977). *Carbohydr. Res.* **57**, 15–23.

Fleming, M. and Manners, D. J. (1966). *Biochem. J.* **100**, 4P–5P.

Flinn, A. M. (1969). Ph. D. Thesis, Queen's University, Belfast.

Forrest, I. S. (1977). *Biochem. Soc. Trans.* **5**, 1154–1156.

Forrest, I. S. and Wainwright, T. (1977). *J. Inst. Brew.* **83**, 279–286.

Fraser, C. G. and Wilkie, K. C. B. (1971). *Phytochemistry* **10**, 1539–1542.

Fulcher, R. G., Setterfield, G., McCully, M. E. and Wood, P. J. (1977). *Aust. J. Plant Physiol.* **4**, 917–928.

Gamborg, O. L. and Eveleigh, D. E. (1968). *Can. J. Biochem.* **46**, 417–421.

Ghaemi, N., Gay, R. and Quittelier, M. (1978). *Bios (Nancy)* **9**, 16–29.

Goldberg, R. (1977). *C. R. Hebd. Seances Acad. Sci., Sec. D* **284**, 1409–1412.

Heiniger, U. and Delmer, D. P. (1977). *Plant Physiol.* **59**, 719–723.

Helsper, J. P. F. G., Veerkamp, J. H. and Sassen, M. M. A. (1977). *Planta* **133**, 303–308.

Henry, R. J. and Stone, B. A. (1982). *Plant Physiol.* **69**, 632–636.

Heyn, A. N. J. (1969). *Arch. Biochem. Biophys.* **132**, 442–449.

Higgins, T. J. V., Zwar, J. A. and Jacobsen, J. V. (1976). *Nature (London)* **260**, 166–169.

Huber, D. J. and Nevins, D. J. (1977). *Plant Physiol.* **60**, 300–304.

Huber, D. J. and Nevins, D. J. (1979). *Plant Cell Physiol.* **20**, 201–212.

Huber, D. J. and Nevins, D. J. (1980). *Plant Physiol.* **65**, 768–773.

Huber, D. J. and Nevins, D. J. (1981). *Planta* **151**, 206–214.

Jones, R. L. (1971). *Plant Physiol.* **47**, 412–416.

Kato, K., Yamada, A. and Noguchi, M. (1973). *Agric. Biol. Chem.* **37**, 1269–1275.

Kato, Y., Katsuhiro, I. and Matsuda, K. (1981). *Agric. Biol. Chem.* **45**, 2737–2744.

Keen, N. T. and Yoshikawa, M. (1983). *Plant Physiol.* **71**, 460–465.

Keen, N. T., Wang, M. C., Bartnicki-Garcia, S. and Zentmyer, G. A. (1975). *Physiol. Plant Pathol.* **7**, 91–97.

Kivilaan, A., Bandurski, R. S. and Schulze, A. (1971). *Plant Physiol.* **48**, 389–393.

Klein, A. S., Montezinos, D. and Delmer, D. P. (1981). *Planta* **152**, 105–114.

Labavitch, J. M. and Ray, P. M. (1978). *Phytochemistry* **17**, 933–937.

Larsen, G. L. and Brummond, D. O. (1974). *Phytochemistry* **13**, 361–365.

Luschinger, W. W. (1962). *Cereal Chem.* **39**, 225–235.

MacLeod, A. M. and McCorquodale, H. (1958). *New Phytol.* **57**, 168–182.

MacLeod, A. M. and Sandie, R. (1961). *New Phytol.* **60**, 117–128.

McNeil, M., Albersheim, P., Taiz, L. and Jones, R. L. (1975). *Plant Physiol.* **55**, 64–68.

Maltby, D., Carpita, N. C., Montezinos, D., Kulow, C. and Delmer, D. P. (1979). *Plant Physiol.* **63**, 1158–1164.

Manners, D. J. and Marshall, J. J. (1969). *J. Inst. Brew.* **75**, 550–561.

Manners, D. J. and Marshall, J. J. (1973). *Phytochemistry* **12**, 547–553.

Manners, D. J. and Wilson, G. (1974). *Carbohydr. Res.* **37**, 9–22.

Manners, D. J. and Wilson, G. (1976). *Carbohydr. Res.* **48**, 255–264.

Manners, D. J., Seiler, A. and Sturgeon, R. J. (1982). *Carbohydr. Res.* **100**, 435–440.

Marchessault, R. H. and Deslandes, Y. (1981). *Carbohydr. Polym.* **1**, 31–38.

Mares, D. J. and Stone, B. A. (1973a). *Aust. J. Biol. Sci.* **26**, 135–150.

Mares, D. J. and Stone, B. A. (1973b). *Aust. J. Biol. Sci.* **26**, 793–812.

Martin, H. L. and Bamforth, C. W. (1980). *J. Inst. Brew.* **86**, 216–221.

Masuda, Y. and Yamamoto, R. (1970). *Dev. Growth Differ.* **11**, 287–296.

Meier, H. and Reid, J. S. G. (1982). *In* "Encyclopedia of Plant Physiology: New Series" (F. A. Loewus and W. Tanner, eds.), Vol. 13A, pp. 418–471. Springer-Verlin, Berlin and New York.

Meier, H., Buchs, L., Buchala, A. J. and Homewood, T. (1981). *Nature (London)* **289**, 821–822.

Meyer, A. M. and Polyakoff-Mayber, A. (1963). "The Germination of Seeds." Pergamon, Oxford.

Miyamoto, C. and Tamari, K. (1973). *Agric. Biol. Chem.* **37**, 1253–1260.

Montague, M. J. and Ikuma, H. (1978). *Plant Physiol.* **62**, 391–396.

Moore, A. E. and Stone, B. A. (1972a). *Biochim. Biophys. Acta* **258**, 238–247.

Moore, A. E. and Stone, B. A. (1972b). *Biochim. Biophys. Acta* **258**, 248–264.

Moore, A. E. and Stone, B. A. (1972c). *Virology* **50**, 791–798.

Morrall, P. and Briggs, D. E. (1978). *Phytochemistry* **17**, 1495–1502.

Nevins, D. J., Huber, D. J., Yamamoto, R. and Loescher, W. H. (1977). *Plant Physiol.* **60**, 617–621.

Nevins, D. J., Yamamoto, R. and Huber, D. J. (1978). *Phytochemistry* **17**, 1503–1505.

Péaud-Lenöel, C. and Axelos, M. (1970). *FEBS Lett.* **8**, 224–228.

Pillonel, C., Buchala, A. J. and Meier, H. (1980). *Planta* **149**, 306–312.

Polyakov, V. A., Pekhtereva, N. T., Popadich, I. A. and Lerner, I. G. (1977). *Prikl. Biokhim. Mikrobiol.* **13**, 750–753 (in Russian).

Preece, I. A. (1957). *R. Inst. Chem. Lect. Monogr.*, Rep. No. 2, pp. 1–24.

Prentice, N. and Faber, S. (1981). *Cereal Chem.* **58**, 77–79.

Prentice, N., Babler, S. and Faber, S. (1980). *Cereal Chem.* **57**, 198–202.

Raymond, Y., Fincher, G. B. and Maclachlan, G. A. (1978). *Plant Physiol.* **61**, 938–942.

Sakurai, N. and Masud, Y. (1979). *Plant Cell Physiol.* **20**, 593–603.

Smith, D. B., Morgan, A. G. and Aastrup, S. (1980). *J. Inst. Brew.* **86**, 277–283.

Smith D. L. (1974). *Protoplasma* **79**, 41–57.

Smith, M. M. and Stone, B. A. (1973a). *Aust. J. Biol. Sci.* **26**, 123–133.

Smith, M. M. and Stone, B. A. (1973b). *Biochim. Biophys. Acta* **313**, 72–94.

Smith, M. M. and Stone, B. A. (1973c). *Phytochemistry* **12**, 1361–1367.

Southworth, D. and Dickinson, D. B. (1975). *Plant Physiol.* **56**, 83–87.

Stinard, P. S. and Nevins, D. J. (1980). *Phytochemistry* **19**, 1467–1468.

Svensson, B. (1978). *Carlsberg Res. Commun.* **43**, 103–115.

Taiz, L. and Honigman, W. A. (1976). *Plant Physiol.* **58**, 380–386.

Taiz, L. and Jones, R. L. (1970). *Planta* **92**, 73–84.

Takeuchi, Y. and Komamine, A. (1981). *Plant Cell Physiol.* **22**, 1585–1594.

Thompson, R. G. and La Berge, D. E. (1977). *Tech. Q. Master Brew Assoc. Am.* **14**, 238–242.

Thompson, R. G. and La Berge, D. E. (1981). *Tech. Q. Master Brew. Assoc. Am.* **18**, 116–121.

Tsai, C. M. and Hassid, W. Z. (1971). *Plant Physiol.* **47**, 740–744.

Tsai, C. M. and Hassid, W. Z. (1973). *Plant Physiol.* **51**, 998–1001.

Valent, B. S., Darvill, A. G., McNeil, M., Robertsen, B. K. and Albersheim, P. (1980). *Carbohydr. Res.* **79**, 165–192.

Van der Woude, W.-J., Lembi, C. A., Morré, J., Kindinger, J. I. and Ordin, L. (1974). *Plant Physiol.* **54**, 333–340.

Varner, J. E. and Mense, R. M. (1972). *Plant Physiol.* **49**, 187–189.

Wada, S. and Ray, P. M. (1978). *Phytochemistry* **17**, 923–931.

Wong, Y.-S. and Maclachlan, G. A. (1979a). *Biochim. Biophys. Acta* **571**, 244–255.

Wong, Y.-S. and Maclachlan, G. A. (1979b). *Biochim. Biophys. Acta* **571**, 256–269.

Wong, Y.-S. and Maclachlan, G. A. (1980). *Plant Physiol.* **65**, 222–228.

Woodward, J. R. and Fincher, G. B. (1982a). *Carbohydr. Res.* **106**, 111–122.

Woodward, J. R. and Fincher, G. B. (1982b). *Eur. J. Biochem.* **121**, 663–669.

Woodward, J. R., Morgan, F. J. and Fincher, G. B. (1982). *FEBS Lett.* **138**, 198–200.

Woolard, G. R., Rathbone, E. B. and Novellie, L. (1976). *Carbohydr. Res.* **51**, 249–252.

Yamamoto, R. and Nevins, D. J. (1978). *Carbohydr. Res.* **67**, 275–280.

Young, D. H. and Pegg, G. F. (1981). *Physiol. Plant Pathol.* **19**, 391–417.

Chapter **7**

Galactomannans

J. S. GRANT REID

Department of Biological Science
University of Stirling
Stirling, Scotland

I. INTRODUCTION

With the obvious exception of starch, the galactomannans are the most extensively studied of the storage polysaccharides of higher plants, and there can be little doubt that the principal reason for this is their commercial importance. Certain plant galactomannans, notably those from carob (*Ceratonia siliqua*) and guar (*Cyamopsis tetragonoloba*) seeds, find widespread application as "industrial gums." Over the past 30 years, numerous reviews on galactomannans have appeared, concentrating mainly on their structures and industrial uses (e.g., Whistler and Smart, 1953; Smith and Montgomery, 1959; Stepanenko, 1960; Glicksman, 1969). One excellent article has analysed their chemistry and interactions—exploring the relationship between their molecular structures and their commercially important interactive properties with

265

other gelling polysaccharides (Dea and Morrison, 1975). The biochemistry of the galactomannans has been reviewed recently with particular emphasis on the types of enzymes which may participate in their catabolism (Dey, 1978). Although essential background information on galactomannan structures and properties is provided, the aim of this chapter is to highlight the more physiological aspects of galactomannan storage in plants—the distribution of galactomannans, the morphology of their storage, the physiology and biochemistry of galactomannan formation and metabolism, and their overall biological function. This approach has been followed in two recent reviews, one dealing with all the nonstarch storage polysaccharides of higher plants (Meier and Reid, 1982), the other more specifically with the cell-wall storage polysaccharides of seeds (Reid, 1984).

It should perhaps be mentioned at the outset that the galactomannans constitute but one of a group of three structurally related types of higher plant storage polysaccharides, the "mannan group" (Meier and Reid, 1982). The other two members of the group, the "pure" mannans and the glucomannans, are reviewed in Chapter 8.

II. STRUCTURES, PROPERTIES, AND DISTRIBUTION

Many higher plant galactomannans have been subjected to structural analysis, and it is generally accepted that they conform closely to the general structure shown in Fig. 1. The D-mannosyl units are linked by β-(1→4) linkages to form a linear "backbone" while the D-galactosyl residues are attached to this by α-(1→6) linkages. Mannose:galactose ratios vary between about 1:1 ($\sim 100\%$ galactose substitution of the backbone) to about 4:1

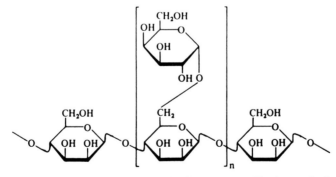

Fig. 1. General structure of leguminous seed galactomannans. The degree of substitution of the β-(1→4)-linked D-mannan backbone by α-D-galactopyranosyl groups varies from about 25% to 100%.

($\sim 25\%$ galactose substitution), depending upon the botanical source of the polysaccharide and, to some extent, on the isolation and purification techniques used to obtain it.

All the higher plant galactomannans have been obtained from seeds, the overwhelming majority of them from the seeds of the Leguminosae. So far, galactomannans have been isolated from the seeds of only three non-leguminous species: *Annona muricata* (Annonaceae) (Kooiman, 1971), *Convolvulus tricolor*, and *Ipomoea muricata* (both Convolvulaceae) (Kooiman, 1971; Khanna and Gupta, 1967).

In the Leguminosae the galactomannans are localised in the endosperm of the seed. The nonendospermic seeds characteristic of several of the tribes of the Leguminosae do not contain galactomannans; if they store poly-saccharide, it takes the form of starch or xyloglucan (amyloid) (Reid, 1984; Chapters 4 and 10, this volume). The nonleguminous seeds that are known to contain galactomannan are endospermic, and it seems reasonable to assume that their galactomannans are localised in the endosperm.

A. Primary Structures

The galactomannan samples that have been subjected to structural analysis have usually been isolated by cold- or hot-water extraction of comminuted seeds. Alkali extraction has been used less frequently. It has been noted that cold-water extraction of carob seeds (*Ceratonia siliqua*) gives a galactom-annan of lower molecular weight and higher galactose content than does a subsequent hot-water extraction (Hui and Neukom, 1964; Dea and Morrison, 1975). A degree of polydispersity and polymolecularity must therefore be assumed.

Galactomannans are normally purified by alcohol precipitation or by their selective complexation by heavy-metal cations such as Cu^{2+} and Ba^{2+}. These procedures are capable of subfractionating a polydisperse and polymolecular native galactomannan: the purified sample could be atypical of the native material in molecular weight and/or structure.

The mannose:galactose ratios of galactomannans have been obtained following total acid hydrolysis, while linkage types have been assigned on the basis of methylation analysis, periodate oxidation, and partial hydrolysis studies. Linkage modes (α or β) were derived by studying optical rotations and by chemical analysis of oligosaccharides liberated on partial hydrolysis.

It should be noted that although the majority of galactomannans conform fully to the structure in Fig. 1 (Dea and Morrison 1975), deviations have been reported. Some galactomannans have been shown to have a small proportion of $(1 \rightarrow 3)$ linkages (Unrau and Choy, 1970a,b; Kapoor and Mukherjee, 1969, 1971) and $(1 \rightarrow 2)$ linkages (Unrau and Choy, 1970a) in the "backbone." One galactomannan is reported to have a proportion of $(1\rightarrow3)$-linked D-mannopyranosyl residues in the backbone and D-galactopyranosyl side chains attached by α-$(1\rightarrow2)$ linkage (Kapoor and Mukherjee, 1971). There are also reports of side branches consisting of short chains of α-$(1\rightarrow6)$-linked D-galactopyranosyl residues (Courtois and Le Dizet, 1963, 1966; Cerezo, 1965).

The principal area of uncertainty with regard to the primary structures of galactomannans is the distribution of D-galactosyl side chains (generally consisting of single units) along the main chains of the incompletely galactose-substituted galactomannans. It is now generally accepted that the distribution is not regular (Dea and Morrison, 1975), but it is probably nonrandom. Degradation of galactomannan by *endo*-β-mannanases has resulted in the production of large mannanase-resistant fragments with mannose:galactose ratios approaching unity (Courtois and Le Dizet, 1968). On the basis of this type of experiment, it has been proposed that galactomannans have a partial "block" structure; areas of low galactose substitution and of high galactose substitution exist in the same molecule (Courtois and Le Dizet, 1968). Periodate oxidation studies (Hoffman *et al.*, 1976) confirm that the distribution of galactose side chains in guar and carob galactomannans is nonrandom. There is evidence that galactomannans with the same mannose:galactose ratio differ with regard to the distribution of their galactosyl substituents (Courtois and Le Dizet, 1966, 1970). Since it appears likely that the distribution of galactosyl substituents is important in determining the industrially important interactions of the galactomannans with other poly-saccharides (Dea and Morrison, 1975), it is important that sensitive techniques to determine and compare distribution patterns be developed. The use of computer modelling in conjunction with a precise knowledge of the action patterns of highly purified *endo*-β-mannanases (McCleary, 1981) shows promise as a fine-structure probe (B. V. McCleary, personal communication).

B. Properties of Galactomannans

The seed galactomannans are hydrophilic substances. In contact with water they hydrate and become mucilaginous, eventually forming a thick paste that is impervious to further water (Dea and Morrison, 1975). This property is the basis of their use commercially for waterproofing explosives (e.g., Yancik *et al.*, 1972), plugging leaking wells (e.g., Walker, 1966) and improving the wet strength of paper (e.g., Chrisp, 1967). The relatively slow hydration of guar-seed galactomannan has been exploited in the manufacture of delayed-release drug preparations (Nürnberg and Rettig, 1974).

When fully dispersed in water, galactomannans afford solutions with very high viscosities; this property has found application in the food industry where galactomannans are used as thickeners for soups, desserts, pie fillings, sauces, and mayonnaise (Glicksman, 1969; Dea and Morrison, 1975). Although galactomannan solutions do not normally form gels, they will do so in the presence of certain ions such as borate (Hart, 1930) and transition-metal cations (Chrisp, 1967), which are capable of cross-linking chains in solution. Galactomannans also enhance synergistically the gelation of other polysaccharides such as the agars and the carrageenans (Dea and Morrison, 1975). The analysis of the molecular interactions responsible for these effects is beyond the scope of this chapter (but see Dea and Morrison, 1975; Morris *et al.*, 1981; Dea *et al.*, 1977).

The significance of the hydrophilic properties of galactomannans in the germinative strategy of the seeds that contain them is discussed in Section V.

C. Chemotaxonomic Significance of Galactomannans

The galactomannans that have been isolated from leguminous seeds span a limited range of mannose:galactose ratios, from approximately 1:1 to just over 5:1. It has been suggested that this ratio may afford a taxonomic character to aid legume classification (Reid and Meier, 1970a; Bailey, 1971; Kooiman, 1972).

The potential chemotaxonomic value of the mannose: galactose ratio cannot be adequately assessed simply by correlating literature values with the accepted taxonomic divisions of the Leguminosae. Such correlations (see Reid and Meier, 1970a, table 1, and Dea and Morrison, 1975, table 1) indicate little more than a tendency for the low-galactose galactomannans to occur in the seeds of the Caesalpinioideae (Caesalpiniaceae) and the highly galactose-substituted galactomannans to be associated with the more advanced tribes of the Faboideae (Fabaceae or Papillionaceae). Dea and Morrison (1975) have correctly pointed out that meaningful comparisons can be made only between galactomannans that have been isolated by methods designed to ensure complete extraction of the polysaccharide from the seed and to avoid the subfractionation of a polymolecular sample during isolation and purification. Where such precautions have been taken (Reid and Meier, 1970*a*; Campbell, 1978), there is a close correlation between mannose:galactose ratio and taxonomic classification. Table I summarises the results of a study of the tribe Trifolieae. The galactomannans from species chosen to represent most of the sections of the tribe display a remarkably constant mannose:galactose ratio, with a single exception—*Trigonella cretica* (section Samaroideae). *Trigonella cretica* seeds from different geographical locations gave the same "deviant" ratio. There may be a case for reexamining the classification of *T. cretica*. A subsequent examination of galactomannans from the tribe Genisteae

Table I

Chemical Compositions of Galactomannans Isolated from the Seeds of Some Species in the Tribe Trifolieae[a]

Species	Section	Galactomannan composition	
		%Galactose	%Mannose
Trifolium incarnatum C. presl.		46	54
Trifolium dubium Sibth.		46	54
Medicago sativa L.		45.5	54.5
Medicago lupulina L.		46	54
Medicago radiata (L.) Heyn		46	54
Melilotus alba medicus		47	53
Melilotus officinalis (L.) pallas		45.5	54.5
Trigonella calliceras Fisch	*Callicerates* Boiss.	47	53
Trigonella corniculata L.	*Falcatulae* Boiss.	46	54
Trigonella hamosa L.	*Falcatulae* Boiss.	46	54
Trigonella caerulea (L.) Ser	*Capitatae* Boiss.	47	53
Trigonella monspeliaca L.	*Reflexae* (Sirj) vass.	48	52
Trigonella polycerata Bieb.	*Bucerates* Boiss.	47	53
Trigonella foenum-graecum L.	*Foenum-graecum* Ser.	47	53
Trigonella cretica (L.) Boiss.	*Samaroideae*[b] (1)	39	61
	(2)	37.5	62.5
	(3)	39	61
	(4)	38	62

[a]After Reid and Meier (1970a).
[b](1) Seeds obtained from Botanical Garden, University of Fribourg, Switzerland.
(2) Seeds obtained from Royal Botanic Garden, Kew, England.
(3) Seeds obtained from Botanical Garden, University of Nancy, France.
(4) Seeds obtained from Botanical Garden, University of Basle, Switzerland.

(Table II) has again revealed a similar constancy of the mannose:galactose ratio.

Only further systematic examination of other leguminous tribes will reveal whether or not the mannose:galactose ratio is a useful taxonomic character.

III.   GALACTOMANNAN FORMATION

A.   Ultrastructural Aspects of Galactomannan Deposition

In the late nineteenth century, Nadelmann (1890) described the deposition of "mucilages," which we now know to be galactomannans, in the developing endosperm of four leguminous species: *Colutea brevialata, Indigofera hirsuta,*

Table II

Compositions of Galactomannans Isolated from the Seeds of some Species
in the Tribe Genisteae, Subtribe Genistinae[a]

Species	Mannose:galactose ratio
Spartium junceum L.	2.2
Genista ovata W.K.	2.0
Genista tinctoria L.	2.1
Genista monosperma Lam.	2.1
Petteria ramentacea (Sieb) Presl.	2.1
Laburnum alpinum (Mill) Presl.	2.1
Laburnum anagyroides medicus	2.2
Ulex europaeus L.	1.9
Cytisus hirsutus L.	2.3
Cytisus supinus L.	2.0
Sarothamnus scoparius (L.) Wimmer ex Koch	2.3

[a] Data from Campbell (1978).

Tetragonolobus purpureus, and *Trigonella foenum-graecum*. He observed that
the mucilage was ultimately deposited in the cell walls of the endosperm, but
that it was formed initially in cytoplasmic vacuoles. Recently Nadelmann's
work has been extended by Reid and Meier (1973a), who used periodate–
Schiff staining and interference contrast microscopy to investigate galactomannan deposition in the developing endosperm of fenugreek (*Trigonella
foenum-graecum*). They confirmed that the galactomannan is a cell-wall
polysaccharide that, in fenugreek, continues to be deposited until almost the
whole of the endosperm is filled with it (Figs. 2, 3, and 4). No evidence for
Nadelmann's (1890) intracellular mucilage vesicles was found using the light
microscope. Curiously, however, electron-microscopic observations (Meier
and Reid, 1977) provide evidence that galactomannan is formed initially
within the cytoplasm, within the cisternae of the rough endoplasmic
reticulum, and then discharged into the wall (Fig. 5). At the height of
galactomannan deposition, the intracisternal material swells to such an extent
that cytoplasmic material becomes "pinched off" to give structures with the
appearance of rough endoplasmic reticulum vesicles with ribosomes on the
inside (Fig. 6). It is possible that the larger intracellular galactomannan
accumulations might have been seen by Nadelmann (1890).

The deposition of galactomannan in the developing fenugreek endosperm
begins next to the embryo. Cells adjacent to the cotyledons are the first to
become "filled" (Fig. 2), while those on the outer periphery of the endosperm
are filled last (Fig. 3). A single outer layer of cells—the aleurone layer—
always remains unfilled (Fig. 4). The aleurone layer is in fact the only layer of
living cells in the mature endosperm, and it subsequently has a key role in the

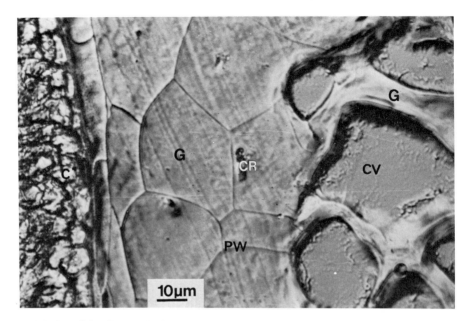

Fig. 2. Light micrograph of part of the endosperm of a developing fenugreek seed, near the beginning of galactomannan deposition. Note that the endosperm cells nearest the cotyledons (C) are already filled with galactomannan (G). CV = cytoplasm and vacuole; PW = primary wall; CR = cytoplasmic remnant. Interference-contrast microscopy.

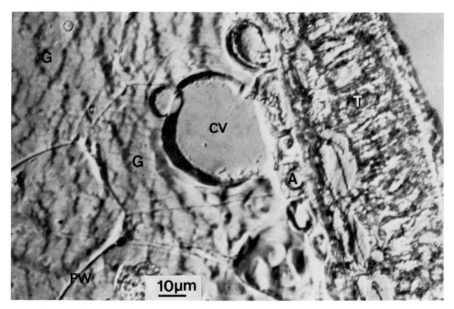

Fig. 3. Light micrograph of part of the endosperm of a developing fenugreek seed nearing the end of galactomannan deposition. Only a few cells next to the aleurone layer (A) remain unfilled with galactomannan (G). CV = cytoplasm and vacuole; PW = primary wall; T = testa. Interference-contrast microscopy.

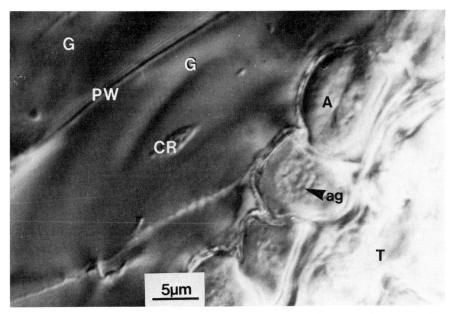

Fig. 4. Light micrograph of part of the endosperm of a developing fenugreek seed after the completion of galactomannan deposition during dehydration. With the exception of the aleurone layer (A), all endosperm cells are filled with galactomannan (G); ag = aleurone grain; PW = primary wall; T = testa; CR = cytoplasmic remnant.

post germinative catabolism of the galactomannan (Reid and Meier, 1972) (see Section IV).

It is implicit in the foregoing description that galactomannan deposition is accompanied by degradative processes that lead to the disappearance of cytoplasm and the cytoplasmic organelles from most of the endosperm cells. The biochemistry of these processes has not yet been studied.

It cannot yet be stated with confidence that the pattern of galactomannan deposition observed in the fenugreek endosperm is universal: there is considerable diversity in the morphology of galactomannan storage (compare Figs. 7 and 12), and no other systems have been studied in recent times.

B. Biochemistry of Galactomannan Synthesis

The only seed system in which the biochemistry of galactomannan synthesis has been studied is the developing fenugreek endosperm, and our knowledge of it is still far from complete.

Before it had been established that the deposition of galactomannan in the fenugreek endosperm is a cell-by-cell process, Reid and Meier (1970b) observed that there is no variation in the mannose:galactose ratio of the

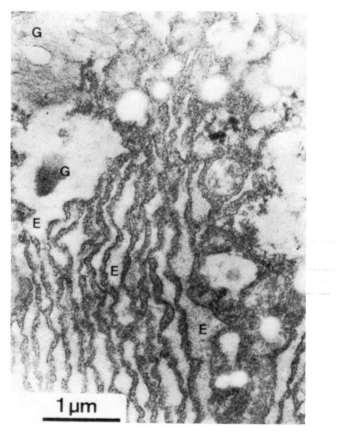

Fig. 5. Electron micrograph showing part of an endosperm cell of fenugreek during galactomannan deposition. The intracisternal space, or enchylema (E), of the endoplasmic reticulum is highly swollen. In the upper left of the photograph, a large ER vesicle containing galactomannan (G) is in connection with the wall. Section contrasted by periodic acid–semicarbazide–silver proteinate. From Meier and Reid (1977).

galactomannan present in the endosperm during its period of formation. This cannot, however, be taken as evidence that the mannose:galactose ratio of a *nascent* galactomannan molecule is constant: the proportion of such molecules at any time must always be small. It has also been shown that the galactosyl-sucrose oligosaccharides of the raffinose–stachyose family (mainly stachyose) are deposited in the endosperm concurrently with galactomannan (Reid and Meier, 1970b), although the final levels of stachyose are very much lower than those of galactomannan (Campbell and Reid, 1982). The parallel accumulation of these two storage carbohydrates led to the suggestion (Reid and Meier, 1970b) that the mode of biosynthesis of the α-D-galactopyranosyl

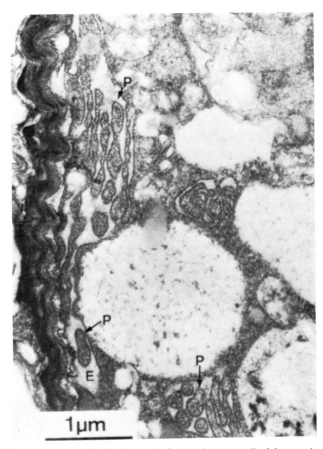

Fig. 6. Electron micrograph showing part of an endosperm cell of fenugreek early during galactomannan deposition. Note "poculiform" ER (P): the intracisternal space has swollen to such an extent that cytoplasmic material has been "pinched off" into "vesicles". E = enchylema. From Meier and Reid (1977).

units in both molecules may proceed by a similar mechanism. Beyond the observation that galactinol, the galactosyl donor in stachyose biosynthesis (Kandler and Hopf, 1982), is present in the developing endosperm, there is no experimental backing for this suggestion. Campbell and Reid (1982) have, however, observed that homogenates prepared from developing endosperms can catalyse the transfer of $\text{D-}[^{14}\text{C}]$galactosyl units from $\text{UDP-D-}[\text{U-}^{14}\text{C}]$ galactose to galactomannan.

The biosynthesis of the mannan "backbone" of the fenugreek galactomannan has been more thoroughly investigated. During, and only during, the period of galactomannan deposition, endosperm homogenates contain an enzyme activity capable of transferring D-[^{14}C]mannosyl units from GDP-D-[U-^{14}C]mannose to a water-soluble product that is indistinguishable from galactomannan (Campbell and Reid, 1982). In the early stages of galacto-mannan accumulation, most of the enzyme activity was associated with a light, particulate fraction (Campbell and Reid, 1982), which behaved like endoplasmic reticulum on density gradient centrifugation (Campbell, 1978). In the later stages, most of the activity was associated with, or occluded within, gross-particulate "cell-wall" material.

IV. CATABOLISM OF GALACTOMANNANS

The disappearance of the endosperm "mucilages" of leguminous seeds following germination was reported by Nadelmann (1890), who concluded that they were reserve substances. His findings have been confirmed and extended in more recent times as a result of physiological studies carried out mainly on three species: fenugreek (*Trigonella foenum-graecum*), carob (*Ceratonia siliqua*), and guar (*Cyamopsis tetragonoloba*). This choice of species was fortunate since their galactomannans have mannose:galactose ratios that typify the two extremes and the middle of the range encountered in nature.

A. Fenugreek

Fenugreek (*Trigonella foenum-graecum*; Leguminosae-faboideae tribe Trifolieae) has a mannose:galactose ratio of ~ 1. The fenugreek seed contains about 30% by weight of a galactomannan the structure of which does not appear to deviate from the established pattern (Fig. 1) (Daoud, 1932; Andrews *et al.*, 1952). Its mannose:galactose ratio is close to unity; thus almost every mannosyl unit in the main chain carries a galactosyl substituent.

Following germination, which is complete at 10 hr under our laboratory conditions (Reid and Bewley, 1979), the amount of galactomannan in the endosperm does not change for about 14 hr; thereafter it declines rapidly to zero (Reid, 1971). This represents a very rapid rate of breakdown relative to the cotyledonary reserves of protein and oil (Leung *et al.*, 1981). The disap-pearance of galactomannan from the endosperm coincides with a large increase in sucrose levels within the embryo, and with the synthesis of transitory starch in the embryo (Reid, 1971). It seems reasonable to associate both transitory starch formation and sucrose accumulation with the influx of galactomannan breakdown products from the endosperm.

It has been shown (Leung *et al.*, 1984) that the *initiation* of starch formation in the cotyledons is independent of a supply of sugars from the endosperm, although the "grand period" of starch accumulation does not occur in the absence of such an influx.

During galactomannan mobilisation there are low but detectable levels of galactose, mannose, and manno-oligosaccharides in the endosperm, suggesting both a hydrolytic mode of breakdown and a rapid uptake of sugars into the embryo (Reid, 1971). The hydrolytic breakdown was confirmed by the observation that galactomannan was broken down and converted quantitatively to galactose and mannose when isolated endosperms were incubated in the absence of the embryo (Reid and Meier, 1972). Three hydrolytic enzymes, together capable of converting galactomannan to galactose and mannose, were subsequently assayed in the endosperm of the germinated seed. Two of them, α-galactosidase and *endo-β*-mannanase, appeared and increased in the endosperm to parallel galactomannan breakdown, while the activity of the third, β-mannosidase, increased sixfold over the same period. Density-labelling and inhibitor studies carried out on isolated endosperms strongly indicate that the α-galactosidase and the β-mannanase are synthesised *de novo* within the endosperm (Reid and Meier, 1973b; Reid *et al.*, 1977; J. S. G. Reid, C. Davies and P. Phizacklea, unpublished). It is not yet clear in the fenugreek system whether or not the β-mannosidase is synthesised *de novo*, but there is excellent evidence that in another leguminous system (see discussion of guar below) this enzyme is present in an active form in the resting endosperm (McCleary, 1983).

The anatomy of the fenugreek seed is illustrated in Fig. 7. The endosperm, which is interposed between the embryo and the seed coat or testa, comprises two distinct zones: a galactomannan-filled area of storage tissue (Fig. 7) and a one-cell aleurone layer (Fig. 4). The storage tissue is essentially a mass of galactomannan (see Section III,A) and is nonliving; the aleurone layer, which is the outermost cell layer of the endosperm, consists of living cells (Reid and Meier, 1972). It is clear that the aleurone layer must be responsible for the *de novo* synthesis of α-galactosidase and endo-β-mannanase following germination. Cytological studies reinforce this conclusion. During galactomannan mobilisation, a dissolution zone appears in the endosperm next to the aleurone layer and increases in size inwards towards the embryo (Figs. 8, 9, and 10) (Reid, 1971). Furthermore, electron-microscopic examination of the aleurone layer prior to and during galactomannan breakdown provides evidence for the intensive synthesis of secretory proteins (Reid and Meier, 1972).

The sugars formed in the endosperm during galactomannan breakdown in intact seeds do not accumulate but are very rapidly taken up by the embryo (Reid, 1971). Neither free galactose nor mannose has been detected in the embryo (Reid, 1971; Uebelmann, 1978); they must be phosphorylated very rapidly and converted into sucrose and starch. Uebelmann (1978) has shown that the uptake, although rapid, occurs by passive diffusion.

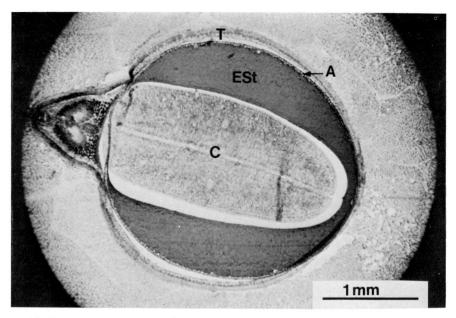

Fig. 7. Cross section of an imbibed fenugreek seed showing cotyledons (C), testa (T), and the endosperm consisting of storage tissue (ESt) and aleurone layer (A). Periodic acid–Schiff staining.

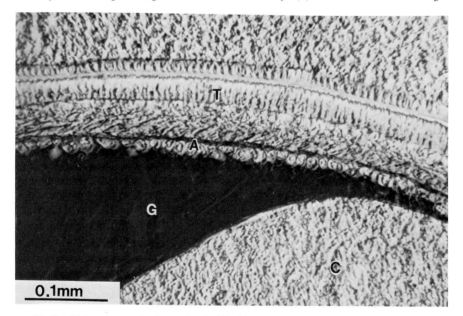

Fig. 8. Light micrograph of part of an imbibed fenugreek seed showing part of cotyledon (C), testa (T), and endosperm. Most of the endosperm is "filled" with dark-staining galactomannan (G), the exception being the one-cell-thick aleurone layer (A). Periodic acid–Schiff staining; Nomarski interference-contrast microscopy.

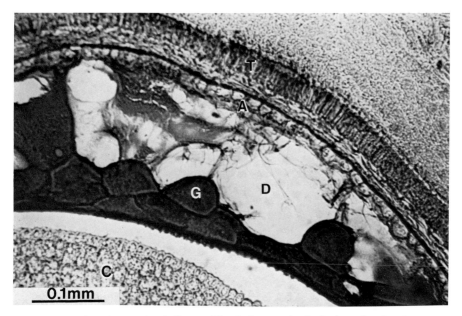

Fig. 9. Light micrograph, similar to Fig. 8, but at the beginning of galactomannan mobilisation. Note the dissolution zone (D) forming in the endosperm next to the aleurone layer (A). T = testa: G = galactomannan. Periodic acid–Schiff staining.

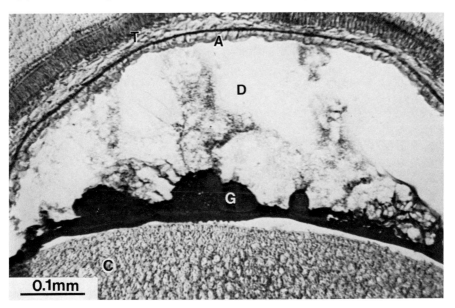

Fig. 10. Light micrograph, similar to Fig. 8, but near the end of galactomannan mobilisation. The size of the dissolution zone (D) has increased. The direction of galactomannan (G) breakdown is from the aleurone layer (A) inwards towards the cotyledons (C). Periodic acid–Schiff staining.

The main physiological events accompanying the mobilisation of galactomannan in the fenugreek seed are illustrated in Fig. 11.

It is tempting, on the basis of Reid and Meier's (1972) observation that
galactomannan breakdown occurs in half endosperms isolated even from dry
seeds, to conclude that the hydrolytic process occurring in the endosperm is
not under embryonic control. Yet this need not be the case. The isolated
endosperms were incubated (Reid and Meier, 1972) under conditions that
allowed the breakdown products to diffuse freely away into the medium.
Halmer and Bewley (1982) have correctly pointed out that under such
conditions, endogenous inhibitors of galactomannan mobilisation might also
diffuse away, thus eliminating the need for any promotive influence from the
embryo. Source–sink regulation is, of course, also possible.

B. Carob

Carob (*Ceratonia siliqua;* Leguminosae-caesalpinioideae) has a
mannose:galactose ratio of ~ 3.5–4. The galactomannan of carob, or locust
"bean," is typical of the low-galactose galactomannans in the seeds of the
Leguminosae-caesalpinioideae (see Section II,C). The seeds contain about
65% by weight of galactomannan and are similar to the fenugreek seed in their
gross anatomy. They differ from fenugreek in the relative size of the

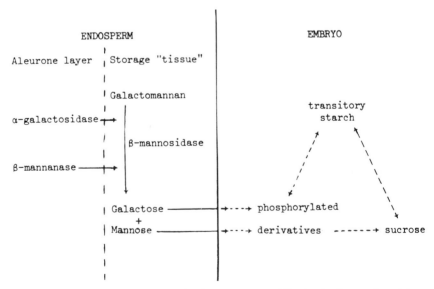

Fig. 11. Schematic representation of galactomannan mobilisation in the germinated fenugreek seed.

endosperm; in carob it accounts for a greater percentage of the seed, and the cotyledons are correspondingly less fleshy (compare Figs. 7 and 12). Galactomannan mobilisation in germinated carob seeds has been investigated by Seiler (1977), who demonstrated that the hydrolytic enzymes are an α-galactosidase, an endo-β-mannanase, and a β-mannosidase, that they are produced within the endosperm, and that at least one of them is synthesised *de novo*. There are, however, several differences between the carob and fenugreek systems. Unlike fenugreek, all the cells of the carob endosperm have living protoplasts, and the galactomannan is stored in the thickened cell walls (Seiler, 1977). There is no specialisation of the carob seed endosperm into storage tissue and aleurone layer.

Following germination, the pattern of galactomannan dissolution differs from that observed in fenugreek. Initially a small amount of wall degradation occurs throughout the endosperm, but this soon ceases and bulk galacto-mannan breakdown then proceeds from the embryo outwards (Fig. 13) (Seiler, 1977). Clearly, all the cells of the endosperm are capable of producing

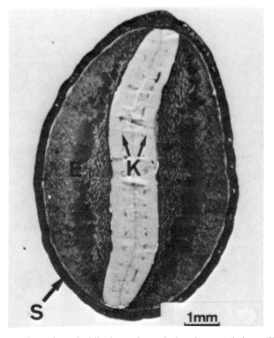

Fig. 12. Cross section of an imbibed carob seed showing cotyledons (K), testa (S) and endosperm (E). Note that there is no clear division of the endosperm into storage tissue and aleurone layer as in fenugreek (Fig. 7). All endosperm cells have living protoplasts. Periodic acid–Schiff staining. After Seiler (1977).

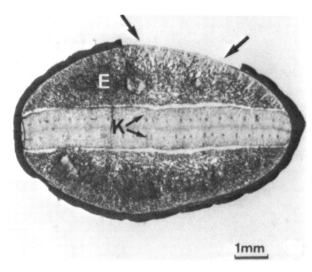

Fig. 13. Cross section of a carob seed during galactomannan breakdown. Obvious break-down occurs next to the cotyledons, except where removal of the testa (between arrows) has allowed diffusion of breakdown products and/or freer diffusion of oxygen. E, endosperm; K, cotyledons. Periodic acid–Schiff staining. After Seiler (1977).

galactomannan-degrading enzymes, but the rate of breakdown is probably controlled by the uptake of breakdown products into the embryo (Seiler, 1977).

C. Guar

Guar (*Cyamopsis tetragonoloba*; Leguminosae-faboideae tribe Indigoferae) has a mannose:galactose ratio of ~2. Guar-seed galactomannan shows no unusual structural features, and its mannose:galactose ratio is about the middle of the naturally occurring range. Scanning electron micrographs (McClendon *et al.*, 1976) indicate that the anatomy of the endosperm is similar to that of fenugreek: "galactomannan-filled" storage tissue is surrounded by an aleurone layer. The aleurone layer of the guar-seed endosperm is, however, several cell layers thick. The physiology of galactomannan mobilisation in guar has not been studied in detail, but it is almost certainly similar to that of fenugreek.

The guar-seed system is of particular interest because it is the only one from which all three galactomannan degrading enzymes have been isolated and purified to homgeneity (McCleary, 1982, 1983). These enzymes have been used *in vitro* to demonstrate their interactions during galactomannan mobilisation *in vivo* (McCleary, 1983).

All three enzymes are acid hydrolases, are monomeric, and do not appear to have cofactor requirements. The α-galactosidase (MW 40,500; pI 3.7; pH optimum 4.5–5.0) and the endo-β-mannanase (MW 41,700; pI 4.8–6.2, 5 isozymes; pH optimum 4–5) are probably synthesised *de novo* (McCleary, 1983). The β-mannosidase is more correctly described as an *exo-β*-mannanase, since manno-oligosaccharides are better substrates than is mannobiose (McCleary, 1982). This enzyme is present in the endosperm even in the resting state and does not increase in activity following germination (McCleary, 1983). The enzyme (MW 59,000; pI 9.4; pH optimum 5–6) is strongly bound to galactomannan and can be quantitatively extracted only by prior depolymerisation of the galactomannan by a fungal endo-β-mannanase (McCleary, 1983).

McCleary (1983) has demonstrated the importance of the three enzymes in galactomannan hydrolysis and sugar uptake by guar cotyledons, by incubating washed, endosperm-free embryos with galactomannan alone and in the presence of all three enzymes singly and in combination. The α-galactosidase alone catalysed the complete removal of the galactose, which was rapidly taken up by the embryo. Wnen α-galactosidase was used together with endo-β-mannanase, the galactomannan was converted to galactose plus a series of manno-oligosaccharides ranging from mannobiose to mannopentaose. The galactose was absorbed rapidly by the cotyledons, but the oligosaccharides were taken up slowly and incompletely. Only in the presence of all three enzymes was the galactomannan hydrolysed completely and the breakdown products absorbed completely within the time scale observed *in vivo*.

V. BIOLOGICAL FUNCTIONS OF GALACTOMANNANS

The doctoral dissertation published by Nadelmann in 1890 had the aim of determining if the "mucilages" of leguminous seed endosperms were reserves as well as water-imbibition substances. The origin of the premise that the mucilages were primarily imbibition substances was not given in the dissertation (Nadelmann, 1890), but may well have been de Vries's (1877) study of the germination of red clover. De Vries (1877) expressed the opinion that the clover-seed endosperm does not supply reserve materials but instead functions as a water store during germination. Nadelmann (1890) concluded that the endosperm mucilages of leguminous seeds were first and foremost reserves, and it has largely been taken for granted since then that leguminous seed galactomannans are carbohydrate reserves. The possibility of a dual function has been admitted (Reid, 1971; Dea and Morrison, 1975; Dey, 1978), but multifunctionality has only recently been demonstrated experimentally (Reid and Bewley, 1979).

Reid and Bewley (1979) have compared the germination of and subsequent seedling development from whole fenugreek seeds and from "naked embryos," i.e. seeds from which the endosperm and its attached testa had been removed. They concluded that there is no purely nutritional reason for some 30% of the seed's reserves to take the form of galactomannan rather than protein, lipid, or even starch. There is, in other words, no clear relationship between the structure and complex hydrophilic properties of the fenugreek galacto-mannan and its biological function as a substrate reserve. On the other hand, these hydrophilic properties were shown to form the basis of a mechanism whereby the fenugreek endosperm can imbibe large quantities of water, distribute it around the embryo, and deploy it to protect or "buffer" the embryo against desiccation during transient periods of water stress following imbibition. The endosperm takes up over 60% of the water imbibed by the seed prior to germination (Fig. 14), and this water is effectively distributed all round the embryo (Fig. 15). Thus the embryo, prior to the emergence of the radicle, cannot be challenged directly by the water potential of the seed's environment—only by that of the endosperm. When the fully imbibed seed is subjected to a highly desiccating external environment, it loses water selectively from the endosperm, which possesses the interesting property of being able to lose most of its water with little change in its water potential (Fig. 16). The germinating embryo, which is challenged only by the water potential of the endosperm, does not lose water until the water content of the endosperm has dropped to 100% of the dry weight (Fig. 16). At 52% relative humidity the embryo is cushioned against desiccation for a 6-hr period; under less extreme conditions, the buffering effect of the endosperm lasts even longer (Reid and Bewley, 1979).

The galactomannan of the fenugreek seed endosperm is, then, a multipur-pose macromolecule. It serves initially in water imbibition. Subsequently it acts as a "water buffer," protecting the germinating embryo against transient periods of water-stress. After germination it is mobilised and used as a substrate reserve. It is interesting to note that fenugreek, and the other species of the tribe Trifolieae, are believed to have radiated out from the semi-arid areas of the Middle East (Hegi, 1953). Perhaps the storage of galactomannan in leguminous seeds is an adaptation to semi-arid conditions.

VI. CONCLUSIONS

Although more is known about the galactomannans than about any other group of nonstarch storage polysaccharides in higher plants, many problems remain. There is still insufficient information concerning the fine structures of galactomannans, particularly the distribution of galactose residues along the

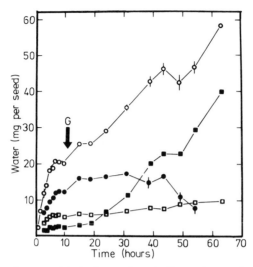

Fig. 14. Pattern of water uptake by fenugreek seeds. Dry seeds were placed on wet cotton at time 0. ○, whole seed; ●, endosperm (+testa); ■, cotyledons; □, axis. G, completion of germination. After Reid and Bewley (1979).

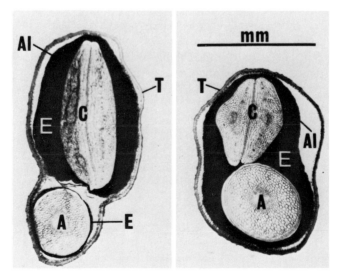

Fig. 15. Sections of fully hydrated fenugreek seeds, periodic acid–Schiff stained. Note that the dark-staining galactomannan completely surrounds the embryo. C, cotyledons; E, endosperm; A, axis; Al, aleurone layer; T, testa. After Reid and Bewley (1979).

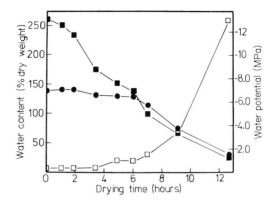

Fig. 16. Water loss at 52% relative humidity from the endosperm testa (■) and from the embryo (○) of the hydrated fenugreek seed in relation to the water potential of the endosperm testa (□, right-hand axis). After Reid and Bewley (1979).

mannan "backbone." A knowledge of fine structure is important to our understanding of the industrially important interactions of galactomannans (Dea and Morrison, 1975). The overall physiology and biochemistry of the postgerminative catabolism of galactomannans are largely understood, but the regulation of the process requires further investigation, particularly in the light of recent information concerning the overall biological function (Reid and Bewley, 1979). Galactomannan biosynthesis is clearly insufficiently understood. There is, above all, a need for continued cooperation between chemists, plant physiologists, and ecologists in the investigation of galacto-mannan structure in relation to biological function.

ACKNOWLEDGEMENT

The author is grateful to the Agricultural and Food Research Council for financial support.

REFERENCES

Andrews, P., Hough, L. and Jones, J. K. W. (1952). *J. Chem. Soc.* pp. 2744–2750.
Bailey, R. W. (1971). *In* "Chemotaxonomy of the Leguminosae" J. B. Harborne, D. Boulter, and B. L. Turner, (eds.), pp. 503–541. Academic Press, New York.
Campbell, J. (1978). Ph.D. Thesis, University of Stirling, Scotland.
Campbell, J. and Reid, J. S. G. (1982). *Planta* **155,** 105–111.
Cerezo, A. S. (1965). *J. Org. Chem.* **30,** 924–927.
Chrisp. J. D. (1967). *Chem. Abstr.* **67,** 92485 U.S. Patent 3301723.

Courtois, J. E. and Le Dizet, P. (1963). *Bull. Soc. Chim. Biol.* **45**, 731–741.
Courtois, J. E. and Le Dizet, P. (1966). *Carbohydr. Res.* **3**, 141–151.
Courtois, J. E. and Le Dizet, P. (1968). *Bull. Soc. Chim. Biol.* **50**, 1695–1710.
Courtois, J. E. and Le Dizet, P. (1970). *Bull. Soc. Chim. Biol.* **52**, 15–22.
Daoud, K. M. (1932). *Biochem. J.* **26**, 255–263.
Dea, I. C. M. and Morrison, A. (1975). *Adv. Carbohydr. Chem. Biochem.* **31**, 241–312.
Dea, I. C. M., Morris, E. R., Rees, D. A., Welsh, E. J., Barnes, H. A. and Price, J. (1977). *Carbohydr. Res.* **57**, 249–272.
de Vries, H. (1877). *Landwirtsch, Jahrb.* **6**, 485–487.
Dey, P. M. (1978). *Adv. Carbohydr. Chem. Biochem.* **35**, 341–376.
Glicksman, M. (1969). "Gum Technology in the Food Industry." Academic Press, New York.
Halmer, P. and Bewley, J. D. (1982). *In* "Encyclopedia of Plant Physiology: New Series" (F. A. Loewus and W. Tanner, eds.), Vol. 13A, pp. 748–793. Springer-Verlag, Berlin, and New York.
Hart, R. (1930) *Ind. Eng. Chem.* **2**, 329–331.
Hegi, G. (1935). "Flora von Mittel-Europa," 2nd ed., Vol. II, Part 3. Lehmann, Munich.
Hoffman, J., Lindberg, B. and Painter, T. (1976). *Acta Chem. Scand Ser. B* **B30**, 365–376.
Hui, P. A. and Neukom, H. (1964). *Tappi* **47**, 39–42.
Kandler, O. and Hopf, H. (1982). *In* "Encyclopedia of Plant Physiology: New Series" (F. A. Loewus and W. Tanner, eds.), Vol. 13A, pp. 348–383. Springer-Verlag, Berlin and New York.
Kapoor, V. P. and Mukherjee, S. (1969). *Can. J. Chem.* **47**, 2883–2887.
Kapoor, V. P. and Mukherjee, S. (1971). *Phytochemistry* **10**, 655–659.
Khanna, S. N. and Gupta, P. C. (1967). *Phytochemistry* **6**, 605–609.
Kooiman, P. (1971). *Carbohydr. Res.* **20**, 329–337.
Kooiman, P. (1972). *Carbohydr. Res.* **25**, 1–9.
Leung, D. M. W., Bewley, J. D. and Reid, J. S. G. (1981). *Planta* **153**, 95–100.
Leung, D. M. W., Reid, J. S. G. and Bewley, J. D. (1984). In preparation.
McCleary, B. V. (1981). *Phytochemistry* **18**, 757–763.
McCleary, B. V. (1982). *Carbohydr. Res.* **101**, 75–92.
McCleary B. V. (1983). *Phytochemistry* **22**, 649–658.
McClendon, J. H., Nolan, W. G. and Wenzler, H. F. (1976). *Am. J. Bot.* **63**, 790–797.
Meier, H. and Reid, J. S. G. (1977). *Planta* **133**, 234–248.
Meier, H. and Reid, J. S. G. (1982). *In* "Encyclopedia of Plant Physiology: New Series" (F. A. Loewus and W. Tanner, eds.), Vol. 13A, pp. 418–471. Springer-Verlag, Berlin and New York.
Morris, E. R., Cutter, A. N., Ross-Murphy, S. B., Rees, D. A. and Price, J. (1981). *Carbohydr. Polym.* **1**, 5–21.
Nadelmann, H. (1980). *Jahrb. Wiss. Bot.* **21**, 1–83.
Nürnberg, E. and Rettig, E. (1974). *Drugs Made Ger.* **17**, 26–31.
Reid, J. S. G. (1971). *Planta* **100**, 131–142.
Reid, J. S. G. (1984). *Adv. Bot. Res.* **11** (in press).
Reid, J. S. G. and Bewley, J. D. (1979). *Planta* **147**, 145–150.
Reid, J. S. G. and Meier, H. (1970a). *Z. Pflanzenphysiol.* **62**, 89–92.
Reid, J. S. G. and Meier, H. (1970b). *Phytochemistry* **9**, 513–520.
Reid, J. S. G. and Meier, H. (1972). *Planta* **106**, 44–60.
Reid, J. S. G. and Meier, H. (1973a). *Caryologia* **25**, *Suppl.*, 219–222.
Reid, J. S. G. and Meier, H. (1973b). *Planta* **112**, 301–308.
Reid, J. S. G., Davies, C., and Meier, H. (1977). *Planta* **133**, 219–222.
Seiler, A. (1977). *Planta* **134**, 209–221.
Smith, F. and Montgomery, R. (1959). "Chemistry of Plant Gums and Mucilages." Van Nostrand-Reinhold, Princeton, New Jersey.
Stepanenko, B. N. (1960). *Bull. Soc. Chim. Biol.* **42**, 1519–1536.

Uebelmann, G. (1978). *Z. Pflanzenphysiol.* **88,** 235–253.

Unrau, A. M. and Choy, Y. M. (1970a). *Carbohydr. Res.* **14,** 151–158.

Unrau, A. M. and Choy, Y. M. (1970b). *Can J. Chem.* **48,** 1123–1128.

Walker, R. E. (1966). *Chem. Abstr.* **64,** 3820e. U.S. Patent 3208524 (1965).

Whistler, R. L. and Smart, C. L. (1953). "Polysaccharide Chemistry." Academic Press, New York.

Yancik, J. J., Schulze, R. E. and Rydlund, P. H. (1972). *Chem. Abstr.* **76,** 101, 828. U.S. Patent 3640784.

Chapter 8

Mannans and Glucomannans

J. Derek Bewley
Plant Physiology Research Group
Department of Biology
University of Calgary
Calgary, Alberta, Canada

J. S. Grant Reid
Department of Biological Sciences
University of Stirling
Stirling, Scotland

I. INTRODUCTION

Although the commercially important galactomannans have been allocated a separate chapter in this volume, it should be borne in mind that they constitute only one member of a trio of plant storage polysaccharides based on the β-$(1\rightarrow4)$-linked β-D-mannosyl structure: the storage polysaccharides of the mannan group (Meier and Reid, 1982). This chapter is concerned with the other two mannan-type storage polysaccharides—the "pure" mannans and the glucomannans. It should preferably be read in conjunction with Chapter 7.

BIOCHEMISTRY OF STORAGE
CARBOHYDRATES IN GREEN PLANTS

II. STRUCTURES AND DISTRIBUTION

A. Mannans and Glucomannans in Seeds

Structurally, the "pure" mannans of seeds are closely related to the galactomannans (see Chapter 7). They comprise a linear "backbone" of $(1 \rightarrow 4)$-linked β-D-mannopyranosyl residues to which is usually attached a small proportion of single-unit $(1 \rightarrow 6)$-linked α-D-galactopyranosyl side chains (compare Fig. 1 with Chapter 7, fig. 1). The seed "mannans" are often defined as those polysaccharides with under 10% galactose substitution of the backbone, while "galactomannans" have 20% substitution and over. Although at first sight arbitrary, these definitions are in fact very useful, since molecular types intermediate between the mannans and galactomannans, as defined, do not appear to exist. Furthermore, the two types are quite distinct in their botanical distribution and in their physical properties. Whereas the galactomannans are highly hydrophilic "seed mucilages," the mannans are water-insoluble crystalline substances that can confer great hardness upon the tissues that contain them (see Section IV).

The seed glucomannans differ from the mannans in that the "backbone" contains $(1 \rightarrow 4)$-linked β-D-glucopyranosyl and D-mannopyranosyl residues (Fig. 2). They are usually lightly galactose-substituted, and in their physical properties they resemble the mannans.

In seeds, mannan-type storage polysaccharides (i.e. mannans, glucomannans, and galactomannans) are restricted to endosperms, as opposed to cotyledons or axes, and they are invariably stored as thickenings of the endosperm cell walls (see Figs. 3 and 6, and Chapter 7, fig. 8). They are all cell-wall storage polysaccharides (Meier and Reid, 1982).

1. Seed Mannans

Mannans have so far been isolated and purified from seeds of species in only three families, one monocotyledonous (the Palmae) and two dicotyledonous (the Umbelliferae and the Rubiaceae). In fact, relatively comprehensive structural studies have been carried out on mannans from the seeds of only

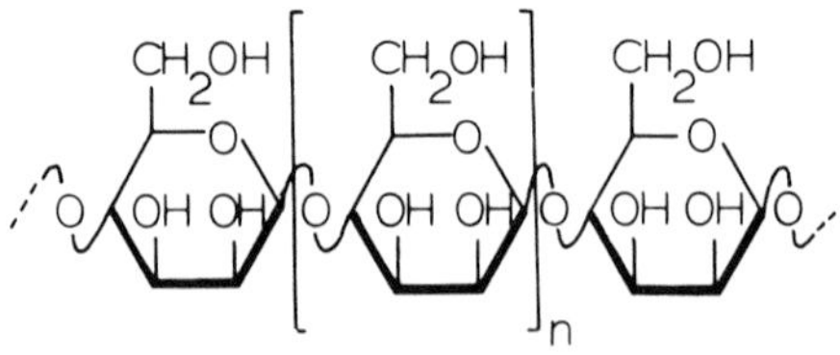

Fig. 1.   General structure of a seed mannan. The linear β-$(1 \rightarrow 4)$-linked chain carries a limited number ($< 10\%$) of D-galactopyranosyl substituents linked α-$(1 \rightarrow 6)$ (see Chapter 7, fig. 1).

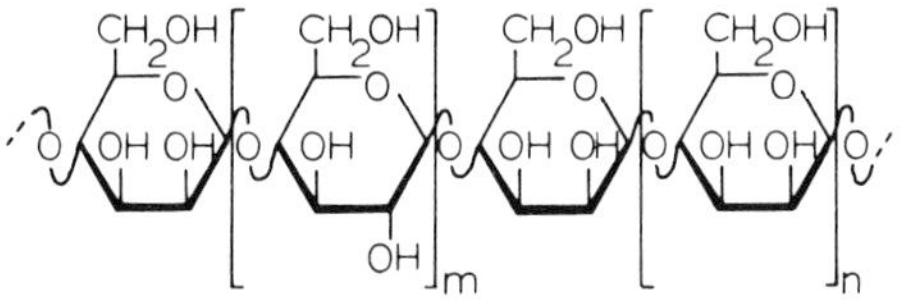

Fig. 2.   General structure of a seed glucomannan. The β-(1→4)-linked chain carries a small number (<10%) of D-galactopyranosyl substituents linked α-(1→6). The distribution of mannosyl and glucosyl units in the chain is uncertain and may be random. The ratio of mannose:glucose is about 1.

four species: *Phoenix dactylifera* (the date palm), *Phytelephas macrocarpa* (the ivory nut palm), *Carum carvi* (caraway, an umbellifer), and *Coffea arabica* (the coffee "bean," Rubiaceae).

Date and ivory nut mannans have been studied over a long period. Both seeds yield two mannan fractions (A and B) differing in their solubilities in aqueous alkali and in cuprammonium solutions (Pringsheim and Seifert, 1922; Lüdtke, 1927). All four are similarly constituted chemically (Klages, 1934a,b; Aspinall *et al.*, 1953, 1958; Meier, 1958), but the more soluble A mannans are of lower molecular weight than B mannans. The number-average degree of polymerisation of date mannan A is 17–21, whereas that of mannan B is about 80 (Meier, 1958). Mannan A from the ivory "nut" released D-mannose (97.1%), D-galactose (1.8%), and D-glucose (0.8%) on complete acid hydrolysis, while mannan B gave D-mannose (98.3%) D-galactose (1.1%), and D-glucose (0.8%). The mannose residues were shown by methylation analysis and partial hydrolysis studies to be linked by β-(1→4) linkages, and the galactose was shown to occupy a terminal nonreducing position. The glucose probably arose from a contaminant (Aspinall *et al.*, 1953, 1958). Although the anomeric configuration of the galactosidic linkage has not been demonstrated in the ivory nut and date mannans, it is probably α, as in the mannan from the endosperm of the palm *Erthyrea edulis* (Robic and Percheron, 1973).

After solvent extraction and sodium chlorite "delignification," the endosperm of the date seed contained 61% mannan A, 31% mannan B, and 8% cellulose (Meier, 1958). Consequently, mannan must constitute the bulk of the massively thickened endosperm cell walls (Fig. 3). Meier (1958) has investigated the structures of date and ivory nut endosperm cell walls by polarised light microscopy, electron microscopy, and X-ray diffraction. He concludes that mannan A is a granular material that is crystalline in the native state and after isolation, while mannan B is microfibrillar, shows birefringence, and is amorphous both before and after isolation. These results suggest that the mannans A and B may be functionally distinct molecular species.

Less comprehensive structural analyses have been carried out on polysaccharides isolated from the seeds of three other palm species: *Erythrea edulis*

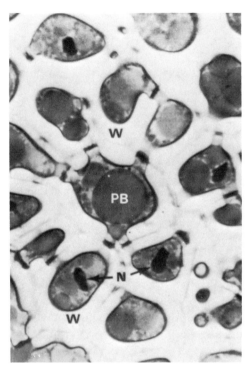

Fig. 3.   Methacrylate-embedded section of the endosperm of *Phoenix dactylifera* stained with toluidine blue. Note the thick cell walls (W). PB, protein body; N, nucleus. From DeMason *et al.* (1983).

(Robic and Percheron, 1973), *Hyphaene thebaica* (El Khadem and Sallam, 1967), and *Cocos nucifera* (Mukherjee and Rao, 1962). These are very similar in structure to the date and ivory nut mannans. Other palm seeds have been shown to release mannose on hydrolysis and to have massively thickened cell walls, and it seems reasonable to assume that the thickened cell walls of palm seed endosperms always contain mannan.

Recently the "reserve cellulose" or cell wall storage polysaccharide in the endosperm of caraway (*Carum carvi*), an umbellifer, has been characterised as a β-(1→4)-linked mannan by comparison of the products formed on total hydrolysis, acetolysis, and enzymatic hydrolysis with those released from date mannan (Hopf and Kandler, 1977). The mannan of the caraway seed is almost totally insoluble in alkali, suggesting that only a single mannan type may be present, resembling date mannan B rather than the alkali-soluble mannan A. Other umbellifers are known to have seeds with thick-walled endosperms (Reiss, 1889). They probably all contain mannans, but further structural studies are necessary before this can be freely assumed.

A β-(1→4)-linked mannan, containing 2% galactose, has been obtained by alkali extraction from green coffee beans (Wolfrom *et al.*, 1961). The bulk of the cell wall, however, was mannose-rich but alkali-insoluble, and its structure was not investigated (Wolfrom and Patin, 1965).

2. Seed Glucomannans

Reasonably comprehensive structural analyses have been carried out on glucomannans obtained by alkali-extraction of seeds of five species belonging to the monocotyledonous families Liliaceae and Iridaceae: *Asparagus officinalis* (Goldberg, 1969; Jakimow-Barras, 1973), *Endymion nutans* (Goldberg, 1969), *Scylla non-scripta* (Thompson and Jones, 1964), *Iris ochroleuca*, and *I. sibirica* (Andrews *et al.*, 1953). All consist of a β-(1→4)-linked backbone of D-mannose and D-glucose residues, carrying a small percentage (3–6%) of single-unit (1→6)-linked D-galactopyranosyl substituents. The mode of the galactosyl linkage has not been demonstrated unequivocally, but it is probably α (Goldberg, 1969). All of the species studied have hard, horny endosperms with thickened cell walls, and there can be no doubt that the glucomannans are localised there. It is known that the seeds of other species from the Liliaceae and Iridaceae store mannose- and glucose-rich polysaccharides (Jakimow-Barras, 1973) or have thick-walled endosperms (Elfert, 1894). The extent of the distribution of glucomannans in seeds is not yet clear. Other seeds contain mannose-rich polysaccharides that cannot yet be classified definitively because insufficient structural information is available. Examples include the storage "mannan" of the lettuce seed (*Latuca sativa*, Compositae) (Halmer *et al.*, 1975), the mannose-rich polysaccharide of *Diospyros kaki* (Ebenaceae) (Ishii, 1895), and *Melampyrum lineare* (Scrophulariaceae) (Curtis and Cantlon, 1966).

B. Mannans and Glucomannans of Bulbs and Tubers

Although galactomannan-type storage polysaccharides appear to be limited to seed endosperms, "pure" mannans and glucomannans have a storage function in the roots, bulbs, and tubers of certain monocotyledonous species in the Lilaceae, Amaryllidaceae, Orchidaceae, and Dioscoraceae.

A "pure" mannan, free of galactose and glucose residues, has been isolated from the tubers of the sweet potato, *Dioscorea batatas* (Misaki *et al.*, 1972). It is predominantly β-(1→4)-linked with a small amount of branching through C-3. Interestingly, the physical properties of this mannan differ greatly from those of the "pure" mannans of seeds. Whereas the seed mannans are highly insoluble, the mannan of *D. batatas* is water-soluble and has been extracted from the plant tissue using water (Misaki *et al.*, 1972). The tuber galactomannan's solubility may be attributed partially to its slightly branched structure and partially to the fact that it is acetylated. It has an acetyl content of 14.2% (Misaki et al., 1972), indicating that on average each D-mannosyl unit carries about 0.65 acetyl groups. It is not known whether the tubers of other species in the Dioscoraceae contain mannans.

Glucomannans have been obtained from the bulbs, tubers, and roots of numerous species in the Liliaceae, Amaryllidaceae, Orchidaceae, and Araceae. All are β-(1→4)-linked, most are slightly branched, and all are water-soluble. They do not carry galactosyl side chains as do the seed glucomannans, but they appear invariably to be acetylated.

In the Lilaceae (Lilium and Eremurus species), the mannose/glucose ratio varies from 1.75 in Lilium maculatum bulb scales (Tomoda and Odaka, 1978) to 3 in Eremurus spectabilis "roots" (Dolvetmuradov, 1970). Acetyl contents when determined are within the range 1.2–5.1%. Branching can be through C-2 of mannose only (Tomoda et al., 1975), C-3 of mannose only (Tomoda and Kaneko, 1976; Kato et al., 1976), C-2 and C-3 of mannose (Tomoda et al., 1976, 1978; Tomoda and Odaka, 1978), or C-2 and C-3 of mannose plus C-3 of glucose (Tomoda and Satoh, 1979).

Glucomannans obtained from bulbs of Narcissus tazetta, Lycoris radiata, and Cooperia pedunculata (Amaryllidaceae) (Kato et al., 1973; Mizuno and Hayashi, 1957; Guess et al., 1960) are similar to the glucomannans of the Liliaceae, except that in the polysaccharide of Lycoris radiata (Higanbanana-mannan) the branching is at C-6 of glucose.

Orchid tubers have long been used in herbal medicine as a source of a mucilaginous drug. Suitable orchid tubers, collectively known as "salep," are also the source of a nutritious drink of the same name (Grieve, 1980). The mucilaginous substance is a glucomannan. That from Orchis morio, the commonest source of "salep," has an acetyl content of 5.3%, a mannose-glucose ratio of 3.3, and a low percentage of branching through C-3 of mannose and glucose (Buchala et al., 1974).

The tubers of several species in the Araceae from the genera Arum and Amorphophallus yield glucomannans. The best studied is that for Amor-phophallus konjac (Konjac-mannan). It has a mannose:glucose ratio of 1.4, is slightly acetylated, and is branched through C-3 of mannose and glucose (Kato and Matsuda, 1973; Shimahara et al., 1975a,b; Kishida et al., 1978; Maekaji, 1978).

The cytology of glucomannan formation and storage in nonseed tissues has been described by Meier and Reid (1982). It is pertinent to note here, however, that the glucomannan is not present in a specific tissue, as in the seeds, but rather in individual specialised cells, or idioblasts, distributed throughout the tissue. The idioblasts appeared to be completely filled with the glucomannan.

III. METABOLISM

A. Biosynthesis

No information is available on the pathways involved in the synthesis of mannans or glucomannans in developing seeds. It has been pointed out by Meier and Reid (1982) that whereas immature seeds of various palm species contain galactomannan, the mature seeds have "pure" mannans as the major storage carbohydrate. Possibly, then, during seed development there is the removal of galactose residues from galactomannan to yield mannans with a low degree of galactose substitution.

Data on mannan and glucomannan synthesis in other organs and tissues of higher plants are equally spartan. Extracts from the shoots of young seedlings of *Phaseolus aureus* (*Vigna radiata*, mung bean) are capable of producing glucomannan *in vitro* when GDP-α-D-mannose and GDP-α-D-glucose are added. In the presence of the former substrate alone, only β-(1→4)-linked mannan is synthesized; in the presence of the latter, β-(1→4)-linked glucan is the product (Villemez, 1971; Heller and Villemez, 1972). For glucosyl-transferase activity to be maintained *in vitro*, there must be the continual production of a mannose-containing acceptor molecule; therefore, this enzyme is dependent on prior or concomitant activity of mannosyltransferase. In contrast, the mannosyltransferase does not require the continual pro-duction of glucose-containing acceptors for its activity. In the presence of GDP-α-D-mannose and GDP-α-D-glucose, the mung bean enzyme extracts preferentially produce a β-(1→4)-linked glucomannan, rather than a β-(1→4)-linked glucan plus a (β-(1→4)-linked mannan. The *in vivo* functions of the mannosyl- and glucosyltransferase remain to be demonstrated, but it is reasonable to expect that they are involved in cell-wall synthesis. Research needs to be initiated to determine if such enzymes exist in developing seeds, tubers, bulbs, etc.

B. Mobilisation

By far the greatest number of studies on the mobilisation of mannan-containing substances have been concerned with galactomannans—most notably those present as cell-wall reserves in the persistent endosperms of legumes (see Chapter 7). Far less information is available on the mobilisation of pure mannans and glucommanans. In fact, only one species of seed storing pure mannan has been studied in any detail—the date (*Phoenix dactylifera*)—and only one species storing glucomannans—asparagus (*Asparagus offici-nalis*). Mobilisation of glucomannans in nonseed tissues has received equally scant attention.

1. Mobilisation of Mannan and Glucomannan in Seeds

a. Date.   That mannan is mobilised from the endosperm of the date seed following germination has been known for a long time (Sachs, 1862). There is, however, only one published report on the morphology and biochemistry of this process—that of Keusch (1968). Mobilisation of the mannans commences when the haustorium, a modified cotyledon, penetrates the endosperm tissue. A dissolution zone is soon observed around the haustorium as the endosperm cell walls are hydrolysed (Figs. 4 and 5), presumably due to the activity of mannanases and β-mannosidase. Both activities have been detected in the endosperm of the germinated date seed. The endo-β-mannanase activity is present almost exclusively in that part of the endosperm nearest to the haustorium, while the β-mannosidase is present throughout the endosperm (D. DeMason, R. E. Sexton and J. S. G. Reid, unpublished data). Keusch's observation that isolated haustoria are capable of degrading a mannan preparation *in vitro* suggests that the haustorium secretes hydrolytic enzymes. Nevertheless, the endosperm cells are living (DeMason *et al.*, 1983) and are capable of enzyme synthesis. Further investigation of the source of mannandegrading activity is clearly necessary. No phosphorolytic activity is present during endosperm hydrolysis. While the mannan in the secondary walls is broken down, the cellulose in the primary cell wall is resistant to hydrolysis and may remain as a thread-like infrastructure.

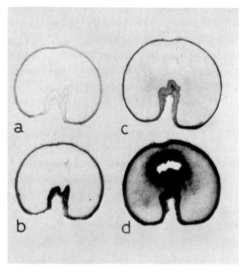

Fig. 4.   Cross section through the endosperm of date seeds in the peripheral regions where there is no mannan hydrolysis (a and b), and in the dissolution zone close to the haustorium, where cell wall breakdown is effected (c and d). From Keusch (1968).

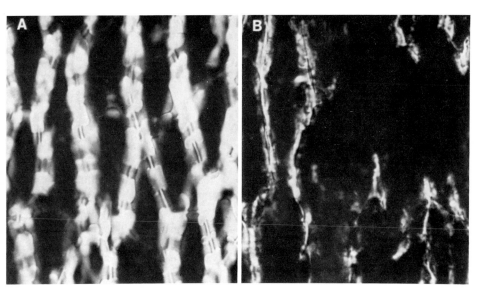

Fig. 5. Cross section through the endosperm of date seeds in the dissolution zone close to the haustorium (B) and the nonhydrolyzed zone away from the haustorium (A). From Keusch (1968).

The major sugar present in the dissolution zone is mannose (Table I); within the haustorium and growing axis of the seedling, however, this sugar does not accumulate. It is probably converted to sucrose immediately upon absorption into the haustorium, and translocated therefrom in this form. The pathway of conversion remains to be determined, but the pathway suggested by Keusch (1968) seems to be reasonable (Fig. 6).

Table I

Percentage of Free Hexoses and Sucrose in Various Parts of the Germinated Data Seed and Seedling during Endosperm Mannan Mobilisation[a]

Tissue	Mannose (%)	Fructose (%)	Glucose (%)	Galactose (%)	Sucrose (%)
Dry embryo	1	1	5	0	93
Endosperm of ungerminated seed	< 1	1	1	< 1	98
Endosperm of imbibed seed	< 1	1	1	< 1	98
Endosperm of germinated seed	1	2	2	< 1	95
Dissolution zone	38	< 1	10	2	50
Haustorium	0	12	13	0	75
Root	0	19	38	0	43
Shoot	0	15	37	0	48

[a]Translated from Keusch (1968).

b. Asparagus. The mode of mobilisation of glucomannans from the endosperm of the asparagus seed resembles, in some ways, that of mannan mobilisation from the date endosperm. In particular, the cotyledon appears to be the site of hydrolytic enzyme production (but see later), and these enzymes are secreted into the surrounding endosperm. Then the hydrolytic products are absorbed back into the cotyledons, and there is no accumulation of mannose therein, suggestive of a rapid conversion to another sugar, probably sucrose. In contrast to the situation in the date seed, the cotyledon does not become haustorial; i.e., it does not penetrate the endosperm *per se*, but simply expands into the space formerly occupied by that tissue.

Following germination, a digestion or dissolution zone appears immediately around the cotyledons (Fig. 7); this gradually spreads throughout the endosperm as the cell walls and other reserves therein are mobilised until, eventually, the endosperm is totally liquified (Goldberg and Roland, 1971). The activity of several enzymes capable of hydrolysing hemicelluloses, including glucomannans, is highest in the digestion zone, especially enzymes with mannosidase-, glucosidase-, and galactosidase-type activity (Table II). In

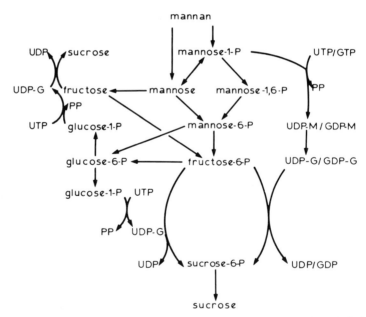

Fig. 6. Proposed scheme for the conversion of mannose, released during the hydrolysis of the endosperm cell walls in date, to sucrose. Since mannose does not accumulate in the haustorium or embryo axes, the conversion probably occurs immediately after absorption. Based on Keusch (1968).

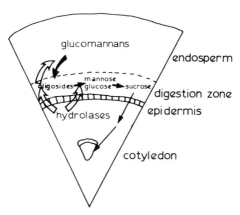

Fig. 7. The dissolution of glucomannans in the endosperm of asparagus, and the absorption of sugars by the cotyledons. After Goldberg and Roland (1971).

that part of the endosperm which is still be hydrolysed, enzyme activity is very low; in contrast, enzyme activity in the cotyledon is quite high. Such an observation is suggestive that the cotyledon is the site of enzyme production. It is possible, though, that because of the juxtaposition of the cotyledon and the dissolution zone, the enzymes are present in the apoplast of the former having diffused there from the endosperm. Hence, it remains to be established unequivocally that the cotyledon is indeed the site of hydrolytic enzyme production. Sucrose synthesis from the hydrolytic products of glucomannan probably occurs within the endosperm itself, and it is in this form that sugar is absorbed into the cotyledon and translocated to the growing axes.

Table II

Enzyme Activity in Various Regions of the Asparagus Seed 16 Days after the Start of Imbibition[a]

	Enzyme activity[b] against				
Region	α-Phenyl glucoside	β-Phenyl glucoside	α-Phenyl mannoside	β-Phenyl mannoside	α-Phenyl galactoside
Endosperm	1.0	1.3	5.5	1.7	3.5
Digestion zone	6.9	3.5	126.2	19.1	29.1
Cotyledon	2.7	1.4	22.9	10.8	29.8

[a]Translated from Goldberg and Roland (1971).

[b]Enzyme activity is expressed on a 100 mg dry weight basis, and is the number of moles of substrate hydrolysed in 20 hr. See Fig. 7 for explanation of the various regions.

2. Mobilisation of Glucomannans in Nonseed Tissues

The tubers of *Orchis* sp. contain a highly viscous water-soluble poly-saccharide, salep mannan (already discussed), which consists of mannose and glucose in a molar ratio of $\sim 3:1$; starch is also present as a storage carbohydrate, but in lesser quantities (Franz, 1979). When growth of the tuber commences in the spring, glucomannan is mobilised first, and starch a considerable time later (Fig. 8a). Accompanying at least the initial breakdown of glucomannans there is an accumulation of free sugars—mannose, mannobiose, glucose, and minor amounts of mannotriose (Fig. 8b). No oligomers containing both glucose and mannose have been detected. Sucrose levels increase and remain high. Presumably this sugar is synthesised by conversions from the hydrolytic products of glucomannan mobilisation, and while some contribution from starch hydrolysis might be expected, this apparently is minor. Sucrose is ultimately transported to the daughter tubers and into the growing vegetative plant.

It is interesting to note that the only glucomannan-degrading enzyme present in the *Orchis* tubers during early hydrolysis of the glucomannan is a mannanase (Fig. 8c). Enzymes capable of mobilising the oligomannans resulting from mannanase activity arise later. Whether or not they are induced or activated in response to prior oligomer production, i.e. by substrate induction, is a matter for speculation.

Glucomannan is also the major reserve polysaccharide in the parent tubers of konjac (*Amorphophallus konjac*) and is mobilised during germination to provide sugars for the newly growing stems, leaves, and young tubers. β-Mannanase activity has been identified as being important (Sugiyama *et al.*, 1973), and, in fact, two mannanases with endoactivity are now known to be involved (Shimahara *et al.*, 1975a,b). Both mannanase I and mannanase II cleave konjac glucomannan to manno-oligosaccharides and glucomanno-oligosaccharides in a random, endo-wise manner, but the randomness of enzyme I, the more active enzyme, is somewhat less than that of enzyme II. Mannosidic bonds linked to glucose residues are attacked slightly more easily than those between mannose residues, in the following order:

$$M\text{-}G > M\text{-}M \ggg G\text{-}M$$

The rate of hydrolysis is greater as the chain length of the substrate increases.

The bulbs of the Easter lily (*Lilium longiflorum*) and of *Ungeria* sp. (*U. sewerzowii* and *U. ferganica*) contain water-soluble polysaccharides, including glucomannan. These are mobilized as the bulb commences growth in the spring (Matsuo and Mizuno, 1974; Sabirova *et al.*, 1977). Glucomannans are found in all bulb scales of the Easter lily, although the outermost ones retain

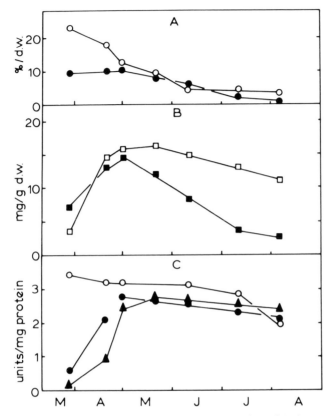

Fig. 8. Hydrolysis of glucomannan and starch in the mother tubers of *Orchis morio* L. during the growing season. (A) Changes in glucomannan (○) and starch (●). (B) Production of sucrose (□) and reducing sugar (■) during polysaccharide reserve mobilisation. (C) Increases in activity of enzymes involved in glucomannan hydrolysis. ●, β-mannanase; ○, β-mannosidase; ▲, β-glucosidase; dw, dry weight. After Franz (1979).

their glucomannan throughout the growing season. In contrast, levels of this polysaccharide decline to zero, with a resultant increase in reducing sugars, including mannose and oligosaccharides (Fig. 9). At least some of the latter could be derived from starch, however, which is present in the bulbs in approximately six times greater amounts and is mobilised along with glucomannans.

Leaves of *Aloe barbardensis* change in their polysaccharide composition with the seasons of the year (Mandal and Das, 1980). While little is known about these changes, the hydrolysis of a glucomannan appears to be one of the events involved.

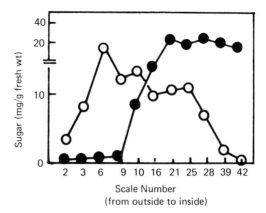

Fig. 9. Polysaccharide and soluble sugar content of the scales of the Easter lily bulb after a low-temperature treatment for 85 days and a growing temperature for 42 days. Note the differences in the distribution between scales of glucomannan (●) and free sugars (○). All scales contained equal amounts of glucomannan and of free sugars at the end of the cold-treatment period. After Matsuo and Mizuno (1974).

IV.   BIOLOGICAL FUNCTIONS

It has been shown that the "storage" galactomannan of the fenugreek seed endosperm is in fact a multipurpose macromolecule with important roles in water-imbibition prior to germination and water-storage during germination, in addition to its storage function following germination (Reid and Bewley, 1979) (see Chapter 7). No experimental evidence is available to suggest that the mannans and glucomannans of seeds are multifunctional. Nevertheless they are stored externally and might be expected to participate in the seed's interaction with its external environment.

Seed mannan and glucomannans are hard, crystalline materials, and they confer hardness on the endosperms which contain them. (Anyone who has accidentally bitten hard on a date seed will appreciate this.) The seed mannans and glucomannans are not hydrophilic, and the seed's hardness is retained after imbibition. Marloth (1883) considered that such endosperms served to protect seeds from "harmful external influences," and it seems reasonable that they would serve as a protection against mechanical damage.

The branched, acetylated mannans and glucomannans of roots, tubers, and bulbs are hydrophilic and may have a function in water storage. It is interesting to note that they are generally stored alongside relatively large quantities of starch, suggesting that their overall function is distinct from that of starch.

V. CONCLUSIONS

At the present time we are faced with insufficient basic information on all aspects of storage mannans and glucomannans. The extent of their distribution is not clear; the regulation of their mobilisation is not understood, and there is a paucity of information available concerning their biosynthesis. Interest in nonstarch storage polysaccharide metabolism has, however, increased in recent years, and it is to be hoped that some of this increased effort will be directed towards the storage mannans and glucomannans.

REFERENCES

Andrews, P., Hough, L. and Jones, J. K. N. (1953). *J. Chem. Soc.* pp. 181–188.

Aspinall, G. O., Hirst, E. L., Percival, E. G. V. and Williamson, I. R. (1953). *J. Chem. Soc.* pp. 3184–3188.

Aspinall, G. O., Rashbrook, R. B. and Kessler, G. (1958). *J. Chem. Soc.* pp. 215–221.

Buchala, A. J. Franz, G. and Meier, H. (1974). *Phytochemistry* **13**, 163–166.

Curtis, E. J. C. and Cantlon, J. E. (1966). *Science* **151**, 580–581.

DeMason, D., Sexton, R. E. and Reid, J. S. G. (1983). *Ann. Bot.* (*London*) [N.S.] **52**, 71–80.

Dovletmuradov, K. (1970). Cited in *Biol Abstr.* **53**, 33637 (1972).

Elfert, T. (1894). *Bibl. Bot.* **30**, 1–25.

El Khadem, H. and Sallam, M. A. E. (1967). *Carbohydr. Res.* **4**, 387–391.

Franz, G. (1979). *Planta Med.* **36**, 68–73.

Goldberg, R. (1969). *Phytochemistry* **8**, 1783–1792.

Goldberg, R. and Roland, J. C. (1971). *Rev. Gen. Bot.* **78**, 75–102.

Grieve, M. (1980). "A Modern Herbal." Penguin Books, London.

Guess, J. W., Hall, N. A. and Rising, L. W. (1960). *J. Am. Pharm. Assoc.* **49**, 102–105.

Halmer, P., Bewley, J. D. and Thorpe, T. A. (1975). *Nature* (*London*) **253**, 716–718.

Heller, J. S. and Villemez, C. L. (1972). *Biochem. J.* **129**, 645–655.

Hopf, H. and Kandler, O. (1977). *Phytochemistry* **16**, 1715–1717.

Ishii, J. (1895). *Landwirtsch. Vers.-Stn.* **45**, 435–436.

Jakimow-Barras, N. (1973). *Phytochemistry* **12**, 1331–1339.

Kato, K. and Matsuda, K. (1973). *Agric. Biol. Chem.* **37**, 2045–2051.

Kato, K., Kawaguchi, Y. and Mizuno, T. (1973). *Carbohydr. Res.* **29**, 469–476.

Kato, K., Yamaguchi, Y., Mutoh, K. and Ueno, Y. (1976). *Agric. Biol. Chem.* **40**, 1393–1398.

Keusch, L. (1968). *Planta* **78**, 321–350.

Kishida, N., Okimasu, S. and Kamata, T. (1978). *Agric. Biol. Chem.* **42**, 1645–1650.

Klages, F. (1934a). *Justus Liebigs Ann. Chem.* **509**, 159–181.

Klages, F. (1934b). *Justus Liebigs Ann. Chem.* **512**, 185–194.

Lüdtke, M. (1927). *Justus Liebigs Ann. Chem.* **456**, 201–224.

Maekaji, K. (1978). *Agric. Biol. Chem.* **42**, 177–178.

Mandal, G. and Das, A. (1980). *Carbohydr. Res.* **86**, 247–257.

Marloth, R. (1883). *Bot. Jahrb. Syst. Pflanzengesch. Pflanzengeogr.* **4**, 225–265.

Matsuo, T. and Mizuno, T. (1974). *Plant Cell Physiol.* **15**, 555–558.

Meier, H. (1958). *Biochim. Biophys. Acta* **28**, 229–240.

Meier, H. and Reid, J. S. G. (1982). *In* "Encyclopedia of Plant Physiology: New Series" (F. A. Loewus and W. Tanner, eds.), Vol. 13A, pp. 418–471. Springer-Verlag, Berlin and New York.

Misaki, A., Ito, T. and Harada, T. (1972). *Agric. Biol. Chem.* **36**, 361–371.

Mizuno, T. and Hayashi, K. (1957). *Nippon Nogei Kagaku Kaishi* **31**, 138–141, *Chem. Abstr* **52**, 18682.

Mukherjee, A. K. and Rao, C. V. N. (1962). *J. Indian Chem. Soc.* **10**, 687–692.

Pringsheim, H. and Seifert, K. (1922). *Hoppe-Seyler's Z. Physiol. Chem.* **123**, 205–212.

Reid, J. S. G. and Bewley, J. D. (1979). *Planta* **147**, 145–150.

Reiss, R. (1889). *Landwirtsch. Jahrb.* **18**, 711–765.

Robic, D. and Percheron, P. (1973). *Phytochemistry* **12**, 1369–1372.

Sabirova, S. K., Rakihimov, D. A., Ismailov, Z. K. and Khamidkhodzhaev, S. A. (1977). *Chem. Nat. Comp. (Engl. Transl.)* **13**, 583–584.

Sachs, J. (1862). *Bot. Ztg.* **20**, 241–246, 250–252.

Shimahara, H., Suzuki, H., Sugiyama, N. and Nisizawa, K. (1975a) *Agric. Biol. Chem.* **39**, 293–299.

Shimahara, H., Suzuki, H., Sugiyama, N. and Nisizawa, K. (1975b). *Agric. Biol. Chem.* **39**, 301–312.

Sugiyama, N., Shimahara, H., Andoh, T. and Takemoto, M. (1973). *Agric. Biol. Chem.* **37**, 9–17.

Thompson, J. L. and Jones, J. K. N. (1964). *Can. J. Chem.* **42**, 1088–1091.

Tomoda, M. and Kaneko, S. (1976). *Chem. Pharm. Bull.* **24**, 2157–2162.

Tomoda, M. and Odaka, C. (1978). *Chem. Pharm. Bull.* **26**, 3373–3377.

Tomoda, M. and Satoh, N. (1979). *Chem. Pharm. Bull.* **27**, 468–474.

Tomoda, M., Kaneko, S. and Nakatsaka, S. (1975). *Chem. Pharm. Bull.* **23**, 430–436.

Tomoda, M., Kaneko, S., Ohmori, C. and Shiozaki, T. (1976). *Chem. Pharm. Bull.* **24**, 2744–2750.

Tomoda, M., Satoh, H. and Ohmori, C. (1978). *Chem. Pharm. Bull.* **26**, 2768–2773.

Villemez, C. L. (1971). *Biochem. J.* **121**, 151–157.

Wolfrom, M. L., Patin, D. L. (1965). *J. Org. Chem.* **30**, 4060–4063.

Wolfrom, M. L., Laver, M. L. and Patin, D. L. (1961). *J. Org. Chem.* **26**, 4533–4535.

Chapter 9

Algal Polysaccharides

ELIZABETH PERCIVAL and RICHARD H. McDOWELL

Department of Chemistry
Royal Holloway College (University of London)
Egham Hill, Egham
Surrey, England

I. GENERAL REMARKS

A. Introduction

At first glance it might seem strange that the diverse forms of plant life making up the algae are all classed as green plants. Some of the green algae found in fresh water could easily be confused with the higher plants growing alongside them, but the red and brown seaweeds and the unicellular organisms, some of which colour hot springs and others that produce sudden "blooms" in fresh or salt water, seem far removed from green plants. The justification for including them is that, with the exception of a few colourless forms, they all contain the photosynthetic pigment chlorophyll *a*.

BIOCHEMISTRY OF STORAGE
CARBOHYDRATES IN GREEN PLANTS

Which of the constituents are storage polysaccharides is a question that at present cannot be given a definite answer. Until much more work has been done on them, the function of many of the polysaccharides of algal cells will remain in doubt. Evidence for the presence of enzymes in the alga that degrade a polysaccharide constituent is one criterion that might be used, but lack of this evidence is no proof that the enzyme is absent, as the extraction of enzymes requires a number of precautions (Marsden *et al.*, 1981) that have not been widely observed in past work.

If the definition of storage products were taken to include only those substances that are broken down to provide energy or building blocks for new growth by respiration or other metabolic processes, only a few algal polysaccharides would be included. On the other hand, most of them, other than the fibrillar cell-wall constituents, may, after a lapse of time, be converted into other materials—even, for example, into extracellular mucilages. Some mention is therefore made in this chapter of most algal polysaccharides, and any known facts about their biosynthesis and degradation are included.

B. Photosynthesis

In addition to chlorophyll *a*, algae contain other photosynthetic pigments (carotenoids and biliproteins, Meeks, 1974; Goodwin, 1974) that aid the collection of light and give the algae their distinctive colours. Using some representatives of different algal divisions (*Codium fragile* and *Ulva lactuca* from Chlorophyta, *Porphyra umbilicalis* and *Chondrus crispus* from Rhodophyta, and *Fucus vesiculosus* from the Phaeophyta), Mishkind and Mauzerall (1980) showed that they all had a common photosynthetic step. The turnover time of the photosynthetic unit (PSU) was 0.5 msec, very close to that of *Chlorella* and higher plants. The size of the PSU is about 2000 chlorophyll units per oxygen molecule evolved in a single flash in *Chlorella* and green benthic algae. It is about half this size in the Rhodophyta and Phaeophyta, the difference being made up by the accessory pigments.

There is now a considerable weight of evidence that the reductive pentose cycle established for *Chlorella* is the principal path of CO_2 fixation in other classes of algae. Kremer and Küppers (1977) carried out short-term photosynthetic experiments on the fixation of $^{14}CO_2$, using a number of red, green, and brown algae. With a period of illumination of 2 to 5 sec, 90% or more of the radioactivity was in phosphate esters, and of this more than 60% was in 3-phosphoglycerate. The proportion of radioactivity in these compounds dropped after 5 sec. In tests for the enzymes that fix the CO_2, ribulose biphosphate carboxylase (RuBP-C) was much the most active in all the algae examined, indicating that the reductive pentose cycle was involved in the greater part of the fixation. In the brown algae there was a high level of phosphoenolpyruvate carboxykinase (PEP-CK), which in the presence of

ADP will fix CO_2 in the dark; the first detectable products are glutamate and aspartate. Much smaller amounts of this enzyme were present in red and green algae. The enzyme phosphoenolpyruvate carboxylase (PEP-C) (responsible for CO_2 fixation in the C4 cycle in higher plants) had negligable activity in any of the algae, contrary to the findings of Kareker and Joshi (1973).

Light-independent CO_2 fixation (LIF) by PEP-CK takes place in the light as well as in the dark. Kremer (1981) suggests that during photosynthesis the substrate PEP is derived from newly synthesised phosphoglycerate but in the dark it is formed by metabolism of mannitol. The products of LIF most commonly found are aspartate, glutamate, and malate: the first product formed is thought to be oxaloacetate, but this is quickly transformed into other substances and has not been detected in algae. The contribution of LIF to the total uptake of CO_2 by brown algae is difficult to quantify, as it depends on the species, part of the plant (Küppers and Kremer, 1978; Kremer, 1980) and the time of year (Küppers and Weidner, 1980). The maximum rate of LIF can be about 10% of photosynthesis in saturated light conditions, so that it forms an important part of carbon fixation in poor light conditions.

The activities of the enzymes RuBP-C and PEP-CK extracted from different parts of plants of *Laminaria* and *Fucus* species and fixation of $^{14}CO_2$ in light and dark by tissue sections were studied by Küppers and Kremer (1978). In the laminarias the maximum activity of RuBP-C and rate of photosynthetic carbon fixation is found in mature parts of the blade, whereas growth takes place at the junction of stipe and blade, products of photosynthesis being translocated to this region. PEP-CK activity and dark fixation are at a maximum in the rapidly growing region, where the enzyme fixes nearly as much CO_2 as does photosynthesis. The activity of this enzyme can be important in the dark days of winter in high latitudes when the growth of laminarias is increasing (Küppers and Weidner, 1980). On the other hand, PEP-CK is less important in the *Fucus* species and RuBP-C and photosynthetic fixation are at a maximum in the growing tips (Küppers and Kremer, 1978).

Although algae do not have the highly developed vascular system found in higher plants, the transport of photosynthate from one part of the plant to another has been demonstrated in a number of species of algae (Schmitz, 1981). The most highly developed system of translocation is in *Macrocystis* species where photosynthetic products are exported from mature blades both to the growing tips and to the base from which new fronds grow (Lobban, 1978).

C. Low Molecular Weight Precursors

In each division of the algae, the carbohydrate metabolism from the pool of sugar phosphates follows a characteristic course. The many low molecular weight sugars, glycosides, polyols, and cyclitols found in algae are listed by

Craigie (1974). Details of such compounds in several algae have been reported by Nagashima and Fukada (1981). Only those that are characteristic of a group of algae or for which metabolic paths are known will be considered in this chapter. Although the amounts of many of these substances are very small, they may be important in metabolism but are rapidly converted into other compounds. This is certainly true of the sugar nucleotides, which are the immediate precursors of polysaccharides and which are considered with their ends products in the appropriate sections.

Sucrose is the main soluble product of photosynthesis in the Chlorophyta (Bidwell, 1958; Percival and Smestad, 1972b; Smestad *et al.*, 1972) and has been found in many other algae, but its occurrence in members of Rhodophyta and Phaeophyta is limited to a few unconfirmed reports.

In the Rhodophyta the most commonly found compound is floridoside (2-*O*-glyceryl-α-D-galactopyranoside), although isofloridoside (1-*O*-glyceryl-α-D-galactoside) is present in larger amounts in some algae (Majak *et al.*, 1966). Isofloridoside has also been found in the Chrysophyte, *Ochromonas malhamensis* (Kauss, 1967). It is considered to be formed by galactosyl transfer from UDP-galactose to glycerol 3-phosphate followed by dephosphorylation (Kauss, 1979). Isofloridoside functions as an osmotic regulator in this alga, which can be grown in fresh and salt water. Nagashima *et al.* (1969) showed that floridoside and trehalose were formed by photosynthesis in the calcareous red alga *Serraticardia maxima*. They appear to be active metabolites.

Algae of the order Ceramiales differ from other red algae in synthesising 2-D-glycerate-α-D-mannopyranoside (digeneaside). When these algae were grown in the presence of $^{14}CO_2$, ^{14}C was found in both the glycerate and mannose components (Kremer, 1978). No floridoside or isofloridoside was found in these algae.

Mannitol is present in all species of Phaeophyceae that have been examined. It shows marked seasonal variations and in the autumn can amount to 25% of the dry weight of some *Laminarias*. It has been found in various other algae, but Kremer (1978) was unable to confirm reports of its presence in red algae. The presence of mannitol in the members of the Prasinophyceae distinguished them from other classes of Chlorophyta. The biosynthesis of mannitol in the brown algae has been extensively studied (Bidwell, 1958; Yamaguchi *et al.*, 1966; 1969; Ikawa *et al.*, 1972; Kremer, 1979). During photosynthesis in the presence of $^{14}CO_2$, the proportion of ^{14}C in this hexitol shows the typical upward curve of accumulated photosynthates, accounting for 30–50% of the radioactivity after 10 min. A corresponding decrease in labelling of sugar phosphates suggests that the mannitol is derived from this pool. Analysis of the intramolecular distribution of ^{14}C in labelled mannitol from *Eisenia bicyclis* also indicates a close relationship with newly photosynthesised hexoses (Yamaguchi *et al.*, 1969). Mannitol-1-phosphate dehydrogenase catalysing the reaction

$$\text{fructose 6-phosphate} \longrightarrow \text{mannitol 1-phosphate}$$

and mannitol-1-phosphatase were first demonstrated in *Spatoglossum pacificum* and *Dictyota dichotoma* by Ikawa *et al.* (1972). Willenbrink and Kremer (1973) showed that mannitol synthesis takes place in the phaeoplasts. These organelles were separated from $^{14}CO_2$-fixing fronds of *Fucus serratus* at different times. After 10 sec photosynthesis, about 80% of the $[^{14}C]$mannitol was in the phaeoplasts, but after 180 sec (when the $[^{14}C]$mannitol had increased 10-fold), most of it was in the rest of the cell.

In addition to mannitol, the seven-carbon polyol volemitol has been found in *Pelvetia canaliculata* (Lindberg and Paju, 1954), and the enzyme catalysing the reduction of sedoheptulose 7-phosphate to this polyol has been demonstrated (Kremer, 1977).

II. NEUTRAL HOMOPOLYSACCHARIDES

A. Glucans

1. (1→4)-α-D-Glucans

a. Natural Occurrence. (1→4)-α-D-Glucans, generally referred to as starches or glycogen, have been shown to be present in species of Cyanophyta, Rhodophyta, Chlorophyta, Cryptophyta, and Dinophyta as particles or granules that stain a purple-brownish or blue colour with iodine. In the Cyanophyta they are very small, the largest size recorded being 31×65 nm (Chao and Bowen, 1971). In the Rhodophyta the granules, which are outside the chloroplasts, vary in size from 1 to 15 μm and have a variety of shapes (Meeuse *et al.*, 1960; Boney, 1975, 1978). Nagashima *et al.* (1971) found that in *Serraticardia maxima* the starch granules were present in inner layers. On the other hand, starch granules of the Chlorophyta are found within the chloroplasts.

In general the amount of starch found in an alga is very small ($< 3\%$ of the dry weight), but larger proportions are sometimes present in spores. Sheath *et al.* (1981) found that 80% of the cell volume of immature seirospores of *Seirospora griffithsiana* was occupied by starch granules of 0.3–1.7 μm diameter. Cultivation of algae under abnormal conditions may also produce an increase in the amount of starch (Antia *et al.*, 1979).

Studies on laboratory cultures of unicellular algae indicate that seasonal variations in starch may be important, although little is known about this in natural conditions. Mangat and Badenhuizen (1971) found that changes in temperature altered the proportions of starch synthesizing enzymes and also the proportions of amylose in *Polytoma uvella*. Sheath *et al.* (1979) found changes in the proportions of these enzymes in *Porphyridium purpureum* during ageing of a batch culture. In those divisions in which starch is present, no essential differences in its structure have been found that can be correlated with a marine or freshwater habitat or indeed with absence of photosynthetic pigment.

    b. Isolation.   In very few cases has it been found possible to obtain sufficient starch free from impurities for examination in the original granular form. Mechanical disruption of the cells is generally a first step, and a mixture of starch and cell debris is obtained from which the starch is separated by solubilisation. Greenwood and Thomson (1961) obtained a small amount of granular starch from *Dilsea edulis*, but for their chemical and physical examination they extracted the starch with hot water under nitrogen. Meeuse *et al.* (1960) obtained granular starch from a number of species, of which *Constantinea subulifera* was the most convenient, but this involved grinding the algae with iodine in potassium iodide solution before differential centrifugation.

    Hot water has been used by many workers to extract the starch, but 30% perchloric acid in the cold (Anderson and King, 1961), 30% chloral hydrate in the cold (Archibald *et al.*, 1960), 0.06 M potassium dihydrogen phosphate at 100°C (Sheath *et al.*, 1979), or dimethyl sulphoxide (Hirst *et al.*, 1972) gave higher yields or purer products in the cases where they were used. These extraction methods give a mixture from which the starch has to be separated. Greenwood and Thomson (1961) used ultracentrifugation to remove galactan. Others have precipitated acidic polysaccharides with quaternary ammonium compounds (Peat *et al.*, 1959a; Ozaki *et al.*, 1967). The starch has been precipitated as the iodine complex in some studies (Anderson and King, 1961; Love *et al.*, 1963). Gradient elution has been used in several cases using a column of DEAE–Sephadex (Turvey and Simpson, 1966) or DEAE–cellulose (Bourne *et al.*, 1974; Carlberg and Percival, 1977).

    c. Constitution and Structure.   i. *General.*   The properties and reactions most commonly studied have been

1. Complete acid or enzymatic (by a mixture of enzymes) hydrolysis and examination of the monosaccharides obtained. Only glucose in almost theoretical yield is to be expected.
2. Optical rotation; a high positive rotation indicating α linkages. The significance of different rotations found for the various starches is not clear.

3. Information on the degree of branching, the linkages involved, and the length of branches has been obtained by:
   a. Methylation and characterisation of the methylated hydrolysis products.
   b. Periodate oxidation.
   c. Partial hydrolysis with acid or with amylolytic enzymes having known action patterns. Bacterial isoamylase, which completely debranches amylopectin and glycogen, is particularly useful (Yokobayashi *et al.*, 1970). The resulting oligosaccharides can be separated by gel permeation and characterised.
   d. Measuring the colour of the complex with iodine, either as the λ_{max} (wavelength of maximum absorption) or the "blue value" (absorption at 680 nm). However, there is no correlation between the degree of branching and the iodine coloration (Archibald *et al.*, 1961).
4. Information on molecular weight is given by viscosity measurements (limiting viscosity number), light scattering, and sedimentation data obtained by ultracentrifugation.

Some of the results and conclusions from these studies are discussed within different sections in this chapter. More details will be found also in Percival and McDowell (1967) and Manners and Sturgeon (1982).

The nomenclature of the $(1\rightarrow4)$-α-D-glucans is based on studies in higher plants and animals and presents some difficulties when applied to algae. The essentially unbranched polymer amylose is sufficiently distinct in its properties to be known by this name in the algae, but the distinction between the branched amylopectin and the more highly branched glycogen is not so clear, and either name has been used by different authors for some of the branched glucans found in algae. Various aspects of the biochemistry of starch, as worked out for land plants, are discussed in more detail in Chapter 4 of this volume.

ii. *Glucans of the Cyanophyta.* The storage polysaccharides of those few blue-green algae that have been examined are $(1\rightarrow4)$-α-D-glucans of the phytoglycogen type. Small iodine-staining bodies had been observed in these algae, but the first to be characterised was that of *Oscillatoria* sp. (Hough *et al.*, 1952). It stained red-brown with iodine and had an average chain length (CL) of 23–26 glucose units, determined by methylation analysis. The glucan from *Oscillatoria princeps* was examined by Fredrick (1953) using periodate oxidation to determine the average CL. The normal alga grown at 25°C had average CL of 19–21, with iodine λ_{max} of 550 nm, but variants appearing in material grown at 5–10°C had CL of 25–29 and λ_{max} of 610 nm.

Chao and Bowen (1971) examined the glucan from *Nostoc muscorum* and found the average CL to be 13 by periodate oxidation. The iodine complex λ_{max} of 410 nm is similar to that of oyster glycogen.

The glucan of *Anacystis nidulans* was debranched by isoamylase, and the resulting mixture of linear maltose oligosaccharides was fractioned by gel permeation (Weber and Wöber, 1975). The average CL was found to be 9 with a spread from 3 to 35, the frequency curve showing a shoulder at 27. (A sample of amylopectin used for comparison had an average CL of 23 with a skew distribution of 7–100 with a second peak at 70.)

iii. *Glucans of the Rhodophyta.*   The $(1\rightarrow4)$-α-D-glucans that have been most studied are those of the red algae, where the granules that stain blue or purple with iodine have been observed for more than a century in many species and have been given the name floridean starch.

Until recently, it was thought that all the red algae produced starches that were entirely branched α-glucans, but now amylose has been found in four marine and three freshwater species of the subclass Porphyridiales and in *Erythrocladia* sp. (Bangiales), a filamentous marine alga (McCracken *et al.*, 1980; McCracken and Cain, 1981). Starch granules and cell debris were obtained from these algae by disrupting the cells and centrifugation, and the starch was brought into solution in water at 100°C. In the case of the Porphyridiales species, the amylose was precipitated by saturating the solution with *n*-butanol. The redissolved amylose gave a blue colour with iodine λ_{max} in the range of 580–620 nm compared with potato amylose at 610–620 nm. It was broken down completely to maltose by β-amylase. The starch from *Erythrocladia* did not give a precipitate with *n*-butanol, but an emulsion layer formed. Retaining the liquid below the emulsion and repeating the treatment gave a solution that showed a blue iodine colour (λ_{max} 580–590) and yielded only maltose on β-amylolysis. Starch solutions from which the amylose had been precipitated gave a brown purple colour with iodine and could not be digested completely with β-amylase, indicating the presence of amylopectin. Sufficient starch to determine the amylose content was obtained only from *Flintiella sanguinaria* Ott (23%) and *Porphyridium aeruginium* Geitley (10%).

Although starch granules have been studied microscopically from very many species of the Rhodophyta, only those from 10 genera have been examined chemically and the absence of amylose has not been demonstrated in all cases. Until many more species have been critically examined, it will not be possible to say how general is the presence of amylose.

A question which has interested many investigators is whether floridean starch should be classed as amylopectin or glycogen. Some of the results that provide evidence on this are given in Table I. Although in general these glucans

Table I

Properties of Some Floridean Starches

Source of starch	$[a]_D$	λ_{max} (nm)	β-Amylolysis (% maltose)	Average chain length	External chain length
Dilsea edulis[a]	173°	530	42	15	—
Dilsea edulis[b]	176°	500	46	9	—
Dilsea edulis[c]	190°	550	49	18.6	!1
Serraticardia maxima[d]		550	60	17	—
Constantinea subulifera[e]	196°	500	50	12	8
Corallina officianalis[f]	171°	500	44	12	7
Rhodymenia pertusa[g]	177°	500	—	12–13	—
Potato amylopectin	195°	530–550	56	20–25	13–16
Glycogen	191°	420–490	40–50	10–14	6–8

[a] Peat *et al.* (1959a).
[b] Fleming *et al.* (1956).
[c] Greenwood and Thomson (1961).
[d] Ozaki *et al.* (1967).
[e] Meeuse *et al.* (1960).
[f] Turvey and Simpson (1966).
[g] Whyte (1971).

are more like amylopectin, the results indicate considerable differences between species. The starch from *Dilsea edulis* has been the most extensively studied. Manners and Wright (1962) found that a purified sample of this starch did not react with concanavalin A, indicating that it is an amylopectin rather than a glycogen. Brunswick and Manners (1981) debranched this starch with isoamylose and found the distribution of the chain length to be nearer to amylopectin than to glycogen.

In ultracentrifuge experiments, Greenwood and Thomson (1961) found that this starch resembled potato amylopectin and differed from glycogen in its variation of sedimentation coefficient with concentration. They also found a weight average molecular weight of 7×10^8 by light scattering and a limiting viscosity number ($[\eta]$, ml/g) of 160, again closer to amylopectin than glycogen.

As well as the presence of amylose in some species, already mentioned, other marked differences between samples should be noted. Peat *et al.* (1959a,b) demonstrated the presence of $(1 \rightarrow 3)$ α links by acid and enzymatic hydrolyses in a sample from *Dilsea edulis*, but $(1 \rightarrow 3)$ links have not been found in other samples. Gelatinisation temperatures have received little attention, but

Meeuse *et al.* (1960) found that starch granules from *Constantinea subulifera* broke down completely only after 6 hr at 100°C, while Greenwood and Thomson (1961) found that *Dilsea edulis* starch granules gelatinised at 45–47°C.

iv. *Glucans of the Chlorophyta.* The division Chlorophyta comprises Chlorophyceae, Prasinophyceae, and Charophyceae. Starch granules have been observed in the chloroplasts of many species, and starches from members of all three classes have been chemically examined. In nearly all cases they have been found to contain both amylose and amylopectin as in higher plants.

Green seaweeds from five genera were examined by Love *et al.* (1963). The amount of starch in these algae was only about 1–2% of dry solids, and much larger quantities of other water-extractable material was present. The starch was separated by iodine precipitation, and amylose was separated from starches of *Enteromorpha compressa*, *Ulva lactuca*, *Cladophora rupestris*, and *Codium fragile*. It was considered that the amylose was destroyed during extraction of the starch from *Chaetomorpha capillaris* and from *Caulerpa filiformis* examined by Mackie and Percival (1960). The amylose contents of the first four seaweed starches were between 16 and 37% (cf. potato, 25%) and had degree of polymerisation (DP) (by viscosity) of about 250–600 (cf. potato, 2220) and λ_{max} of iodine colour at 610–636 nm (640 for potato). The λ_{max} of the six amylopectins ranged from 540 to 565 nm (560 nm for potato). The structures of *Cladophora* and *Enteromorpha* amylopectins, as essentially (1→4)-α-D-glucans with (1→6) branches, were confirmed by methylation analysis, and the average CL was found to be 27.

Other marine filamentous algae that have been examined are *Urospora penicilliformis* (Bourne *et al.*, 1974), *U. wormskoldii*, and another stage in the life cycle of the latter, *Codiolum pusillum* (Carlberg and Percival, 1977). Starches were found in each species, but no fractionation was attempted. The presence of amylose was proved in *U. penicilliformis* and amylopectin in *U. wormskoldii* and *Codiolum*.

Although the fructan in *Acetabularia* is thought to be the main storage polysaccharide (see Section II,B), Werz and Clauss (1970) have shown that starch granules are present in the chloroplasts of *A. mediterranea* and *A. crenulata*.

Although an insoluble (1→4)-α-D-glucan was found to be the major labelled product when $^{14}CO_2$ was incorporated photosynthetically by *Caulerpa simpliciuscula* (Howard *et al.*, 1977) water-soluble D-glucans were also synthesised. These were examined by Hawthorne *et al.* (1979) and found to be (1→3)-β-glucan and (1→4)-α-glucan in the proportions 3:1, and with an approximate molecular weight of 4000 (see Section II,A,2).

Unicellular algae have the advantage that pure strains can be cultured in quantity in the laboratory and have been very widely studied in metabolic experiments, but the starch has not been examined in many of them. Although much of the work on incorporation of CO_2 into carbohydrates was done with *Chlorella vulgaris*, the only detailed study on the starch from this organism has been made in the abnormal condition of heterotrophic growth at different temperatures (Kobayashi *et al.*, 1978). The amylose content remained low at about 7%, although the phosphorus content and swelling power of the granule was changed by the temperature of growth. The degree of branching of the amylopectin was not recorded.

Chlorella pyrenoidosa was examined by Olaitan and Northcote (1962), who found that only about 8% of the alkali-extractable carbohydrate was starch, which contained 7% amylose. The CL of the whole starch found by periodate oxidation was 9, about half that found in most amylopectins. Duynstee and Schmidt (1967) examined a high-temperature strain of *C. pyrenoidosa* and found an amylose content of 30% measured by absorption of the iodine complex at 600 nm.

There has been some controversy about the relation between *Chlorella* and the colourless organism *Prototheca*. Conte and Pore (1973), on the basis of cell-wall constituents and other criteria, concluded that *Prototheca* was closely related to *C. protothecoides* but not to other species of *Chlorella*. Manners *et al.* (1973) examined the water-soluble carbohydrates from *P. zopfii* and found that the major component was a galactan. The glucan present was of the glycogen type having an average CL of 14 glucose residues and with an iodine complex of λ_{max} 485 nm. No amylose was found. On the other hand, Fredrick (1979) found that the α-glucan from *P. zopfii* contained both amylose and amylopectin, the latter having an iodine complex with λ_{max} 580 nm. Details of solution and fractionation of the glucan were not given. Examination of the starch of *C. protothecoides* and reexamination of that from other *Chlorella* species by modern methods would be useful.

In some members of the Volvocales, there is no striking distinction between the starches of photosynthetic and colourless forms. *Dunalliella bioculata* (Eddy *et al.*, 1958), a halophilic species, was grown in double-strength seawater with a good supply of CO_2 and in bright light. The starch contained 12–14% amylose, and the β-amylolysis limit of the amylopectin was 60%. The starch of the related colourless form *Polytomella coeca* (Bourne *et al.*, 1950) grown on an acetate medium contained 13–16% amylose, and the amylopectin had a β-amylolysis limit of 48%.

The starch of the colourless freshwater alga *Polytoma uvella*, also grown with acetate as carbon source (Manners *et al.*, 1965), contained 15% amylose. However, Mangat and Badenhuizen (1970, 1971) found that the amylose content could vary from 5 to 13.5% depending on temperature and the stage of growth. The starch of the related photosynthetic *Chlamydomonas* has not been examined. Another freshwater member of the Volvocales, *Haematococcus pluvialis*, contains starch with 22% amylose, both starch components being similar to those from other green algae (Hirst *et al.*, 1972). *Tetraselmis carteriiformis*, a unicellular photosynthetic marine alga of the class Prasino-phyceae, produced a starch of similar structure with 20–21% amylose when grown in the laboratory (Hirst *et al.*, 1972).

Anderson and King (1961) examined the starch of the filamentous freshwater alga *Nitella translucens* (Characeae) and again found a starch similar to others from green algae. Amylose content was 12% and amylopectin average CL 19.

The present evidence is that α-D-glucans of the Chlorophyta do not differ fundamentally from one another. It is indeed possible that the differences found are more a consequence of the stage of growth at which the samples were collected or, in the case of laboratory-grown samples, the culture conditions, rather than the species from which the starch was extracted.

v. *Glucans from other Algal Divisions.* Very few samples have been examined; these have all been from laboratory cultures. *Chilomonas para-mecium*, a freshwater alga included in the Cryptophyceae, was examined by Archibald *et al.* (1960). The starch had an amylose content of 45%, and the amylopectin component had λ_{max} at 540 nm and average CL of 22. *Chroomo-nas salina*, a marine member of the Cryptophyceae, was grown by Antia *et al.* (1979) in seawater containing 0.25 M glycerol under illumination. Giant cells formed with overproduction of starch, and they burst after 50 days culture. However, 49% of the dry matter was wax. The separated and purified starch gave an iodine complex with λ_{max} at 605 nm and blue value of 0.36, from which an amylose content of 30% was calculated. Methylation analysis gave an average CL of 15 for the whole starch.

The dinoflagellate *Thecadinium inclinatum* (Dinophyceae) was grown in the laboratory by Vogel and Meeuse (1968), and the starch was extracted from the sonicated cells with chloral hydrate solution. Hydrolysis with acid or enzyme gave only glucose, and the $[\alpha]_D$ +173° indicated α linkages. No separation into amylose and amylopectin fractions was attempted, but the λ_{max} of the iodine complex at 620–630 nm suggests a high amylose content. On the other hand, a β-amylolysis limit of 45% indicates that much of the starch must be highly branched.

Prochloron, a green prokaryotic alga that is symbiotic with an ascidian (Lewin, 1976), synthesises a starch of which the iodine complex shows two absorption peaks, at 420 and 610 nm, suggesting a mixture of amylose and glycogen. The two components were separated by electrophoresis on cellulose acetate membrane in borate buffer, after complexing with Procian Red dye. Comparison with glucans from *Nostoc* and *Chlorella* agreed with this conclusion (Fredrick, 1980b).

d. Biosynthesis. Much of the work on the synthesis of glucans in the algae has been to determine the evolutionary relationship of prokaryotic and eukaryotic algae and higher plants (Fredrick, 1981). As the enzymes involved catalyse the same reactions as those in higher plants, only the differences found between the various classes of algae will be considered in this chapter.

Two or perhaps three groups of glucosyltransferases, all of which occur in multiple forms, are involved in the formation of the α-glucans. They are the synthases, the branching enzymes, and perhaps the phosphorylases. Synthases that transfer glucose from a nucleoside diphosphate glucose to a primer chain of $(1\rightarrow4)$-α-D-glucose residues have been detected in all the main divisions of the algae, but few detailed studies of their action have been made. Preiss and Greenberg (1967) purified starch synthase from *Chlorella pyrenoidosa* and found that only ADPG and to a much smaller extent deoxy-ADPG could serve as a substrate. Sanwal and Preiss (1967) found that ADPG pyrophosphorylase from *Chlorella* was activated by 3-phosphoglycerate and inhibited by inorganic phosphate and ADP, showing that the starch synthesis is regulated at the pyrophosphorylase level. Nagashima *et al.* (1971) prepared extracts from the red alga *Serraticardia maxima* and found that the synthetic activity was associated with the starch granules, not with the chloroplasts. ADPG was the most effective doner, but UDPG and GDPG could be utilised less efficiently. Similarly, McCracken and Badenhuizen (1970) found that with a synthase from *Polytoma uvella* starch synthesis was seven to eight times faster with ADPG than with UDPG.

A second group of transferases is the phosphorylases, which, before the discovery of the synthases, were thought to be the enzymes involved in chain building. Although their main function is probably the breakdown of glucans, their synthetic function may also be important. As it has been found that some phosphorylases (a_2), considered to be glycoproteins, initiate glucan synthesis from glucose 1-phosphate without the need for a primer (Fredrick, 1971), this activity may make the presence of a phosphorylase essential for initiating starch synthesis. Fredrick (1975) showed that the non-primer-requiring phosphorylases in *Oscillatoria princeps* can be complexed with concanavalin A, whereas that requiring a primer (a_1), cannot, confirming the glycoprotein nature of the a_2 type. Further evidence for this is that the a_2 type is converted to a_1 by digestion with α-amylase, and that when the original extract from *O. princeps* was incubated with UDPG only the a_2 type of phosphorylase could be separated by electrophoresis on a polyacrylamide gel (Fredrick, 1973b). Kashiwabara *et al.* (1965) found that an extract from the green alga *Ulva pertusa* liberated phosphate from glucose 1-phosphate more rapidly in the presence of starch than without it.

Steup and Melkonian (1981) isolated phosphorylases from the unicellular green alga *Eremospheara viridis* and showed by electrophoresis that they existed in three forms, all of which had both phosphorolytic activity on starch in the presence of phosphate and synthetic activity dependent on glucose 1-phosphate being present. The latter activity did not need a primer, although the rate of synthesis was greater when starch, glycogen, or dextrin was present.

It is in the branching enzymes that the major differences between the different divisions of the algae are found. They can transfer amylose chains to one another by $(1 \rightarrow 6)$ α links; they are of two types, the Q-enzymes, which can convert amylose into amylopectin, and branching enzymes (BE), which will introduce more branches into amylopectin, giving glycogen-like molecules. Q-Enzyme was first found in *Polytomella coeca* by Bebbington *et al.* (1952), who used it to convert amylose to amylopectin. The occurrence of the various branching enzymes in representatives of the different divisions has been most extensively studied by Fredrick (1973a, 1980a, 1981). The general conclusion is that Cyanophyta have branching enzymes only of the BE type, while the Rhodophyta have both BE and Q-enzymes and the Chlorophyta only Q-enzymes.

Fredrick (1973a) has postulated that the first primitive cells to store glucose in a polymeric form contained an enzyme that combined the chain-forming and branching functions and that in the course of evolution these functions were taken over by separate but, closely related enzymes. In support of this theory, Fredrick (1977) found that in *Oscillatoria princeps* the two phosphorylases and the two synthases showed immunological cross reactions with each other, and weak cross reactions with the branching enzymes. From immunological studies, Fredrick (1980a) demonstrated the close affinity of the branching enzymes from different divisions of the algae. The Q-enzyme from *Rhodymenia pertusa* appeared to be more closely related to the BE enzymes of the blue-green *Anacystis nidulans* than to the Q-enzyme of *Chlorella*. However, as the enzyme systems of only a small proportion of the algae have been studied, much more work is required to understand the way in which the different glucans are synthesised.

e. Degradation.   The most energy-efficient breakdown of $(1 \rightarrow 4)$-α-D-glucans to D-glucose is by phosphorylation (see Chapter 3), and phosphorylases appear to be universally present in the algae. No information is available on the regulation of their action that would lead to degradation rather than synthesis. Debranching enzymes will be necessary to degrade the branched glucans. None have yet been found in algae, although several different types have been isolated from higher plants and from bacteria.

α-Amylase is widely distributed in algae and could provide an alternative hydrolytic pathway for glucan degradation.

2. *(1→3)-β-ᴅ-Glucans and Related Polysaccharides*

a. Natural Occurrence. The (1→3)-linked β-ᴅ-glucans occur widely in algae. They are present as laminaran in all the Phaeophyta (Percival and McDowell, 1967; Manners and Sturgeon, 1982) and occasionally in the Chlorophyta (Carlberg and Percival, 1977; Hawthorne *et al.*, 1979), as leucosin, chrysolaminaran, and paramylon in Bacillariophyceae (Chryso-phyta) (Beattie *et al.*, 1961; Ford and Percival, 1965b; Handa, 1969; Allan *et al.*, 1972), in small marine flagellates also belonging to the Chrysophyta (Kreger and Van der Veer, 1970), and in the Eugleenophyta (Clarke and Stone, 1960; Leedale *et al.*, 1965).

Laminaran-containing vesicles have been demonstrated in the gameto-phytes of the Laminariales. They were identified in cells of species of *Alaria, Agarum, Laminaria,* and *Nerocystis* as dense granules 100–200 Å in diameter in the vesicles (Rusanowski and Vadas, 1981). Paramylon is laid down in *Euglena spirogyra* as large oval rings anterior and posterior to the nucleus. It appears as granules with slight indication of helical organisation under the electron microscope (Leedale, 1966).

Fucus serratus, F. spiralis, Laminaria digitata, L. hyperborea, and L. saccharina have been investigated (Küppers and Kremer, 1978) for the distribution of enzymatic CO_2 fixation capacities and of laminaran in different regions of the thallus. In the three *Laminaria* species there are marked differences in the distribution of the enzyme activities, whereas the highest concentration of laminaran is in the actively growing region of the frond in all three species. In contrast, the distribution of enzymatic activity is the same in the two *Fucus* species, and the highest concentration of laminaran is found in the tips of the fronds, which are also the growing regions. A similar study (Kremer, 1980) with the three *Laminaria* species but by transverse profiles as distinct from the longitudinal profiles of the earlier studies reported the highest concentration of laminaran in the meristoderm in the stipe and in the cortex in the blade. It should, however, be remembered that this tissue in the stipe is quantitatively very small. The values for the carbox-ylating enzymes correspond well with the earlier finding from the appro-priate thallus regions.

Seasonal variation in the amount of laminaran occurs in all species of Phaeophyta, but the extent of variation is very dependent on the species, being less in the Fucales than in the Laminariales. In the autumn and winter between 20 and 36% of the dry weight of the frond of *Laminarias* can be laminaran (Black and Dewar, 1949), whereas in the Fucales it may only vary from 2 to 5%. The significance of these variations in the Laminariales is somewhat questionable in view of the following results. In short-term photoassimilation experiments in $^{14}CO_2$ using *Laminaria digitata* and *L. saccharina*, where care was taken to ensure all the natural conditions were observed, it was found that there was little loss (1–3%) of assimilated labelled material (Ferrier *et al.*, 1981). However, in seasonal studies (Johnston *et al.*, 1977), it was found that the high concentration of laminaran synthesised during the summer is to a large extent lost in autumn–winter due to the loss of laminaran-containing tissue from the distal end of the frond into the sea. This laminaran is not therefore available to support growth of the plants in the following spring. Similar results were obtained for *Laminaria longicruris* by Chapman and Craigie (1978).

Handa (1969) showed that the glucan from *Skeletonema costatum* was a storage polysaccharide. This diatom was one of a number cultured by Allan *et al.* (1972) in enriched seawater from which they isolated hot-water-soluble polysaccharides characterised by them as $(1 \rightarrow 3)$-linked glucans similar to that separated from *Phaeodactylum tricornutum* by Ford and Percival (1965b).

Myklestad (1974, 1977) cultured nine different species of marine diatoms and analysed them for, among other things, their acid-soluble glucans. All species tended to accumulate the latter in the stationary growth phase. This accumulation was observed under low dilution rates in continuous culture. The glucan contents for stationary phase cultures of *Skeletonema costatum* and *Chaetoceros curvisetas* were 81 and 25% of the organic dry matter, respectively. A similar glucan from *Chaetoceros affinis* was shown by partial acid hydrolysis studies and comparison with laminaran (Myklestad and Haug, 1972) to be a $(1 \rightarrow 3)$-β-D-glucan. The high glucan content in the stationary phase appeared to coincide with depletion of nitrate from the medium. The $(1 \rightarrow 3)$-β-D-glucan contents of phytoplankton collected from the Trondheim Fjord (Norway) over a period of some months varied from 7.7 to 36.5% of the organic dry matter (Haug *et al.*, 1973).

A glucan consisting of a number of $(1 \rightarrow 4)$- and $(1 \rightarrow 3)$-linked β-glucose units has been isolated from *Monodus subterraneus* (Xanthophyceae) (Beattie and Percival, 1962; Ford and Percival, 1965a).

b. Isolation. Most laminarans of the Phaeophyta can readily be isolated by aqueous extraction of the weed, "soluble" laminaran by cold water and "insoluble" by hot water. Unfortunately, other polysaccharides such as "fucans" and alginates may be present in the aqueous extracts. However, extraction with 2% aqueous calcium chloride prevents a solubilisation of the alginates, and contaminating "fucan" can be separated on DE52–cellulose columns (Jabbar Mian and Percival, 1973a) (see also Section IV,B,1). Chemically the two forms of laminaran are very similar.

$(1\to3)$-β-Laminarans can be isolated from diatoms by cold-water extraction after removal of colouring matter, and can be purified by fractionation on DEAE–cellulose columns (Ford and Percival, 1965b) or by hot-water extraction after removal of an extracellular mucilage (Allan *et al.*, 1972)

Two methods for separation of the paramylon granules from the Euglenaphyta have been reported, from *Euglena gracilis* by Clarke and Stone (1960) and from a number of species by Leedale *et al.*, (1965).

c. Constitution and Structure. All laminarans are polysaccharides of comparatively small molecular size, ranging from 20 to 60 units. The majority of the laminarans of the Phaeophyta consist of linear chains of $(1\to3)$-linked β-D-glucose units. Partial acid hydrolysis studies by Peat *et al.* (1958) on insoluble laminaran from *Laminaria hyperborea* and soluble laminaran from *L. digitata* led to the separation from each of small quantities of mannitol and 1-O-β-glucosylmannitol. At the same time, they also separated gentiobiose and two isomeric trisaccharides, each containing $(1\to3)$ and $(1\to6)$ linkages. The proportion of mannitol and $(1\to6)$ β-D-glucosidic linkages was somewhat higher in the soluble laminaran. From these results it was concluded that a proportion of the molecules were terminated at the reducing end by a molecule of mannitol (M-chains) linked through one of the two primary groups and the rest (G-chains) terminated by reducing glucose residues linked through C-3. Periodate oxidation analyses (Annan *et al.*, 1965a,b) confirmed that the mannitol was monosubstituted and that the $(1\to6)$ linkages were interchain linkages. This latter conclusion had previously been arrived at by Hirst *et al.* (1958) from Barry degradation of laminaran. Similar laminarans have been isolated from *Desmarestia aculeata* (Percival and Young, 1974). It follows that laminaran consists of chains of $(1\to3)$-linked β-D-glucose residues with a low degree of branching by $(1\to6)$ β-D-glucosidic linkages and with a proportion of the chains terminated by mannitol. The mannitol content of different laminarans varied between 2 and 4%. However, the laminaran from *Himanthalia*, *Bifurcaria* and *Padina* appeared to be devoid of mannitol (Jabbar Mian and Percival, 1973a)

From periodate oxidation analysis it was established that soluble laminaran was more highly branched than the insoluble form (Fleming *et al.*, 1966). The former would therefore have a lower degree of linear orientation and intermolecular bonding, with resulting greater solubility. Enzymatic degradation, before and after Smith degradation of laminaran, by an exo-$(1\to3)$-glucanase (Nelson and Lewis, 1974) provides evidence of side chains containing only a single glucose residue. For details of the laminarans of individual species, see Percival and McDowell (1967).

In contrast to the above, laminarans of *Eisenia bicyclis*, a member of the Laminariales, and *Ishige okamurai*, belonging to the Chondariales, both consist of linear chains containing (1→3) and (1→6) linkages in the ratio of (1→6) to (1→3) of 1:3 and 1:4, respectively. Furthermore the laminaran from *Eisenia* appears to be devoid of branch points, and both were devoid of mannitol (Maeda and Nisizawa, 1968). The term "laminaran" should be regarded not as a unique biological entity but as a class of β-glucans.

Although the presence of (1→3)-linked β-D-glucans have been reported in a number of Chlorophyta (Percival and McDowell, 1967), the only one isolated in a pure state and investigated structurally is that from *Caulerpa simpliciuscula* (Hawthorne *et al.*, 1979). It comprises ~30 glucose units, of which 27 are linked by (1→3) β linkages, and there are between one and two (1→6)-linked branches per molecule with a maximum of four glucose units per side chain. Since this β-glucan coexists with soluble starch in the ratio of 3 moles of β-(1→3) form to 1 mole of starch, their metabolic role is not so clear-cut, and these authors suggest that it is a specific reserve for the synthesis of cell-wall material that is laid down during the final stages of wound healing (see Section II,A,1).

The chrysolaminaran metabolised by diatoms closely resembles that of the Phaeophyta in that it has a small proportion of (1→6) interchain linkages, an average of one branch point per molecule, and a DP of ~20 (Beattie *et al.*, 1961; Ford and Percival, 1965b). *Skeletonema costatum* produces a glucan with M_n 13,000. Methylation analysis and periodate oxidation studies (Smestad-Paulsen and Myklestad, 1978) showed the polysaccharide to be a (1→3)-linked β-D-glucan with branches at positions 2 and 6 and an average CL of 11. This is the first time (1→2) linkages have been reported in algal reserve polysaccharides.

The glucan of the Euglenophyta (paramylon) consists of chains of (1→3)-linked β-D-glucose residues (Kreger and Meeuse, 1952; Clarke and Stone, 1960). No evidence of branching or for the presence of any nonglucose units have been found. For a detailed physical study, see Marchessault and Deslandes (1979).

d. Biosynthesis. Few biosynthetic studies on (1→3)-β-D-glucans in algae have been reported (Nisizawa, 1981), except for paramylon in *Euglena gracilis*. Using ^{14}C-labelled nucleotide glucoses with extracts from dark-cultured cells of *Euglena gracilis*, Goldemberg and Marechal (1963) found that a (1→3)-β-glucan is synthesised. Of those tested, UDPG was the most efficient. The synthesis took place with or without the addition of the primer paramylon. The enzymatic transformation of the UDPG to the (1→3)-β-glucan attained 20–30% yields (Marechal and Goldemberg, 1964). The enzyme that carried out the syntheses of the paramylon was a UDPG glucotransferase.

When species of *Fucus*, *Laminaria*, or *Eisenia* were allowed to photosynthesise in the presence of ^{14}C]bicarbonate, large amounts of ^{14}C]mannitol were rapidly produced (Bidwell, 1958; Yamaguchi *et al.*, 1969) and contained as much as 65% of the total radioactivity. Enzymes such as aldolase, hexose diphosphatase, mannitol-1-phosphatase, and glucose phosphate isomerase have been detected in five species of brown algae (Yamaguchi *et al.*, 1969). These could form a possible biosynthetic pathway for mannitol. The ^{14}C from bicarbonate is subsequently incorporated into laminaran and other heteropolysaccharides.

e. Degradation. Degradation is apparently taken over by other enzymes such as $(1 \rightarrow 3)$-β-glucan phosphorylase (Dwyer *et al.*, 1970) and a laminarase, $(1 \rightarrow 3)$-β-glucanase (Dwyer and Smillie, 1970). A similar phosphorylase has been isolated from the golden-brown flagellate *Ochromonas malhamensis* (Chrysophyceae), which will act on laminaran or chrysolaminaran (Kauss and Kriebitzsch, 1969). The enzyme is stimulated by a low concentration of AMP and inhibited by a high concentration (Albrecht and Kauss, 1971). Chrysolaminaran is an assimilatory product during photosynthesis and a respiratory substrate in darkness (Kauss, 1962).

During photoassimilation of $^{14}CO_2$, ^{14}C is rapidly incorporated into mannitol and laminaran in a seawater culture of *Eisenia bicyclis* and there is a rapid decrease in both these carbohydrates in the dark (Yamaguchi *et al.*, 1966). These authors consider that the carbohydrates play the same role in the alga as starch and sucrose in higher plants. As the laminaran is broken down in the dark, fucoidan is synthesised.

Quatrano and Stevens (1976) suggest that laminaran present in the unfertilized eggs of species of *Fucus* is the source of carbon for the cell-wall polysaccharides as well as providing energy. They obtained an enzyme preparation from *Fucus distichus* that is probably an exo-$(1 \rightarrow 3)$ β-glucanase. The only radioactive product released from ^{14}C-labelled laminaran by this enzyme was glucose.

B. Fructans

1. Natural Occurrence

Meeuse (1962) and Craigie (1974) have reviewed the earlier literature on the occurrence of fructans in various algae. Fructans appear to be restricted to members of the Dasycladales and to some other members of the Chlorophyta. In addition to *Acetabularia* sp. (Bourne *et al.*, 1972), three marine species of *Cladophora* and two of *Rhizoclonium* contained low-molecular-weight fructans, but these polymers appeared to be absent from two species of *Chaetomorpha* and a species of *Urospora* (Percival and Young, 1971). No variation in the fructan could be detected between samples collected in March and in September or from localities as far apart as Nova Scotia, the south of England, or northwestern Scotland. These fructans can be detected as spherocrystals, characteristic of inulin and practically indistinguishable from those of *Dahlia variabilis*, in alcohol-treated plants of *Acetabularia mediterranea* (Rubat du Merac, 1953; Meeuse, 1963). They had $[a]_D$ $-44°$ and gave X-ray powder photographs similar to those of inulin.

In *Acetabularia crenulata*, the caps of the cells contained 25 times as much hot-water-soluble fructan as the stalks (Bourne *et al.*, 1972). Winkenbach *et al.* (1972) studied chloroplast preparations from *A. mediterranea* and reached the conclusion that while sucrose may be synthesised in the chloroplasts, subsequent metabolism to monosaccharides and fructan requires the organisation of whole cells and that synthesis of the latter and its storage is associated with the vacuole. In subsequent studies, Shephard and Bidwell (1973) concluded that the bulk of the carbon fixation occurs through the Calvin cycle, together with some β-carboxylation in the chloroplasts, and that in intact cells the carbon flows into fructans that are formed in the cytoplasm. This is supported by the observation of Rubat du Merac (1953) that much of the precipitated fructan appears to be attached to the cytoplasmic layer along the side of the cell wall.

2. Isolation

Rubat du Merac (1953) extracted *Acetabularia mediterranea* with boiling water and precipitated the fructan with alcohol as a white powder. From *Cladophora rupestris*, from two other species of *Cladophora* and two of *Rhizoclonium* (Percival and Young, 1971), and from *Acetabularia crenulata* (Bourne *et al.*, 1972), 80% ethanol extracted an homologous series of fructose-containing oligosaccharides from the trisaccharide upwards. Subsequent treatment with cold water followed by hot water in an atmosphere of carbon dioxide extracted fructans of higher molecular weight from the last weed. Each of the aqueous extracts were separated into neutral fructans and acidic polysaccharides (see Section IV,B,2) on a DEAE–cellulose column from which the fructans were eluted with water. After dialysis and concentration, they were freeze-dried to white powders. A somewhat more complicated method of extraction and purification of the fructan from *Batophora oestedi* was adopted by Meeuse (1963).

3. Constitution and Structure

These algal fructans have $[a]_D$ $\sim -40°$ and on hydrolysis yield 90% fructose and a small proportion of glucose. Methylation analyses of fructans from *Cladophora* (Percival and Young, 1971) and from *Acetabularia* (Bourne *et al.*, 1972) established the presence of linear chains of $(2\rightarrow 1)$-linked fructofuranose units terminated at the reducing end by a molecule of glucopyranose, linked through C-1 to C-2 of fructose—in other words, terminated by a molecule of sucrose. The fructan is therefore of the inulin type. An estimation of the glucose: fructose ratio in the *Acetabularia* fructan gave 1:33 and hence an average CL of 34 for the fructan.

4. Biosynthesis

The metabolic role of these fructans has not been demonstrated in algae, but they are presumed to be photosynthetic reserves. Prolonged light is reported to favour the accumulation of inulin in *Acetabularia* (Vanden Driessche, 1969).

Sucrose is not a storage product in *Acetabularia* (Smestad *et al.*, 1972), as the radioactivity it acquires in culture containing [14C]bicarbonate decreases rapidly after transfer to nonradioactive medium. It is an active metabolite and is converted into other products by various routes. It is most probably the precursor in the synthesis of the homologous series of fructose-containing oligomers found in *A. mediterranea* in the same way that it is utilized in the synthesis of a similar homologous series of fructans in *Helianthus tuberosum* (Edelman and Jefford, 1968). These authors observed that fructofuranose is transferred from one sucrose unit to the fructose moiety of another sucrose unit. One enzyme system, sucrose–sucrose 1-fructosyltransferase, was shown to be responsible for the formation of the trisaccharide, and another, β-(2→1′)-fructan 1-fructosyltransferase, for the formation of higher oligomers. The glucose released from the sucrose residues was believed to be transformed into sucrose again. That a similar series of transfers occurs in *Acetabularia mediterranea* cells is supported by several studies with 14C. The major products of 10-min photosynthesis with $^{14}CO_2$ on *A. mediterranea* were sucrose and glucose (Shepard *et al.*, 1968; Bidwell, 1970). In later experiments, smaller amounts of fructose and water-soluble fructose-containing oligosaccharides were also detected (Winkenbach *et al.*, 1972).

From culture experiments with *A. mediterranea* in medium containing [14C]bicarbonate followed by transfer to nonradioactive medium (Smestad *et al.*, 1972), it was deduced that a pattern of fructan synthesis similar to that of land plants is followed. Fructotriose is transformed into higher homologues. Part of the sucrose is hydrolysed to glucose and fructose, both from the synthesis of the oligofructans and from the hydrolysis of sucrose. The change in radioactivity of the glucose followed almost the same pattern as sucrose, which suggests that the glucose is phosphorylated and transformed into sucrose. The small amount of free glucose found in this alga (Bourne *et al.*, 1972) supports this suggestion. The radioactivity in fructose increased throughout all experiments, indicating that fructose is probably not metabolised further. In support of this, *Acetabularia* was shown to contain large amounts of free fructose (Bourne *et al.*, 1972).

C. Xylans

1. Natural Occurrence

Although xylose has been reported as a constituent of a number of heteropolysaccharides from Phaeophyta, no pure xylans have been separated. In contrast, pure xylans have been isolated from a number of Rhodophyta species and from various species of the Chlorophyta (Percival and McDowell, 1967; Manners and Sturgeon, 1982). It is not clear whether xylans have a reserve or a structural role in the alga. It is the major polysaccharide in *Rhodymenia palmata* (recently changed to *Palmaria palmata*) (Percival and Chanda, 1950) and in *Chaetangium fastigiatum* (Cerezo *et al.*, 1971). The solubility in water of these two xylans indicated that they are probably reserve polysaccharides. However, further extraction of the residual weed (*R. palmata*) with dilute alkali (Turvey and Williams, 1970) gave a second xylan of different structure (to be discussed later), and it may be that xylans fulfil more than a single function in the same alga. That they definitely form part of the cell wall—for example, in *Porphyra umbilicalis* and *Bangia fuscopurpurea*—was shown when Frei and Preston (1964b) mechanically separated xylans from the cell walls and showed they comprised a network of microfibrils. Turvey and Williams (1970) have surveyed the xylans from several species of red algae and concluded that two distinct types of structure exist, one of which is present as a component of the cell wall and another that does not serve as skeletal material.

Although classed as cell-wall xylan by Wurtz and Zetsche (1976), it is worth commenting on the fact that the haploid gametophyte *Halicystis ovalis* produces a xylan while the diploid form of this alga *Derbesia marina* replaces this by a mannan.

2. Isolation

Although xylans have been extracted from algae with water, acid, alkali, and chlorite (for details see references in Percival and McDowell, 1982), only those that are soluble in water are considered to function as reserve polysaccharides.

3. Constitution and Structure

Work on the xylan of *Rhodymenia palmata* (see Percival and McDowell, 1967) and on *Chaetangium* (Cerezo *et al.*, 1971) indicated that they consist of linear chains of $(1 \rightarrow 3)$- and $(1 \rightarrow 4)$-linked β-D-xylose units in the proportion of 1:3 with a small degree of branching. Björndal *et al.* (1965) examined xylans from *R. palmata* extracted with water and then with 0.2 *M* hydrochloric acid and concluded that the only difference between the two extracts was a difference in the proportions of the two types of linkage. Enzymatic degradative studies by these authors indicated a random distribution of the two linkages along the chains.

From a survey of xylans from species of red algae, Turvey and Williams (1970) concluded that xylans containing $(1 \to 3)$ and $(1 \to 4)$ linkages are present as separate xylans or combined in a heteropolysaccharide, neither of which is skeletal material, whereas linear xylans containing either $(1 \to 4)$ or $(1 \to 3)$ linkages are components of the cell wall. The Japanese workers Iriki *et al.* (1960) and Miwa *et al.* (1961) have investigated a number of xylans from species of the Chlorophyta, all of which they consider to be cell wall polysaccharides.

Alkaline extraction of the green seaweed *Caulerpa filiformis* gave a $(1 \to 3)$-β-D-xylan (Mackie and Percival, 1959), but it is unlikely that this fulfils a reserve function. It should be noted that the Chlorophyta xylans are closely associated with glucans of the laminaran type. Frei and Preston (1964a) consider that the glucan forms an encrusting material in which the microfibrils of the xylans are embedded.

4. Biosynthesis

Wurtz and Zetsche (1976) found that the enzymes phosphoglucoisomerase, phosphoglucomutase, and UDPG pyrophosphorylase had a higher activity in *Halicystis*, the form of the life history that synthesises the xylan, than in *Derbesia*, the form that synthesises a mannan. They did not advance any theory as to what part, if any, these enzymes played in the synthesis of xylan. In fact, the mechanism of the biosynthesis of these xylans is unknown. Extracts from various seaweeds have shown hydrolytic activity to *R. palmata* xylan (Manners and Sturgeon, 1982).

D. Mannans

Mannans have never been found in the Phaeophyta, although mannose is often a constituent of a number of heteropolysaccharides, the so-called "fucans" (see Section IV,B,1).

The cuticles of *Porphyra umbilicalis* and *Bangia fuscopurpurea* (Rhodophyta) were shown by Frei and Preston (1964b) to contain a mannan.

$(1 \to 4)$-β-D-Mannans are found in the Chlorophyta (Percival and Mc-Dowell, 1982), but these are considered to be cell-wall polysaccharides. A water-soluble $(1\rightarrow3)$-α-D-mannan is synthesised by *Urospora penicilliformis* (Bourne *et al.*, 1974) and may possibly serve as a storage polysaccharide.

The cell wall of the diatom *Phaeodactylum tricornutum* consists of a branched $(1\rightarrow3)$-β-D-mannan, the branches containing both mannose and glucuronic acid (Ford and Percival, 1965c).

Precipitation with Cetavlon of a sulphated D-mannan in yield greater than 30% of the dry weight of *Nemalion vermiculare* from a hot-water extract by Usov *et al.* (1973) was followed by desulphation with methanolic hydrogen chloride and methylation. Analysis of the derived methylated mannoses showed them to be $(1\rightarrow3)$-linked in the polysaccharide. An $[\alpha]_D$ $+59°$ indicated α linkages. A $(1\rightarrow4)$-linked β-D-mannan branched at C-6 and sulphated at C-2 has been isolated from *Codiolum pusillum* by Carlberg and Percival (1977).

No direct evidence that any of these mannans serve as storage poly-saccharides has so far been obtained.

III. SULPHATED GALACTANS

A. Agar, Carrageen, and Related Polysaccharides

1. Natural Occurrence

The galactans are the major polysaccharides of the Rhodophyta. They occur in the cell walls and in the intercellular matrix, and consist essentially of linear chains of alternating $(1\rightarrow3)$-β-D-galactose and $(1\rightarrow4)$-α-L-galactose in the agars and related polysaccharides and D-galactose in the carrageenans, furcellarans, and similar polysaccharides. Many of these residues are masked by substitution with half ester sulphate and with methoxyl groups and pyruvic acid groups (as the 4,6-O-1-carboxy-ethylidene derivative) and by modifica-tion to the 3,6-anhydro sugar (Fig. 1). The range of galactan sulphates occurring even within a single species is wide and probably consists of a continuous spectrum of structural types (Duckworth and Yaphe, 1971). Extracts from particular genera differ in the proportion of D- and L-galactose and in the degree of substitution and modification of the residues, and it is these differences that determine the overall shape or conformation of the macromolecules and hence the physical properties of the polysaccharides. There are extracts that give stiff gels in dilute solution and others that have no gelling properties. Since these polysaccharides are not considered to be storage polysaccharides, their chemical constitution and structure will not be dealt with further, but there are excellent reviews of their chemistry, properties, and commercial uses (see Rees, 1972; Percival, 1978a; Turvey, 1978; McCandless, 1981).

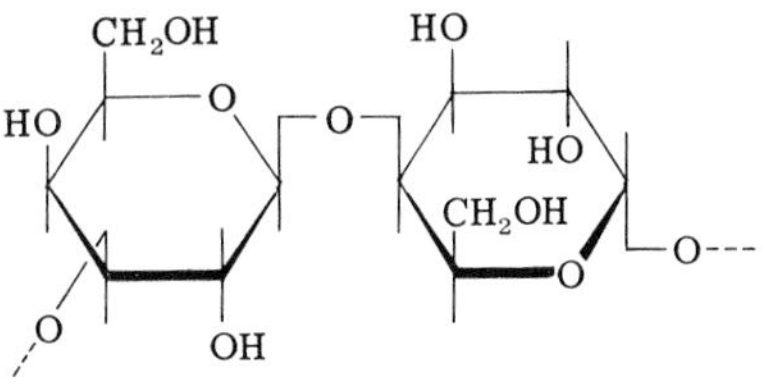

L-Galactose

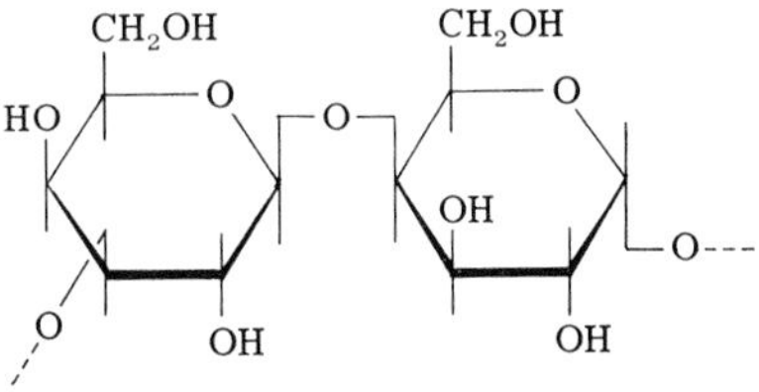

D-Galactose

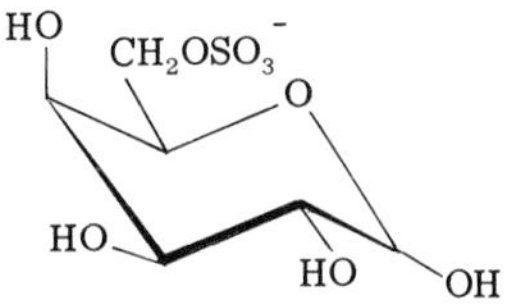

D-Galactose
6-sulphate

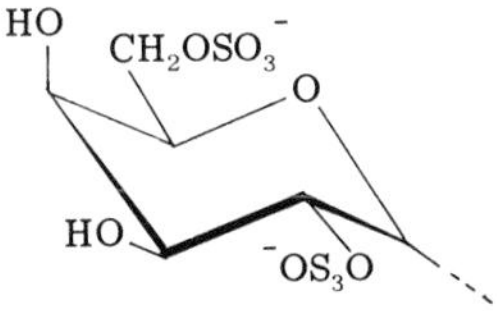

D-Galactose
2,6-disulphate

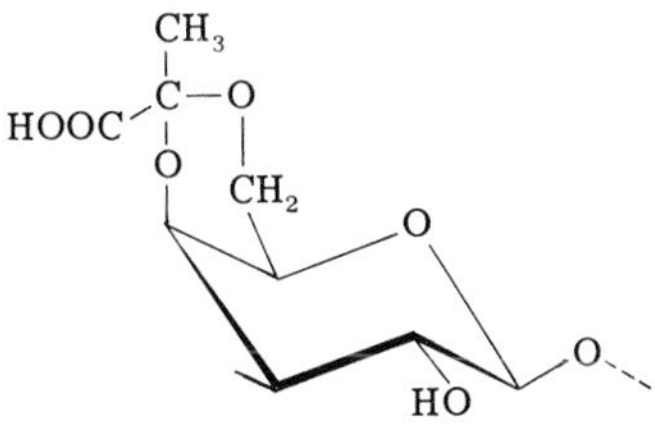

Galactose 4,6- *O*-1-
carboxylethylidene

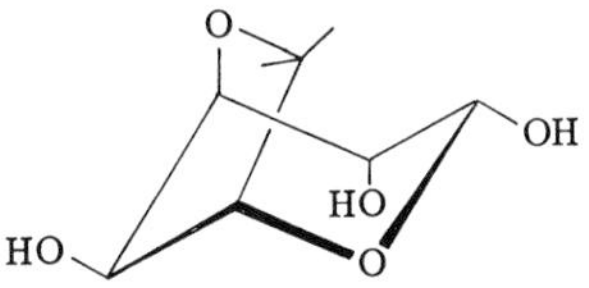

3,6-Anhydro-α-L-galactose

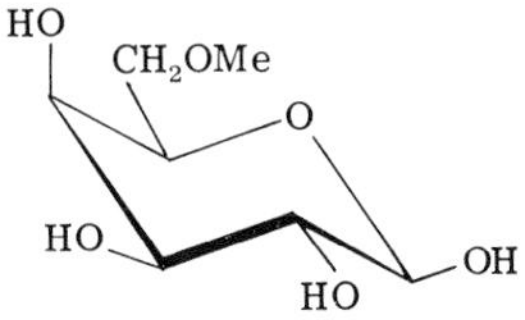

6- *O*-Methyl-D-galactose

Fig. 1. Structural units of sulphated galactans.

2. Biosynthesis

Su and Hassid (1962) found GDP-L-galactose and UDP-D-galactose in *Porphyra*. They considered that the former was probably derived from GDP-D-mannose. Although inositol derivatives act as precursors of sugars in higher plants (Loewus, 1971), this possibility has not been investigated in algae, but the idea is attractive since *O*-methyl ethers of inositols could be the precursors of the methylated galactans. On the other hand, no *O*-methylated sugar nucleotides have been isolated from algae, and it has been suggested that methylation takes place at the polysaccharide level by transfer from *S*-adenosyl-L-methionine (Turvey, 1978). Soluble galactose-6-sulphate alkyltransferases have been separated from *Porphyra* by Rees (1961) and from *Gigartina* by Lawson and Rees (1970). The enzymes from the two sources appear to be different from each other since they have pH

optima at 7.0 and 6.2, respectively. They require a thiol compound for stabilisation. Wong and Craigie (1978) have isolated similar enzymes, the sul-
phohydrolyses. Carbon-14 tracer techniques have shown that the galactans
are not derived from galactitol or glucitol, the major photosynthetic products in *Bostrychia scorpioides*, but are synthesised in parallel with these
compounds (Kremer, 1976).

Little is known with certainty about the pathway of synthesis of these
polysaccharides. It almost certainly occurs in Golgi vesicles, where sulphation
occurs, and is followed by secretion (Peyriere, 1970; Ramus, 1973; Evans *et al.*,
1973).

Craigie and Wong (1979) have put forward likely pathways from biological
and chemical knowledge and identified potential control points in the
biosyntheses of the different carrageenans. These involve sulphotransferase
action on the precursor, followed by sulphohydrolyase to convert highly
sulphated units, where appropriate, to anhydrogalactose.

Lestang-Bremond and Quillet (1981a,b) postulate that there is a turnover of
sulphate in the carrageenan in *Catanella*, the sulphate being carried by methyl
cytosine diphosphate sulphate.

IV. HETEROPOLYSACCHARIDES

A. Nonsulphated

1. Natural Occurrence

Complex water-soluble heteropolysaccharides comprising a number of
monosaccharides have been separated from *Tribonema aequale* (Cleare and
Percival, 1972), a freshwater member of the Xanthophyceae. Heteropolysaccharides have also been separated from another freshwater species of the
Xanthophyceae, *Monodus subterraneus* (Beattie and Percival, 1962). Members
of the Chlorophyta, *Mougeotia* sp. and *Microspora* sp. (Megarry, 1973), and
Chroomonas, a member of the Cryptophyta (Beattie, 1961), synthesise water-
soluble heteropolysaccharides containing a number of monosaccharides.
Polysaccharides of the Cyanophyta appear to be extremely complex. Of those
investigated, *Nostoc* by Hough *et al.* (1952), *Anabeana cylindrica* by Bishop
et al. (1954), and *Phormidium* sp. by Matulewicz *et al.* (1984) all synthesise
polysaccharides composed of at least five monosaccharides and uronic acids.
Microscopic examination of *Anabaena* led to the suggestion that the
polysaccharide may be derived from the mucilaginous envelopes surrounding
the trichomes. A different polysaccharide has been isolated from the envelopes
of heterocysts and spores of *Anabaena cylindrica* (Cardemil and Wolk, 1976).

A complex heteropolysaccharide has been extracted from *Emiliana huxleyi*, a member of cocoolithophorideae (Chrysophyta) (Fichtinger-Schepman *et al.*, 1980).

2. Isolation

The majority of these polysaccharides are isolated, after cold- or hot-water extraction and dialysis, by precipitation with acetone or ethanol. However, the isolation of a colourless polysaccharide from *Nostoc* involved a lengthy procedure of isolation and purification (Hough *et al.*, 1952). The polysaccharide from *Anabaena* was isolated from the cultured organism after removal of noncarbohydrate material by treatment with hot dilute alkali (Bishop *et al.*, 1954). From a cultured sample of *Monodus subterraneous*, hot-water extraction gave a complex polysaccharide containing a number of monosaccharides, while further extraction with aqueous chlorite gave a polysaccharide consisting of glucose and xylose (Beattie and Percival, 1962).

3. Constitution and Structure

Most of these polysaccharides have defied fractionation when subjected to the standard techniques of polysaccharide fractionation. Apart from their constituent monosaccharides, little is known about the structure of many of them. The polysaccharides from *Monodus* consist of glucose, mannose, galactose, xylose, and a trace of fucose from the hot-water extraction and glucose and xylose in a subsequent aqueous chlorite extract (Beattie and Percival, 1962). *Tribonema* polysaccharide consists of glucose, galactose, mannose, rhamnose, and fucose, and traces of arabinose and xylose (Cleare and Percival, 1972). *Mougeotia* polysaccharide contains galactose, glucose, mannose, arabinose, xylose, fucose, rhamnose, and uronic acids, and *Microspora* water-soluble polysaccharide is very similar in its monosaccharide content (Megarry, 1973). Partial hydrolysis of the polysaccharide from *Mougeotia* led to the isolation and characterisation of D-glucuronosyl-$(1\rightarrow3)$-galactose, an aldobiouronic acid that must be a structural feature of this polysaccharide.

Anabaena cylindrica polysaccharide consists of glucose, xylose, glucuronic acid, galactose, rhamnose, and arabinose in a molar ratio of 5:4:4:1:1:1 (Bishop *et al.*, 1954). The polysaccharides of the heterocysts and spore envelopes of *A. cylindrica* each contain glucose, mannose, galactose, and xylose, and the two polysaccharides have identical linkages (Cardemil and Wolk, 1976). They consist of repeating units containing one mannosyl and three glucosyl residues, all linked by $(1\rightarrow3)$-β-glycosidic bonds with glucose, xylose, galactose, and mannose present as side chains. From enzymatic degradative studies, Cardemil and Wolk (1979) advance a subunit structure for the polysaccharide.

Nostoc polysaccharide contains glucose, xylose, galactose, rhamnose, and glucuronic and galacturonic acids (Hough *et al.*, 1952). *Phormidium* sp. water-soluble polysaccharide contains rhamnose, fucose, arabinose, xylose, mannose, galactose, and glucose in the approximate molar proportions of 2.5:2.0:1.0:1.5:2.0:2.0:4.0, together with 9% uronic acid comprising both glucuronic and galacturonic acids. Periodate oxidation, methylation, and partial hydrolysis studies (Matulewicz *et al.*, 1984) provide evidence for a highly branched molecule with (1→3)-linked glucose and (1→4)-linked galactose as the dominant interchain units. After partial hydrolysis, ∼30% of the starting material was recovered as a polymer. It contained 75% carbohydrate, of which 18% was uronic acid and 9.4% was protein. The neutral monosaccharides in a total hydrolysate of this polymer were glucose, mannose, galactose, rhamnose, fucose, arabinose, and xylose in the approx-imate percentages of 45, 30, 10, 5, 3, 2, and 5, respectively. Comparison of these figures with the percentages of the sugars in the initial polysaccharide indicates a preferential loss of rhamnose, arabinose, and xylose on partial hydrolysis, indicating that these sugars are probably on the periphery of the macromolecule.

The polysaccharide from *Emiliana huxleyi* contains D-galacturonic acid, mannose (D and L), L-rhamnose, 6-*O*-methylmannose (D and L), 2,3-di-*O*-methyl-L-rhamnose, 3-*O*-methyl-D-xylose, D-ribose, L-galactose, L-arabinose, D-glucose, and D-xylose. Partial hydrolysis and methylation of the derived oligosaccharides resulted in the characterisation of nine oligosaccharides (Fichtinger-Schepman *et al.*, 1980).

4. Biosynthesis

Cardemil and Wolk (1979) suggest that the subunits detailed by them for the polysaccharides of the envelopes of the heterocysts and spores of *Anabaena cylindrica* are synthesised first and then assembled as prefabricated units in the elongated envelope polysaccharides.

It is impossible to say on the basis of present knowledge what function this wide variety of heteropolysaccharides play in the various algae.

B. Sulphated

Since the function of the complex water-soluble sulphated heteropoly-saccharides of the Phaeophyta and of the Chlorophyta is largely unknown, it is appropriate that they should be included briefly in this chapter.

1. "Fucans"

a. Natural Occurrence. The so-called fucans occur as matrix polysaccharides in all members of the Phaeophyta investigated. They can be found in the literature under various names: fucoidin, fucoidan, ascophyllan, sargassan, pelvetian, glucuronoxylofucan, and fucans. For simplicity, in this chapter, they will all be designated "fucans." Extensive purification in the early studies was carried out in an attempt to isolate a polysaccharide containing only fucose residues. This was never achieved completely, and clearly much of the native polysaccharide was lost in the process of purification. More recent studies have shown (Larsen *et al.*, 1970) that they all comprise a family of polydisperse heteromolecules based on fucose, xylose, glucuronic acid, together in some species with galactose and mannose and in all species with half ester sulphate. Attempts to separate completely protein from these molecules have proved unsuccessful, indicating that they may be proteoglycans. Tentative evidence for linkage to the protein through serine and threonine for the "fucan" from *Ascophyllum nodosum* has been advanced by Medcalf and Larsen (1977). It is suggested by Larsen *et al.* (1970) that in its native state fucan is mostly present as a building element of a much more complex macromolecule and that the isolation of fucan entails chemical degradation as well as physical separation. This helps to explain the conflicting reports concerning the monomeric composition of fucans. When small proportions of algin and protein are removed from "fucan" extracts, the mucilage loses its colloidal properties.

The proportion of fucan in brown seaweeds varies both seasonally and from species to species. It comprises from 18 to 24% of the dry weight of *Pelvetia canaliculata*, but in *Durvillea potatorum* it is never more than 2% (Madgwick and Ralph, 1969). For the fucans isolated from a large number of different species of brown seaweeds, see Percival and McDowell (1982).

Fucans are present in the intercellular tissues and in the mucilage that exudes from the surface of fronds (Doner and Whistler, 1973). Light-microscope histochemistry by Evans and Callow (1974) and electron-microscope X-ray analysis by Callow and Evans (1976) have located the "fucan" in secretory canals, the middle lamella of cell walls, and on the thallus surface.

b. Isolation. A certain proportion of the fucans present in the alga can readily be extracted by cold or hot water, preferably in the presence of Ca^{2+} ions to insolubilise any alginate that might otherwise be extracted. After dialysis, the extract (the fucan) can be isolated as a white powder by freeze-drying. With many brown weeds, to avoid contamination with brown phenolic substances, preliminary treatment with formalin is recommended

(Jabbar Mian and Percival, 1973a). This polymerises the phenolic compounds and renders them insoluble in water. It should be pointed out that the residual algal material still contains fucan that can only be extracted by more drastic methods. Many authors use various concentrations of hydrochloric acid to extract the fucan. Purification and fractionation is often achieved with DEAE–Sephadex (Mori and Nisizawa, 1982).

c. Constitution and Structure.   The fucans are a family of complex, highly branched polysaccharides consisting of fucose, xylose, glucuronic acid together, in some species, with galactose and mannose. Methylation, periodate oxidation, and partial hydrolysis studies have revealed the essential structural similarity of the fucans separated from *Himalthalia lorea, Bifurcaria bifurcata, Fucus vesiculosus,* and *Ascophyllum nodosum,* namely the presence of $(1\rightarrow2)$- and $(1\rightarrow3)$-linked L-fucose residues with sulphate at C-4. The glucuronic acid and xylose units are not sulphated and appear to be on the periphery of the molecules (Percival and McDowell, 1981). Partial hydrolysis of fucans from a number of genera of brown seaweeds revealed D-glucuronosyl-$(1\rightarrow3)$-L-fucose (Percival, 1967; Jabbar Mian and Percival, 1973b) and β-D-xylopyranosyl-$(1\rightarrow3)$-fucose in the fucan from *Ascophyllum nodosum* (Larsen, 1967) as structural features of the macromolecule. The galactose present in the fucan from *Desmarestia aculeata* appears to be present as end-group and $(1\rightarrow3)$-linked units (Percival and Young, 1974). As many as six contiguous galactose units are present in the fucan from *Lessonia nigrescens,* with mainly $(1\rightarrow3)$ and $(1\rightarrow6)$ linkages (Percival *et al.,* 1983; 1984).

In the fucans that contain appreciable proportions of mannose—for example, those from *Sargassum pallidum* (Khomenko *et al.,* 1971), *Pelvetia wrightii* (Solov'eva *et al.,* 1973), *Sargassum linifolium* (Abdel-Fattah *et al.,* 1974a,b), and from *Lessonia nigrescens* (Percival *et al.,* 1984)—the mannose is mainly linked in linear chains of alternating units of mannose and glucuronic acid. A variation that incorporates glucose is reported by Hussein *et al.* (1980) for the water-soluble fucan from *Padina pavonia* in which the inner part of the molecule is composed of $(1\rightarrow4)$-linked β-D-glucuronic acid, $(1\rightarrow4)$-linked β-D-mannose, and $(1\rightarrow4)$-linked β-D-glucose. Also in this polysaccharide are $(1\rightarrow4)$-linked galactose, galactose sulphate. $(1\rightarrow3)$-linked β-D-xylose, and the usual α-$(1\rightarrow2)$-linked L-fucose 4-sulphate units. Methylation of the fucans from *Sargassum pallidum* and *Pelvetia wrightii* (Pavlenko *et al.,* 1974) established the linkages of the monosaccharides and the essential similarity of the two "fucans."

Lestang and Quillet (1981) consider that the core of *Pelvetia* fucan is made up of a helicoidal chain of sulphated fucose units linked through C-1 → C-2, six of them coiling spontaneously without any tension. The sulphate groups point outward in a particularly reactive position, while the oxygen atoms gather all around the axis, making a sort of tube that is able to complex Ca^{2+}.

d. Functions. Lestang and Quillet (1974, 1981a) put forward a number of functions for the fucans; including prevention of poisoning by excess Na^+ ions by means of rapid turnover of sulphate and the bonded Na^+, during which inorganic SO_4^{2-} reesterified the polysaccharide. These authors showed that *Pelvetia canaliculata* can survive 8 days of immersion twice a month. Using pulse–chase experiments with $^{35}SO_4$ on whole living thalluses immersed in seawater Lestang-Bremond and Quillet (1976) demonstrated a rapid turnover of the sulphate groups on the periphery of the globular molecules, the renewal time being approximately 20 min. An enzymatic process continuously breaks the saturated sulphates, which return to the sea. Cytosine–ribose diphospho-sulphate (CDPS) was identified as the activator to provide the energy for sulphation.

A second function is protection against desiccation. Lestang and Quillet (1981) demonstrated the presence of a much higher proportion of magnesium to sodium associated with the fucan of *Pelvetia canaliculata* than found in seawater, where there is considerable excess of sodium over magnesium. The magnesium in solution is always coordinated with six water molecules, and the authors suggest that the coordinated magnesium protects the plant against desiccation and high salt concentrations. This is supported by the very low "fucan" content of weeds that are permanently submerged.

e. Biosynthesis. Autoradiography experiments with $^{35}SO_4$ by Evans *et al.* (1973) showed that in *Pelvetia* the sulphated fucan is synthesised by all cell types, particularly the epidermal cells. However, in *Laminaria* spp. this activity appears to be confined to specialised secretory cells that discharge into mucilage canals. In both genera, sulphation seems to occur in the Golgi-rich perinuclear region (see also Quatrano *et al.*, 1979b). The sulphated material is secreted through the thallus into mucilaginous canals (Evans and Callow 1974, which empty their contents onto the surface.

In an attempt to determine the metabolic pathway of fucans in *Fucus vesiculosus*, the fronds were allowed to assimilate $^{14}CO_2$ for various periods of time; in a second experiment, after 10 min assimilation in $^{14}CO_2$ the fronds were transferred to fresh medium containing $^{12}CO_2$. The "fucans" were extracted sequentially by dilute acid and alkali from the fronds after various periods of time and the radioactivity in each extract was measured (Bidwell *et al.*, 1972). Fractionation of the extracts resulted in the separation of three "fucans" that differed in the relative proportions of their constituent monosaccharides. In addition the supernatant from the acid extracts contained highly labelled oligosaccharides consisting of fucose and galactose.

Measurement of the radioactivity after various periods of time showed that all these materials were active metabolites, and enabled the authors to postulate that the oligosaccharides are first metabolised as a pool of carbohydrate material and are then transferred to the respective polysaccharides.

Various enzymes have been shown to be associated with Golgi-rich fractions from the vegetative thallus of *F. serratus*, including UDP-galactose: L-fucose β-galactosyltransferase (Coughlan and Evans, 1978), which transfers D-galactose from UDP-galactose to both L-fucose and "fucan." It seems likely that the sulphotransferase enzymes also occur in the Golgi cisternae.

Extensive studies on the "fucans" synthesised by the zygotes and young embryos of *Fucus distichus* (= *gardneri*), *F. vesiculosus*, and *F. inflatus* have been reported (Quatrano *et al.*, 1979a). Fucans at different stages of growth of the zygote are described. Similar studies by Callow *et al.* (1978) also report the detection of ^{35}S-APS (adenosine 5'-phosphosulphate) and ^{35}S-PAPS (adenosine 3'-phosphate 5'-phosphosulphate) at an early stage in the zygote growth, which suggests that SO_4^{2-} is activated by these high-energy phosphorylated compounds before polysaccharide sulphation can occur.

2. Heteropolysaccharides of the Chlorophyta

a. Natural Occurrence.   Starch-type or inulin-type polysaccharides have been separated from the extracts of all the Chlorophyta, but in addition to these storage polysaccharides, water-soluble sulphated heteropolysaccharides have been found in all the green seaweeds. Although a proportion of the polysaccharides is readily extracted with cold water, it is impossible to remove them completely from the alga even with hot water or acid. Only after chlorite treatment and alkaline extraction is the residual material free from these polysaccharides. They appear to comprise the matrix material of the cell wall and are present in the plant as a gel. It is possible that they fulfil the same functions as the "fucans" in the Phaeophyta. Haug (1976) found that the glucuronoxylorhamnan in *Ulva lactuca* was only retained in the live fronds if they were immersed in a solution of pH 8 containing the amount of borate and calcium normally found in seawater. Haug postulated that the cis hydroxyl groups on C-2 and C-3 of rhamnose in the polysaccharide complexed with borate ion, which was then stabilised by the Ca^{2+}. In the native polysaccharide, a number of the hydroxyl groups on C-2 of rhamnose are replaced by sulphate groups (Percival and Wold, 1963), which would prevent those particular residues complexing with borate. It seems that by the extent of sulphation the alga can regulate the gelling properties of its polysaccharide. The sulphated arabinoxylogalactan type polysaccharide present in *Cladophora rupestris* only requires Ca^{2+} ions to form a stiff gel in a 1% solution.

Many of these polysaccharides fall into two main groups: glucuronoxylorhamnans with a negative specific rotation, found typically in the Ulvales, and arabinoxylogalactans with a positive rotation, as in the Cladophorales. At the same time, there are other polysaccharides that differ somewhat from either of these groups (Percival and Smestad, 1972a). Classical structural studies have revealed that they are all highly branched heteropolysaccharides. Many of the linkages between the monosaccharides and the more dominant structural features have been determined (see Percival, 1978a).

b. Biosynthesis. In $^{14}CO_2$ studies by Bidwell (1967) and Craigie *et al.* (1966), on *Ulva lactuca*, *Cladophora* sp., *Chaetomorpha melagonium*, *Monostroma fusium*, and *Enteromorpha intestinalis* and members of the Volvcales, both found that sucrose was the most highly labelled low-molecular-weight carbohydrate. Similar studies on *Ulva lactuca* by Percival and Smestad (1972b) confirmed this, but these authors found that glucose, fructose, xylose, and *myo*-inositol also incorporated radioactivity. Both sets of workers found that constituents insoluble in 80% ethanol also became highly radioactive. The sulphated polysaccharide in *U. lactuca* was 24 times more radioactive after 3 hr culture than after 10 min culture. The activity in the individual sugars in the water-soluble sulphated glucuronoxylorhamnan was also measured. From the results, the last authors postulated that a rhamnan is built up first and the glucuronic acid and xylose units are added later. UDP-L-rhamnose has been found in *Chlorella* (Barber and Chang, 1967), and this may be the precursor of the rhamnan. Oligosaccharides containing L-rhamnose have been found in ethanol extracts of *Codiolum pusillum*, and in neutral polysaccharides isolated from aqueous extracts of *Urospora wormskioldii* and *C. pusillum* (Carlberg and Percival, 1977).

V. POLYURONIDES

A. Alginic Acid

1. Natural Occurrence

Alginic acid has been found in all species of the Phaeophyta that have been examined, and recently small amounts have been detected in various species of the Corallinaceae (Okazaki *et al.*, 1982), but none was detected in several non-Corallinacean red algae examined. No alginate has been found in higher plants, but the extracellular mucilage formed by certain bacteria (Linker and Jones, 1964; Gorin and Spencer, 1966) differs from algal alginate only in being acetylated.

The proportion of alginate in brown algae varies widely from one species to another, for example, from 50% of dry weight in some samples of *Durvillea antarctica* (South, 1979) to 13% in *Padina pavonia* (Jabbar Mian and Percival, 1973a). Changes in alginate content with season and location of growth have been reported (Percival and McDowell, 1967). It is normally present as a mixed salt of sodium, calcium, and magnesium.

In view of the changes in alginate content and the presence of alginate degrading enzymes in algae (Madgwick *et al.*, 1973a; Shiraiwa *et al.*, 1975), it is possible that alginate can be, to some extent, a storage product as well as a cell-wall constituent.

2. Isolation

The alginate can be extracted from the algae by ion exchange (more commonly with the free acid as an intermediate) to the sodium salt and dissolution in water. The fact that alginate can be extracted by such relatively mild methods indicates that it is not covalently linked to insoluble cell-wall constituents. There does, however, seem to be a close association with sulphated heteropolysaccharides (Section IV,B,1), which are extracted with it.

3. Constitution and Structure

Alginic acid is a linear polymer of $(1 \rightarrow 4)$-linked β-D-mannuronic acid (M), and α-L-guluronic acid (G) residues (Fig. 2). Isolated material commonly has a DP in the range 1,000–10,000, and presumably has a higher DP than this in plant tissues. Studies over the last two decades have shown that the M and G residues are not arranged in a completely random manner but that there are sequences of G residues (G blocks) and of M residues (M blocks). There are also regions in which the residues are in a nearly random sequence (MG blocks) (Percival and McDowell, 1982).

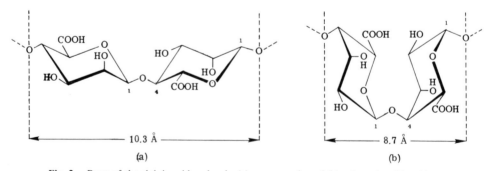

Fig. 2.   Parts of the alginic acid molecule: (a) mannuronic and (b) guluronic acid residues.

The use of enzymes with specific degrading activities (Min *et al.*, 1977; Boyd and Turvey, 1978) together with nuclear magnetic resonance (NMR) (Grasdalen *et al.*, 1979a,b) to characterise the derived oligosaccharides is probably the most promising method of studying the fine structure of alginate molecules, although an obstacle to obtaining a complete picture of the sequences of residues is that all the alginate-cleaving enzymes so far found are eliminases giving fragments with a 4,5-unsaturated uronic acid at their nonreducing ends. Identical unsaturated acids are given by mannuronic and guluronic acids.

Using this approach (Currie, 1983) separated, on Biogel P resin, the products obtained by the action of the guluronolyase from *Klebsiella* on a commercial alginate, and examined the fractions with ^{1}H- and ^{13}C-NMR. They found that many of the oligosaccharides up to hexasaccharides contained G residues, and that those made up entirely of M residues had a range of chain length from 1 to 14 with a main peak at DP 5–6 and a smaller one at DP 10–12. A preliminary examination of some other commerical alginates indicated a similar pattern, but the mannuronic-rich alginate from *Ascophyllum* fruiting bodies gave higher proportions of DP 8–13. Similar studies with a mannuronolyase from an unidentified *Flavobacterium* suggest that the G blocks in these samples have a similar range of sizes. These block sizes are considerably smaller than was thought previously, and more work on these lines using alginates from known species and different algal tissues would be very valuable.

The conformation of the M and G residues (Atkins *et al.*, 1970, 1971) (see Fig. 2) is such that sequences of G residues have a strong affinity for calcium (Haug and Smidsrod, 1967; Smidsrod and Haug, 1968), leading to aggregation of alginate chains (Smidsrod and Haug, 1972; Morris *et al.*, 1973, 1978) and gel formation (Grant *et al.*, 1973).

The significance of the variation in alginate composition in different parts of a large seaweed can be considered in the light of these ideas. The most obvious difference is between the mucilage that can be squeezed out of the fruiting bodies of members of the Fucales and alginates extracted from other algal tissues (Haug *et al.* 1969). The former consists of a viscous solution of alginates with a mannuronic content of over 90%, and can be mixed with seawater without gel formation. All other alginates examined contained at least 28% guluronate, some with up to 70%. Other variations in composition are reviewed by Percival and McDowell (1967, 1982). Although, in the light of the recent work mentioned above, the proportions of the residues in homopolymeric sequences may be somewhat lower than was thought previously, the general conclusions remain that the highest proportion of M sequences are present in fruiting bodies and young tissues, and that the stiffest parts of the plants have most G sequences. The high proportion of G sequences in the holdfasts of *Alaria esculenta* (about 60%) (Stockton *et al.*, 1980a) and in *Laminaria digitata* (about 50%) (Stockton *et al.*, 1980b) suggests that this provides firm adhesion to the substrate.

4. Biosynthesis

The biosynthesis of alginate has recently been reviewed by Larsen (1981). Although knowledge of the pathway is far from complete, a number of possible steps have been elucidated. Lin and Hassid (1966a) isolated the nucleotides GDP-mannuronic acid and GDP-guluronic acid from *Fucus gardneri* (= *distichus*) and carried out a series of experiments on the same alga (Lin and Hassid; 1966b) to elucidate the pathway of formation of these nucleotides and their incorporation into alginate. They first showed that slices of the alga would incorporate radioactivity from both ^{14}C-labelled glucose and ^{14}C-labelled mannose into alginic acid, mannose giving a somewhat higher proportion. They then investigated the effect of a particulate preparation from the alga on a number of reactions and showed that enzymes were present that would catalyse the reactions

$$\text{D-Mannose} + \text{ATP} \xrightarrow{\text{hexokinase}} \text{D-mannose 6-phosphate} + \text{ADP}$$

$$\text{D-Mannose 6-phosphate} \xrightarrow{\text{phosphomannomutase}} \text{D-mannose 1-phosphate}$$

$$\text{D-Mannose 1-phosphate} + \text{GTP} \xrightarrow{\text{guanyltransferase}} \text{GDP-D-mannose} + \text{PP}_i$$

$$\text{GDP-D-mannose} + \text{NaDP}^+ \xrightarrow{\text{dehydrogenase}} \text{GDP-D-mannuronic acid} + \text{NADPH}$$

$$\text{GDP-D-Mannuronic acid} \xrightarrow{\text{transferase}} \text{alginic acid}$$

They also speculated that GDP-D-mannuronic acid was epimerised to GDP-L-guluronic acid, which was also transferred to alginic acid, but they did not detect the necessary enzymatic pathway.

Pinder and Bucke (1975) showed that similar enzymatic reactions took place in the biosynthesis of alginate from sucrose by *Azotobacter vinlandii*. Reaction steps from mannose 6-phosphate onwards to alginic acid were the same as found for *Fucus*, but as no trace of monomeric GDP-L-guluronic acid was found, although the alginate polymer contained guluronic residues, they conclude that the epimerisation took place at the polymer level. The necessary polymannuronate C-5 epimerase for this conversion was discovered in *A. vinlandii* by Haug and Larsen (1971). The mechanism of the reaction was studied by Larsen and Haug (1971) using tritiated water. They found that tritium was incorporated in the guluronic acid formed by the enzymatic action, suggesting that a hydrogen atom adjacent to the carboxyl group was removed and replaced with inversion.

A similar epimerase was found in *Pelvetia canaliculata* (Madgwick *et al.*, 1973b, 1978) and shown to have the same reaction mechanism. Ishikawa and

Nisizawa (1981) have detected polymannuronic acid 5-epimerase in five species of brown algae growing in Japanese waters, the enzyme from *Spatoglossum pacificum* being the most active. In the case of *Eisenia bicyclis* preparations from the freshly growing regions of the frond were much more active than from other parts of the plant. The activity of the enzyme is increased by magnesium and manganese ions. The action of calcium is more complex, as it alters the course of the reaction as well as increasing the rate. Using an extracellular enzyme from *Azotobacter vinlandii* and calcium ion concentrations of 0.85–6.8 mM, Ofstad and Larsen (1981) found that at lower concentrations of calcium the enzyme epimerised preferentially those M residues next to G residues, while at higher concentrations those next to another M residue were attacked. As there was some difference in results with different enzyme preparations, it is suggested that at least two epimerases with different action pattern are present. Purified enzymes are needed to check these conclusions.

There is growing evidence that the M residues are epimerised to G residues only after they are present in the polymer. Hellebust and Haug (1972) carried out experiments on the photosynthetic and dark incorporation of $^{14}CO_2$ into whole plants and selected tissues of *Laminaria digitata* and *L. hyperborea* and found that in the early stages much more radioactivity was present in the M and MG blocks than in G blocks. Similar results were obtained by Abe *et al.* (1973) using *Ishige okamurai*. The increase in the proportion of G residues with advancing age of the tissues of some algae has already been mentioned.

Although several studies have been made on the incorporation of radioactive carbon into alginate during photosynthesis (Yamaguchi *et al.*, 1966; Bidwell, 1967; Hellebust and Haug, 1972; Abe *et al.*, 1973), the course of reactions from the initial products of photosynthesis to the immediate alginate precursors is still not known. Although Lin and Hassid (1966b) showed that mannose and glucose could be used in alginate synthesis, it has not been proved that either of these sugars is present in the plant in natural conditions. Mannitol is the most abundant low-molecular-weight compound in brown algae and, although mannitol-1-phosphate dehydrogenase is present in them (Ikawa *et al.*, 1972; Kremer, 1977), there is considerable doubt about its role as an alginate precursor (Bidwell, 1967; Hellebust and Haug, 1972). The latter authors suggest that amino acids could be precursors of alginate, as they accounted for more radioactivity than mannitol after photosynthesis in the presence of $^{14}CO_2$ and lost it more rapidly during subsequent growth periods in $^{12}CO_2$.

On the other hand, Quatrano and Stevens (1976) produced strong evidence that laminaran was the substrate for cellulose in the cell walls of developing *Fucus* zygotes and suggested that it was also the precursor of the alginate in the wall. From the varying results obtained by others using different seaweeds, it is possible that the contribution of one substance or another to alginate synthesis could vary from species to species.

5. Degradation

Alginate-degrading enzymes have been found in many bacteria and marine animals (Percival and McDowell, 1967). The use of some of these alginate lyases in the determination of alginate structure has been discussed above. Madgwick *et al.* (1973a) showed that an extract from *Laminaria digitata* stipes reduced the viscosity of an alginate solution and produced products with unsaturated end groups, characteristic of the action of an alginate lyase. Shiraiwa *et al.* (1975) obtained crude enzyme preparations from eight seaweeds from Japanese waters and demonstrated lyase activity in all of them. More than one enzyme seems to be involved, as the relative attack on M-rich and G-rich alginate was different from species to species. There was more activity in old than in young fronds. The authors suggest that the alginate may be metabolised to provide energy for developing reproductive bodies.

No information is available on the breakdown of alginates beyond the stage of monomeric unsaturated uronic acids by algal enzymes, but Preiss and Ashwell (1962) demonstrated that a *Pseudomonas* could break these down to pyruvate and triose phosphate.

REFERENCES

Abdel-Fattah, A. F., Hussein, M. M.-D. and Salem, H. M. (1974a). *Carbohydr. Res.* **33**, 9–17.
Abdel-Fattah, A. F., Hussein, M. M.-D. and Salem, H. M. (1974b). *Carbohydr. Res.* **33**, 1924.
Abe, K., Sakamoto, T., Saski, S. F. and Nisizawa, K. (1973). *Bot. Mar.* **16**, 229–234.
Albrecht, G. I. and Kauss, H. (1971). *Phytochemistry* **10**, 1293–1298.
Allan, G. G., Lewin, J. and Johnson, P. G. (1972). *Bot. Mar.* **15**, 102–108.
Anderson, D. M. W. and King, N. J. (1961). *J. Chem. Soc.* pp. 2914–2919.
Annan, W. D., Hirst, E. L. and Manners, D. J. (1965a). *J. Chem. Soc.* pp. 220–226.
Annan, W. D., Hirst, E. L. and Manners, D. J. (1965b). *J. Chem. Soc.* pp. 885–891.
Antia, N. J., Cheng, J. Y., Foyle, R. A. J. and Percival, E. (1979). *J. Phycol.* **15**, 57–62.
Archibald, A. R., Hirst, E. L., Manners, D. J. and Ryley, J. F. (1960). *J. Chem. Soc.* pp. 556–560.
Archibald, A. R., Fleming, I. D., Liddle, A. M., Manners, D. J., Mercer, G. and Wright, A. (1961).
     J. Chem. Soc. pp. 1183–1190.
Atkins, E. D. T., Mackie, W. and Smolko, E. E. (1970). *Nature (London)* **225**, 626–628.
Atkins, E. D. T., Mackie, W., Parker, K. D. and Smolko, E. E. (1971). *Polym. Lett.* **9**, 311–316.
Barber, G. A. and Chang, M. T. Y. (1967). *Arch. Biochem. Biophys.* **118**, 659–663.
Beattie, A. C. (1961). Ph. D. Thesis, University of Edinburgh.
Beattie, A. and Percival, E. (1962). *Proc. R. Soc. Edinburgh* **60**, 171–185.
Beattie, A., Hirst, E. L. and Percival, E. (1961). *Biochem. J.* **79**, 531–537.
Bebbington, A., Bourne, E. J., Stacey, M. and Wilkinson, I. A. (1952). *J. Chem. Soc.* pp. 240–245.
Bidwell, R. G. S. (1958). *Can. J. Bot.* **36**, 337–349.

Bidwell, R. G. S. (1967). *Can. J. Bot.* **45**, 1557–1565.

Bidwell, R. G. S. (1970). *Plant Physiol* **45**, 70–75.

Bidwell, R. G. S., Percival, E. and Smestad, B. (1972). *Can. J. Bot.* **50**, 191–197.

Bishop, C. T., Adams, G. A. and Hughes, F. O. (1954). *Can. J. Chem.* **32**, 999–1004.

Björndal, H., Eriksson, K. E., Garegg, P. J., Lindberg, B. and Swan, B. B. (1965). *Acta Chem. Scand.* **19**, 2309–2315.

Black, W. A. P. and Dewar, E. T. (1949). *J. Mar. Biol. Assoc. U. K.* **28**, 673–699.

Boney, A. D. (1975). *Nova Hedwigia* **26**, 269–273.

Boney, A. D. (1978). *Nova Hedwigia* **29**, 565–567.

Bourne, E. J., Stacey, M. and Wilkinson, I. A. (1950). *J. Chem. Soc.* pp. 2694–2698.

Bourne, E. J., Percival, E. and Smestad, B. (1972). *Carbohydr. Res.* **22**, 75–82.

Bourne, E. J., Megarry, M. L. and Percival, E. (1974). *J. Carbohydr., Nucleosides Nucleotides* **1**, 235–264.

Boyd, J. and Turvey, J. R. (1978). *Carbohydr. Res.* **66**, 187–194.

Brunswick, P. and Manners, D. J. (1981). *Biochem. Soc. Trans.* **9**, 290P.

Callow, M. E. and Evans, L. V. (1976). *Planta* **131**, 155–157.

Callow, M. E., Coughlan, S. J. and Evans, L. V. (1978). *J. Cell Sci.* **32**, 337–356.

Cardemil, L. and Wolk, C. P. (1976). *J. Biol. Chem.* **251**, 2967–2975.

Cardemil, L. and Wolk, C. P. (1979). *J. Biol. Chem.* **254**, 736–741.

Carlberg, G. E. and Percival, E. (1977). *Carbohydr. Res.* **57**, 223–234.

Cerezo, A. S., Lezerovich, R., Labriola, R. and Rees, D. A. (1971). *Carbohydr. Res.* **19**, 289–296.

Chao, L. and Bowen, C. C. (1971). *J. Bacteriol.* **105**, 331–338.

Chapman, A. R. O. and Craigie, J. S. (1978). *Mar. Biol.* **46**, 209–213.

Clarke, A. E. and Stone, B. A. (1960). *Biochim. Biophys. Acta* **44**, 161–163.

Cleare, M. and Percival, E. (1972). *Br. Phycol. J.* **7**, 185–193.

Conte, M. V. and Pore, R. S. (1973). *Arch. Mikrobiol.* **92**, 227–233.

Coughlan, S. J. and Evans, L. V. (1978). *J. Exp. Bot.* **29**, 55–68.

Craigie, J. S. (1974). *In* "Algal Physiology and Biochemistry" (W. D. P. Stewart, ed.), pp. 206–235. Blackwell, Oxford.

Craigie, J. S. and Wong, K. F. (1979). *Proc. Int. Seaweed Symp., 9th, 1977* pp. 369–377.

Craigie, J. S., McLachlan, J., Majak, W., Ackman, R. G. and Tocher, C. S. (1966). *Can. J. Bot.* **44**, 1247–1254.

Currie, A. J. (1983). Ph. D. Thesis, University of Wales.

Doner, L. W. and Whistler, R. L. (1973). *In* "Industrial Gums" (R. L. Whistler and J. Bemiller, eds.) 2nd ed., pp. 115–121. Academic Press, New York.

Duckworth, N. and Yaphe, W. (1971). *Carbohydr. Res.* **16**, 189–197.

Duynstee, E. E. and Schmidt, R. R. (1967). *Arch. Biochem. Biophys.* **119**, 382–386.

Dwyer, M. R. and Smillie, R. M. (1970). *Biochim. Biophys. Acta* **216**, 392–401.

Dwyer, M. R., Smydzuk, J. and Smillie, R. M. (1970). *Aust. J. Biol. Sci.* **23**, 1005–1013.

Eddy, B. P., Fleming, I. D. and Manners, D. J. (1958). *J. Chem. Soc.* pp. 2827–2830.

Edelman, J. and Jefford, T. G. (1968). *New Phytol.* **67**, 517–531.

Evans, L. V. and Callow, M. E. (1974). *Planta* **117**, 93–95.

Evans, L. V., Simpson, M. and Callow, M. E. (1973). *Planta* **110**, 237–252.

Ferrier, N. C., Davies, J. M., Sneddon, I. M. and Johnston, C. S. (1981). *Proc. Int. Seaweed Symp., 8th, 1974* pp. 172–175.

Fichtinger-Schepman, A. M. J., Kamerling, J. P., Versluis, C. and Vliegenthart, J. F. G. (1980). *Carbohydr. Res.* **86**, 215–225, and ref. cited therein.

Fleming, I. D., Hirst, E. L. and Manners, D. J. (1956). *J. Chem. Soc.* pp. 2831–2836.

Fleming, M., Hirst, E. L. and Manners, D. J. (1966). *Proc. Int. Seaweed Symp., 5th, 1965* pp. 255–260.

Ford, C. W. and Percival, E. (1965a). *J. Chem. Soc.* pp. 3014–3016.
Ford, C. W. and Percival, E. (1965b). *J. Chem. Soc.* pp. 7035–7041.
Ford, C. W. and Percival, E. (1965c). *J. Chem. Soc.* pp. 7042–7046.
Fredrick, J. F. (1953). *Physiol. Plant.* **6**, 100–105.
Fredrick, J. F. (1971). *Physiol. Plant.* **25**, 32–34.
Fredrick, J. F. (1973a). *Ann. N. Y. Acad. Sci.* **210**, 254–264.
Fredrick, J. F. (1973b). *Plant Sci. Lett.* **1**, 457–462.
Fredrick, J. F. (1975). *Plant Sci. Lett.* **5**, 131–135.
Fredrick, J. F. (1977). *Phytochemistry* **16**, 55–57.
Fredrick, J. F. (1979). *Phytochemistry* **18**, 1823–1825.
Fredrick, J. F. (1980a). *Phytochemistry* **19**, 539–542.
Fredrick, J. F. (1980b). *Phytochemistry* **19**, 2611–2613.
Fredrick, J. F. (1981). *Ann. N. Y. Acad. Sci.* **361**, 426–434.
Frei, E. and Preston, R. D. (1964a). *Proc. R. Soc. London, Ser, B* **160**, 293–313.
Frei, E. and Preston, R. D. (1964b). *Proc. R. Soc. London, Ser, B* **160**, 314–327.
Goldemberg, S. H. and Marechal, L. R. (1963). *Biochim. Biophys. Acta* **71**, 743–744.
Goodwin, T. W. (1974). *In* "Algal Physiology and Biochemistry" (W. P. D. Stewart, ed.), pp. 176–205. Blackwell, Oxford.
Gorin, P. A. J. and Spencer, J. F. T. (1966). *Can. J. Chem.* **44**, 993–998.
Grant, T., Morris, E. R., Rees, D. A., Smith, J. C. and Thom, D. (1973). *FEBS Lett.* **32**, 195–198.
Grasdalen, H., Larsen, B. and Smidsrod, O. (1979a). *Proc. Int. Seaweed Symp., 9th, 1977* pp. 309–317.
Grasdalen, H., Larsen, B. and Smidsrod, O. (1979b). *Carbohydr. Res.* **68**, 23–31.
Greenwood, C. T. and Thomson, J. (1961). *J. Chem. Soc.* pp. 1534–1537.
Handa, N. (1969). *Mar. Biol.* **4**, 208–214.
Haug, A. (1976). *Acta Chem. Scand., Ser. B* **30**, 562–566.
Haug, A. and Larsen, B. (1971). *Carbohydr. Res.* **17**, 297–308.
Haug, A. and Smidsrod, O. (1967). *Nature (London)* **215**, 1167–1168.
Haug, A., Larsen, B. and Baardseth, E. (1969). *Proc. Int. Seaweed Symp., 6th, 1968* pp. 443–451.
Haug. A., Myklestad, S. and Sakshaug, E. (1973). *J. Exp. Mar. Biol. Ecol.* **11**, 15–26.
Hawthorne, D. B., Sawyer, W. H. and Grant, B. R. (1979). *Carbohydr. Res.* **77**, 157–167.
Hellebust, J. A. and Haug, A. (1972). *Can. J. Bot.* **50**, 169–176, 177–184.
Hirst, E. L., O'Donnell, J. J. and Percival, E. (1958). *Chem. Ind. (London)* p. 834.
Hirst, E. L., Manners, D. J. and Pennie, I. R. (1972). *Carbohydr. Res.* **22**, 5–11.
Hough, L., Jones, J. K. N. and Wadman, W. H. (1952). *J. Chem. Soc.* pp. 3393–3399.
Howard, R. J., Grant, B. R. and Fock, H. (1977). *J. Phycol.* **13**, 340–345.
Hussein, M. M.-D., Abdel-Aziz, A. and Salem, H. M. (1980). *Phytochemistry* **19**, 2133–2135.
Ikawa, T., Watanabe, T. and Nisizawa, K. (1972). *Plant Cell Physiol.* **13**, 1017–1029.
Iriki, Y., Suzuki, T., Nisizawa, K. and Miwa, T. (1960). *Nature (London)* **187**, 82–83.
Ishikawa, M. and Nisizawa, K. (1981). *Bull. Jpn. Soc. Sci. Fish.* **47**, 889–893, 895–899.
Jabbar Mian, A. and Percival, E. (1973a). *Carbohydr. Res.* **26**, 133–146.
Jabbar Mian, A. and Percival, E. (1973b). *Carbohydr. Res.* **26**, 147–161.
Johnston, C. S., Jones, R. G. and Hunt, R. D. (1977). *Helgol. Wiss. Meeresunters* **30**, 527–545.
Kareker, M. D. and Joshi, G. V. (1973). *Bot. Mar.* **16**, 216–220.
Kashiwabara, Y., Suzuki, H. and Nisizawa, K. (1965). *Plant Cell Physiol.* **6**, 537–546.
Kauss, H. (1962). *Vortr. Gesamtgeb. Bot.* **1**, 129–132.
Kauss, H. (1967). *Z. Pflanzenphysiol.* **56**, 453–465.
Kaus, H. (1979). *Ber. Dtsch. Bot. Ges.* **92**, 11–22.
Kauss, H. and Kriebitzsch, C. (1969). *Biochem. Biophys. Res. Commun.* **35**, 926–930.

Khomenko, V. A., Pavlenko, A. F., Solov'eva, T. F. and Ovodov, Yu. S, (1971). *Khim. Prir. Soedin.* pp. 393–396.

Kobayashi, T., Inone, M., Tanabe, I., Onishi, H. and Fuksui, S. (1978). *J. Jpn. Soc. Starch Sci.* **25**, 186–192.

Kreger, D. R. and Meeuse, B. J. D. (1952). *Biochim. Biophys. Acta* **9**, 699–700.

Kreger, D. R. and Van der Veer, J. (1970). *Acta Bot. Neerl.* **19**, 401–402.

Kremer, B. P. (1976). *Planta* **129**, 63–67.

Kremer, B. P. (1977). *Z. Pflanzenphysiol.* **81**, 68–73.

Kremer, B. P. (1978). *Can. J. Bot.* **56**, 1655–1659.

Kremer, B. P. (1979). *Proc. Int. Seaweed Symp., 9th, 1977* pp. 421–428.

Kremer, B. P. (1980). *Mar. Biol.* **59**, 95–103.

Kremer, B. P. (1981). *Proc. Int. Seaweed Symp., 10th, 1980* pp. 437–442.

Kremer, B. P. and Kuppers, U. (1977). *Planta* **133**, 191–196.

Küppers, U. and Kremer, B. P. (1978). *Plant Physiol.* **62**, 49–53.

Küppers, U. and Weidner, M. (1980). *Planta* **148**, 222–230.

Larsen, B. (1967). *Acta Chem. Scand.* **21**, 1395–1396.

Larsen, B. (1981). *Proc. Int. Seaweed Symp., 10th, 1980* pp. 7–34.

Larsen, B. and Haug, A. (1971). *Carbohydr. Res.* **20**, 225–232.

Larsen, B., Haug, A. and Painter, T. (1970) *Acta Chem. Scand.* **24**, 3339–3352.

Lawson, C. J. and Rees, D. A. (1970). *Nature (London)* **227**, 392–393.

Leedale, G. F. (1966). *Adv. Sci.* **23**, 22–37.

Leedale, G. F., Meeuse, B. J. D. and Pringsheim, E. G. (1965). *Arch. Mikrobiol.* **50**, 133–155.

de Lestang, G. and Quillet, M. (1974). *Physiol. Veg.* **12**, 199–227.

de Lestang, G. and Quillet, M. (1981). *Proc. Int. Seaweed Symp. 8th, 1974* pp. 200–204. Marine Science Labs, Menai Bridge.

de Lestang-Bremond, G. and Quillet, M. (1976). *Physiol. Veg.* **14**, 259–269.

de Lestang-Bremond, G. and Quillet, M. (1981a). *Proc. Int. Seaweed Symp., 10th, 1980* pp. 449–454.

de Lestang-Bremond, G. and Quillet, M. (1981b). *Proc. Int. Seaweed Symp., 10th, 1980* pp. 503–507.

Lewin. R. A. (1976). *Nature (London)* **261**, 697.

Lin. T. Y. and Hassid, W. Z. (1966a). *J. Biol. Chem.* **241**, 3283–3293.

Lin. T. Y. and Hassid, W. Z. (1966b). *J. Biol. Chem.* **241**, 5284–5297.

Lindberg, B. and Paju, J. (1954). *Acta Chem. Scand.* **8**, 817–820.

Linker, A. and Jones, R. S. (1964). *Nature (London)* **204**, 187–188.

Lobban, C. S. (1978). *Plant Physiol.* **61**, 585–589.

Loewus, F. A. (1971). *Annu. Rev. Plant Physiol.* **22**, 337–364.

Love, J., Mackie, W., McKinnell, J. W. and Percival, E. (1963). *J. Chem. Soc.* pp. 4177–4182.

McCandless, E. L. (1981). *In* "The Biology of Seaweeds" (C. S. Lobban and M. J. Wynne, eds.), pp. 550–588. Blackwell, Oxford.

McCracken, D. A. and Badenhuizen, N. P. (1970). *Staerke* **22**, 289–291.

McCracken, D. A. and Cain, J. R. (1981). *New Phytol.* **88**, 67–71.

McCracken, D. A., Nadakavukaren, M. J. and Cain, J. R. (1980). *New Phytol.* **86**, 39–44.

Mackie, I. M. and Percival, E. (1959). *J. Chem. Soc.* pp. 1151–1156.

Mackie, W. and Percival, E. (1960). *J. Chem. Soc.* pp. 2381–2384.

Madgwick, J., Haug, A. and Larsen, B. (1973a). *Acta Chem. Scand.* **27**, 711–712.

Madgwick, J., Haug, A. and Larsen, B. (1973b). *Acta Chem. Scand.* **27**, 3592–3594.

Madgwick, J., Haug, A. and Larsen, B. (1978). *Bot. Mar.* **21**, 1–3.

Madgwick, J. C. and Ralph, B. J. (1969). *Proc. Int. Seaweed Symp., 6th, 1968* pp. 539–544.

Maeda, M. and Nisizawa, K. (1968). *Carbohydr. Res.* **7**, 97–99.
Majak, W., Craigie, J. S. and McLachlan, J. (1966). *Can. J. Bot.* **44**, 541–549.
Mangat, B. S. and Badenhuizen, N. P. (1970). *Staerke* **22**, 329–333.
Mangat, B. S. and Badenhuizen, N. P. (1971). *Can. J. Bot.* **49**, 1787–1792.
Manners, D. J. and Sturgeon, R. J. (1982). *In* "Encyclopedia of Plant Physiology" (F. A. Loewus and W. Tanner, eds.) vol. 13A, pp. 472–515. Springer-Verlag, Berlin and New York.
Manners, D. J. and Wright, A. (1962). *J. Chem. Soc.* pp. 4592–4595.
Manners, D. J., Mercer, G. A., Stark, J. R. and Ryley, J. F. (1965). *Biochem. J.* **96**, 530–532.
Manners, D. J., Pennie, I. R. and Ryley, J. F. (1973). *Carbohydr. Res.* **29**, 63–77.
Marchessault, R. H. and Deslandes, Y. (1979). *Carbohydr. Res.* **75**, 231–242.
Marechal, L. R. and Goldemberg, S. H. (1964). *J. Biol Chem.* **239**, 3163–3167.
Marsden, W. J. N., Callow, J. A. and Evans, L. V. (1981) *Mar. Biol Lett.* **2**, 353–362.
Matulewicz, C., Percival, E. and Weigel, H. (1984). *Phytochemistry* **23**, 103–105.
Medcalf, D. G. and Larsen, B. (1977). *Carbohydr. Res.* **59**, 531–537, 539–546.
Meeks, J. C. (1974). *In* "Algal Physiology and Biochemistry" (W. P. D. Stewart, ed.), pp. 161–175. Blackwell, Oxford.
Meeuse, B. J. D. (1962). *In* "Physiology and Biochemistry of Algae" (R. A. Lewin, ed.), pp. 289–313. Academic Press, New York.
Meeuse, R. J. D. (1963). *Acta Bot. Neerl* **12**, 315–318.
Meeuse, B. J. D., Andries, M. and Wood, J. A. (1960). *J. Exp. Bot.* **11**, 129–140.
Megarry, M. L. (1973). Ph. D. Thesis, University of London.
Min, K. H., Sasaki, S. F., Kashiwabara, Y., Umekawa, M. and Nisizawa, K. (1977). *J. Biochem.* (*Tokyo*) **81**, 555–562.
Mishkind, M. and Mauzerall, D. (1980). *Mar Biol.* **56**, 261–265.
Miwa, T., Iriki, Y. and Suzuki. T. (1961). *Colloq. Int. C. N. R. S.* **103**, 135–144.
Mori, H. and Nisizawa, K. (1982). *Bull Jpn. Soc. Sci. Fish.* **48**, 981–986.
Morris, E. R., Rees, D. A. and Thom, D. (1973). *Chem. Commun.* pp' 245–246.
Morris, E. R., Rees, D. A., Thom, D. and Boyd, J. (1978). *Carbohydr. Res.* **66**, 145–154.
Myklestad, S. (1974). *J. Exp. Mar. Biol. Ecol.* **15**, 261–274.
Myklestad, S. (1977). *J. Exp. Mar. Biol. Ecol.* **29**, 161–179.
Myklestad, S. and Haug, A. (1972). *J. Exp. Mar. Biol. Ecol.* **9**, 125–136.
Nagashima, H. and Fukada, I. (1981). *Phytochemistry* **20**, 439–442.
Nagashima, H., Ozaki, H. and Nisizawa, K. (1969). *Bot. Mag.* **82**, 462–473.
Nagashima, H., Nakamura, S., Nisizawa, K. and Hori, T. (1971). *Plant Cell Physiol.* **12**, 243–253.
Nelson, T. E. and Lewis, B. A. (1974). *Carbohydr. Res.* **33**, 63–74.
Nisizawa, K. (1981). *In* "Handbook of Biosolar Resources" (O. R. Zaborsky, ed.), Vol. 1, Part 1, pp. 373–378. CRC Press, Boca Raton, Florida.
Ofstad, R. and Larsen, B. (1981). *Proc. Int. Seaweed Symp., 10th, 1980* pp. 485–493.
Okazaki, M., Furaya, K., Tsukayama, K. and Nisizawa, K. (1982). *Bot. Mar.* **25**, 123–131.
Olaitan, S. A. and Northcote, D. H. (1962). *Biochem. J.* **82**, 509–519.
Ozaki, H., Maeda, M. and Nisizawa, K. (1967). *J. Biochem.* (*Tokyo*) **61**, 497–503.
Pavlenko, A. F., Kurika, A. V., Khomenko, V. A. and Ovodov, Yu. S. (1974). *Khim. Prir. Soedin* pp. 142–145.
Peat, S., Whelan, W. J. and Lawley, H. G. (1958). *J. Chem. Soc.* pp. 729–737.
Peat, S., Turvey, J. R. and Evans, J. M. (1959a). *J. Chem. Soc.* pp. 3223–3227.
Peat, S., Turvey, J. R. and Evans, J. M. (1959b). *J. Chem. Soc.* pp. 3341–3344.
Percival, E. (1967). *Chem. Ind.* (*London*) p. 511.
Percival, E. (1978a). *ACS Symp. Ser.* **77**, 203–212.
Percival, E. (1978b). *ACS Symp. Ser.* **77**, 213–224.

Percival, E. and McDowell, R. H. (1967). "Chemistry and Enzymology of Marine Algal Polysaccharides." Academic Press, New York.

Percival, E. and McDowell, R. H. (1981). *In* "Encyclopedia of Plant Physiology: New Series" (W. Tanner and F. A. Loewus, eds.), Vol. 13B, pp. 277–316. Springer-Verlag, Berlin and New York.

Percival, E. and Smestad, B. (1972a). *Carbohydr. Res.* **25**, 199–312.

Percival, E. and Smestad, B. (1972b). *Phytochemistry* **11**, 1967–1972.

Percival, E. and Wold. J. K. (1963). *J. Chem. Soc.* pp. 5459–5468.

Percival, E. and Young, M. (1971). *Phytochemistry* **10**, 807–812.

Percival, E. and Young, M. (1974). *Carbohydr. Res.* **32**, 195–201.

Percival, E., Venegas Jara, M. and Weigel, H. (1983). *Phytochemistry* **22**, 1429–1432.

Percival, E., Venegas Jara, M. and Weigel, H. (1984). *Carbohydr. Res.* **125**, 283–290.

Percival, E. G. V. and Chanda, S. K. (1950). *Nature (London)* **166**, 787.

Peyriere, M. (1970). *C. R. Hebd, Seances Acad, Sci., Ser, D* **270**, 2071–2074.

Pinder, D. F. and Bucke, C. (1975). *Biochem. J.* **152**, 617–622.

Preiss, J. and Ashwell, G. (1962). *J. Biol. Chem.* **237**, 309–316; 317–321.

Preiss, J. and Greenberg, E. (1967). *Arch. Biochem, Biophys.* **118**, 702–708.

Quatrano, R. S. and Stevens, P. T. (1976). *Plant Physiol.* **58**, 224–231.

Quatrano, R. S., Hogsett, W. E. and Roberts, M. (1979a). *Proc. Int. Seaweed Symp., 9th, 1977* pp. 113–123.

Quatrano, R. S., Brawley, S. H. and Hogsett, W. E. (1979b). *In* "Determination of Spatial Organization" (S. Subtelny and J. R. Konisberg, eds.), pp. 77–95. Academic Press, New York.

Ramus, J. (1973). *In* "Biogenesis of Plant Cell Wall Polysaccharides" (F. A. Loewus ed.), pp. 333–359. Academic Press, New York.

Rees, D. A. (1961). *Biochem. J.* **81**, 347–352.

Rees, D. A. (1972). *Biochem. J.* **126**, 257–273.

Rubat du Merac, M. L. (1953). *Rev. Gen. Bot.* **60**, 689–705.

Rusanowski, P. C. and Vadas, R. L. (1981). *Proc. Int. Seaweed Symp., 8th, 1974* pp. 232–243.

Sanwal, G. G. and Preiss, J. (1967). *Arch. Biochem. Biophys.* **119**, 454–469.

Schmitz, K. (1981). *In* "The Biology of Seaweeds" (C. S. Lobban and M. J. Wynne, eds.), pp. 534–558. Blackwell, Oxford.

Sheath, R. G., Hellebust, J. A. and Sawa, T. (1979). *Phycologia* **18**, 149–163.

Sheath, R. G., Hellebust, J. A. and Sawa, T. (1981). *Phycologia* **20**, 292–297.

Shephard, D. C. and Bidwell, R. G. S. (1973). *Protoplasma* **76**, 289–307.

Shephard, D. C., Levin, W. B. and Bidwell, R. G. S. (1968). *Biochem. Biophys. Res. Commun,* **32**, 413–420.

Shiraiwa, Y., Abe, K., Sasaki, S. F., Ikawa, T. and Nisizawa, K. (1975). *Bot. Mar.* **18**, 97–104.

Smestad, B., Percival, E. and Bidwell, R. G. S. (1972). *Can. J. Bot.* **50**, 1357–1361.

Smestad-Paulsen, B. and Miklestad, S. (1978). *Carbohydr. Res.* **62**, 386–388.

Smidsrod, O. and Haug, A. (1968). *Acta. Chem. Scand.* **22**, 1989–1997.

Smidsrod, O. and Haug, A. (1972). *Acta. Chem. Scand.* **26**, 2063–2074.

Solov'eva, T. F., Pavlenko, A. F., Bondareva, T. V. and Ovodov, Yu. S. (1973). *Khim. Prir. Soedin* pp. 145–148.

South, G. R. (1979). *Proc. Int. Seaweed Symp., 9th, 1977* pp. 133–142.

Steup, M. and Melkonian, M. (1981). *Physiol. Plant* **51**, 343–348.

Stockton, B., Evans, L. V., Morris, E. R. and Rees, D. A. (1980a). *Int. J. Biol. Macromol.* **2**, 176–178.

Stockton, B., Evans, L. V., Morris, E. R. Powell, D. A. and Rees, D. A. (1980b). *Bot Mar.* **23**, 563–567.

Su, J. C. and Hassid, W. Z. (1962). *Biochemistry* **1,** 474–480.
Turvey, J. R. (1978). *Int. Rev. Biochem.* **16,** 153–177.
Turvey, J. R. and Simpson, P. R. (1966). *Proc. Int. Seaweed Symp., 5th, 1965* pp. 323–327.
Turvey, J. R. and Williams, E. L. (1970). *Phytochemistry* **9,** 2383–2388.
Usov, A. I., Adamyants, K. S., Yarotsky, S. V., Anoshina, A. A. and Kochetkov, N. K. (1973). *Carbohydr. Res.* **26,** 282–283.
Vanden Driessche, T. (1969) *Prog. Photosynth. Res., Proc. Int. Congr.* [*1st*], *1968* pp. 450–457.
Vogel, K. and Meeuse, D. J. B. (1968). *J. Phycol.* **4,** 317–318.
Weber, M. and Wöber, G. (1975). *Carbohydr. Res.* **39,** 295–302.
Werz, G. and Clauss, H. (1970). *Planta* **91,** 165–168.
Whyte, J. N. C. (1971). *Carbohydr. Res.* **16,** 295–302.
Willenbrink, J. and Kremer, B. P. (1973). *Planta* **113,** 173–178.
Winkenbach, F., Parthasarathy, M. V. and Bidwell, R. G. S. (1972). *Can. J. Bot.* **50,** 1367–1375.
Wong, F. and Craigie, J. S. (1978). *Plant Physiol.* **61,** 663–666.
Wurtz, M. and Zetsche, K. (1976). *Planta* **129,** 211–216.
Yamaguchi, T., Ikawa, T. and Nisizawa, K. (1966). *Plant Cell Physiol.* **7,** 217–229.
Yamaguchi, T., Ikawa, T. and Nisizawa, K. (1969). *Plant Cell Physiol.* **10,** 425–440.
Yokobayashi, K., Akira, M. and Harada, T. (1970). *Biochim, Biophys. Acta* **212,** 458–469.

Chapter 10

Polysaccharides Containing Xylose, Arabinose, and Galactose in Higher Plants

K. BRINSON and P. M. DEY

Department of Biochemistry
Royal Holloway College (University of London)
Egham Hill, Egham
Surrey, England

I. INTRODUCTION

The term *storage carbohydrate* is somewhat difficult to define precisely. In higher plant seeds the term has classically been used to describe those polysaccharides of endosperm or cotyledons whose mobilisation has been clearly linked to germination, whereas in vegetative tissues the term has

BIOCHEMISTRY OF STORAGE
CARBOHYDRATES IN GREEN PLANTS

usually been confined to the description (apart from starch) of mannans and fructans.

However, it is now generally accepted that in growing plant tissues, and even in senescent tissues such as fruits at late stages of ripeness, there occurs considerable turnover of certain cell-wall polysaccharides (Labavitch, 1981). For example, turnover of xyloglucan has been demonstrated during elongation growth of pea stem sections (Labavitch and Ray, 1974a,b), and in the developing strawberry, turnover of wall-bound galactose was noted accompanying cell expansion growth, followed by loss of both galactose and arabinose from the wall during the subsequent cell separation that forms part of the ripening process (Knee et al., 1977). In several instances the liberated monosaccharides were shown to be incorporated into polysaccharide fractions. Cell-wall polysaccharides may function as reserve substances in vegetative tissues, the catabolism of such polysaccharides contributing to the processes such as resynthesis of cell-wall components and respiration. For certain wall polysaccharides, especially in the nonlignified tissues of storage organs such as fruits, tubers, and bulbs, distinguishing between exclusively "structural" and exclusively "storage" roles may not prove a straightforward exercise (see also Chapters 6 and 9, this volume).

Accordingly, the treatment of polymers of xylose, arabinose, and galactose adopted here incorporates discussion of both plant-storage polysaccharides and polymers usually ascribed structural roles in the primary cell wall. In order to assist the reader's appreciation of the possible location of constituent polysaccharides within the cell wall, one tentative model of the primary wall of dicotyledenous plants is presented in Fig. 1.

II. XYLOGLUCANS

A. Seed "Amyloids"

1. Occurrence

These polysaccharides, which stain blue with iodine/potassium iodide reagent (this starch-like reaction accounting for their early name) were first reported in seeds by Vogel and Schleiden (1839). All known seed sources of these storage xyloglucans come from dicotyledonous species, of which more than 230 are known. The most comprehensive catalogue of amyloid-containing seeds was composed by Kooiman (1960) on the basis of applying the amyloid reaction with iodine/KI to endosperm and/or embryo tissue of 2600 species. The polysaccharide was located in the cell walls of these tissues.

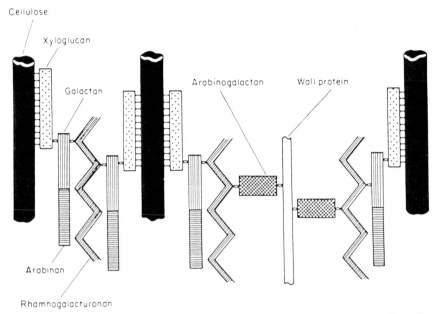

Fig. 1. Suggested scheme for the structure of the primary wall of sycamore callus cells. Redrawn by Robinson (1977) from Keegstra *et al.* (1973) and Albersheim (1976). Reproduced with permission from Robinson (1977). Copyright: Academic Press Inc. (London) Ltd.

2. Structure

The seed sources of xyloglucan from which the polysaccharide has been isolated and characterised include *Tamarindus indica* (Kooiman, 1961; Courtois and Le Dizet, 1974a), *Tropaeolum majus* (Hsu and Reeves, 1967; Courtois and Le Dizet, 1974a), and *Brassica campestris* (Siddiqui and Wood, 1971, 1977). The xyloglucans in all of the above-mentioned seeds possess a backbone of β-$(1\rightarrow4)$-linked D-glucosyl residues with single D-xylosyl groups linked by α-$(1\rightarrow6)$ linkages to the glucan backbone. Some of the xylosyl groups are terminal, while others are substituted with β-$(1\rightarrow2)$-linked D-galactosyl groups. It is now recognised that these polymers are galactoxyloglucans (the general structure shown in Fig. 2). However, the glycosyl composition varies somewhat with the source. The ratio D-galactose:D-xylose:D-glucose is 1:2:3 for the "amyloids" of *Tamarindus indica* and *Tropaeolum majus*, 1:2:4–5 for those from *Impatiens balsamina* (Courtois and Le Dizet, 1974b), and 1:1:4 for *Annona muricata* "amyloids" (Kooiman, 1967). Courtois *et al.* (1976) have speculated on the possible inclusion of β-$(1\rightarrow3)$-linkages and of branching in the glucan backbone, although the β-$(1\rightarrow3)$-linked glucosyl residues might be part of a contaminating β-$(1\rightarrow3)$-D-glucan.

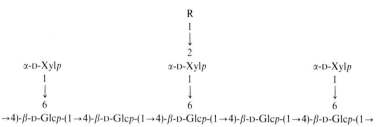

Fig. 2. General structure of xyloglucans. Glc*p*, glucopyranose; Xyl*p*, xylopyranose; R, β-(1→2)-linked D-galactopyranose or *O*-L-fucopyranosyl D-galactopyranose unit.

3. Biosynthesis and Mobilisation

Nothing has been published on the biosynthesis of xyloglucans in seeds. Early studies with *I. balsamina* (Heinricher, 1888), *T. majus*, and *Cyclamen europaeum* (Reiss, 1889) described mobilisation, following germination, of "amyloids" from cotyledon cell walls. A more recent study by Gould *et al.* (1971) noted the disappearance of xyloglucan from the cotyledons of white mustard (*Sinapis alba*) during germination, although nothing is known about the enzymology of xyloglucan mobilisation in this species or in any other seed that contains xyloglucan as a major cell-wall polysaccharide. β-Galactosidases, α-xylosidases, and endoglucanases are the likely enzymes involved in the degradation process.

B. Primary Cell-Wall Xyloglucans

1. Occurrence

Xyloglucans that probably originate from the primary cell wall have been isolated from the media of suspension-cultured cells of *Phaseolus vulgaris* (Wilder and Albersheim, 1973), *Rosa glauca* (Barnoud *et al.*, 1977), and *Acer pseudoplatanus* (Aspinall *et al.*, 1969; Bauer *et al.*, 1973). It is now believed that xyloglucans may be widely distributed in the hemicellulosic wall fractions of dicotyledonous species. In the intact primary wall of sycamore (*A. pseudoplatanus*) cells, the polymer is believed to be hydrogen-bonded to cellulose and covalently bonded at its reducing end to arabinan/galactan (Bauer *et al.*, 1973; Wilder and Albersheim, 1973; Darvill *et al.*, 1980). Xyloglucan accounts for 19% of the wall material in the cells. A xyloglucan similar to the sycamore polymer has been located in the primary wall of pear parenchyma tissue (Ahmed, 1979), and analysis of apple fruit cortex cell walls (Knee, 1973) has strongly suggested the presence of a similar polysaccharide.

There is some evidence, based largely upon the isolation of (4→6)-linked glucosyl residues, for the presence or xyloglucan in monocot primary cell walls

(Burke *et al.*, 1974; Darvill *et al.*, 1977; Labavitch and Ray, 1978), although xyloglucan accounts for only about 2% of the wall material. In addition, oligosaccharides containing xylosyl residues, $(1 \rightarrow 6)$-linked to $(1 \rightarrow 4)$-linked glucosyl residues, have been isolated following treatment of oat coleoptile walls with a crude preparation of endo-β-$(1 \rightarrow 4)$-D-glucanase (Labavitch and Ray, 1978).

2. Structure

The primary cell-wall xyloglucans that have been partially characterised possess structures similar to seed "amyloids." The wall xyloglucans from cultured sycamore (Bauer *et al.*, 1973) and bean (*P. vulgaris*) (Wilder and Albersheim, 1973) cells contain α-L-fucosyl residues probably linked to C-2 of D-galactose, whereas the seed polymers lack these terminal L-fucosyl residues (Kooiman, 1961; Hsu and Reeves, 1967; Gould *et al.*, 1971; Aspinall *et al.*, 1977; Siddiqui and Wood, 1971, 1977). As seed tissue is likely to contain secondary walls, it has been postulated (Darvill *et al.*, 1980) that fucose may be removed from the primary wall xyloglucan during secondary differentiation of the wall. The sycamore and bean primary wall xyloglucans also contain minor amounts of arabinopyranose, which is probably linked to C-2 of a few D-glucosyl residues of the polymer backbone (Bauer *et al.*, 1973; Wilder and Albersheim, 1973; Darvill *et al.*, 1980). A proposed partial structure for the sycamore primary wall xyloglucan is shown in Fig. 3.

3. Biosynthesis and Mobilisation

As with other noncellulosic cell-wall polysaccharides, our present understanding of xyloglucan biosynthesis is very limited. However, enzymes from *Phasealus aureus* and *Pisum sativum* seedlings incorporate D-xylose from UDP-xylose into a polysaccharide with some of the characteristics of xyloglucan, this incorporation being stimulated by UDP-glucose in the incubation medium (Villemez and Hinman, 1975). Ray (1975) has shown that in pea tissue UDP-xylose:xylosyltransferase and UDP-glucose:β-$(1 \rightarrow 4)$-glucan glucosyltransferase are located in a Golgi dictyosome fraction, the proposed site of xyloglucan synthesis. Several leading authors have proposed involvement of the Golgi apparatus and, perhaps, the endoplasmic reticulum in noncellulosic polysaccharide biosynthesis and secretion (Ray, 1973a,b, 1980; Northcote, 1974; Bowles and Northcote, 1976; Robinson *et al.*, 1976; Kawasaki, 1981), and recently Fincher *et al.* (1981) have proposed routes for synthesis/secretion located in the Golgi apparatus and endoplasmic reticulum with perhaps a third site for location of polysaccharide synthetases in the plasma membrane. Reivewing the present state of knowledge in this field, these last authors concluded that data currently available are consistent with a

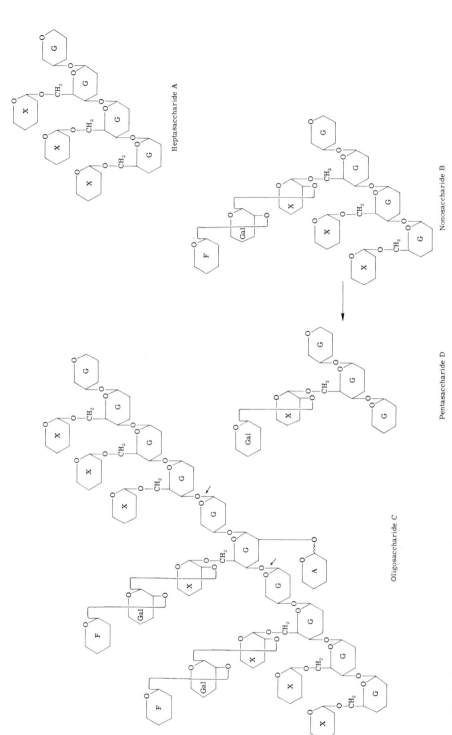

Fig. 3. Proposed structure of a portion of the hemicellulosic xyloglucan of the primary cell wall of dicots. Heptasaccharide A and nonasaccharide B are derived from oligosaccharide C by the action of endo-(1→4)-β-D-glucanase at the bonds indicated by arrows. Pentasaccharide D is derived from B by the combined action of α-L-fucosidase, α-D-xylosidase, and β-D-glucosidase. A, L-arabinopyranose; F, L-fucose; G, D-glucose; Gal, D-galactose; X, D-xylose.

role of dolichol phosphate sugars as intermediates between nucleotide sugars and completed glycoproteins in glycoprotein synthesis in the endoplasmic reticulum and Golgi, but whether these lipid-linked sugars are involved in noncellulosic polysaccharide synthesis at these locations or at plasma membrane sites is unknown at present.

Labavitch and Ray (1974a,b) have demonstrated auxin-induced turnover of primary wall xyloglucan during elongation growth of pea stem sections. There was loss of ^{14}C-labelled xylose and glucose from the wall, and these liberated monosaccharides could be recovered in ethanol-precipitable fraction (i.e. polymeric fraction) from tissue homogenates. These authors, as well as Loescher and Nevins (1972) and Albersheim (1978), have suggested that covalent bonds between xyloglucan and neutral pectic polymers (possibly galactan) are broken and resynthesised during such elongation growth. It is possible that when bonds between xyloglucan and neutral pectic polymers are degraded, this allows removal of xyloglucan from the wall and simultaneous slippage of cellulose fibrils, facilitating wall extension that is "fixed" by reinsertion of newly synthesised xyloglucan into the extended wall.

A xyloglucan similar to the sycamore polymer has been located in the primary wall of pear parenchyma tissue (Ahmed, 1979), and analysis of apple fruit cortex cell walls (Knee, 1973) has strongly suggested the presence of a similar polysaccharide. However, apart from increases in free, wall-derived glucose and xylose in the strawberry (Knee et al., 1977) and in free xylose in the mango (Sarkar, 1963), there is little evidence in the literature to suggest mobilisation of xyloglucans during fruit ripening.

III. XYLANS

A. Occurrence

Xylans constitute the major components of the hemicellulose fractions of most angiosperms, where they account for 20–30% of the dry weight of woody tissues; in gymnosperms they are less prominent (8%). They comprise 20–30% of the dry weight of agricultural residues such as cereal straws and grain hulls and, apart from cellulose, are the main polysaccharide components (15–20%) of the cell walls of grasses (Aspinall, 1970a,b, 1980).

Xylans are mainly secondary cell wall constituents, but in monocotyledon-ous plants they are also found in the primary walls of suspension-cultured cells (Burke et al., 1974; Darvill et al., 1977; Wada and Ray, 1978), young internode tissue (Joseleau and Barnoud, 1974), and coleoptiles (Wada and Ray, 1978; Labavitch and Ray, 1978; Darvill et al., 1978). The first glucuronoarabinoxy-lan to be clearly established as a constituent of a dicot primary cell wall (comprising 5% of the wall) was extracted from suspension-cultured sycamore cells (Darvill et al., 1980).

Gould et al. (1971) and Smith (1974) have proposed that in some seeds minor cell-wall storage polysaccharides, including arabinoxylans, are present in small amounts alongisde much larger quantities of major storage substances such as starch and protein. Such minor storage polysaccharides are probably of widespread occurrence but remain largely uncharacterised, with the exception of the nonstarch storage polysaccharides of barley grain (see also Chapter 6, this volume).

Fincher (1975) has studied the cell walls of the starchy endosperm of barley grain. These walls contained approximately 5% insoluble "microfibrillar" material, the remainder being water-extractable "matrix" material. The matrix contained 25% arabinoxylan and 75% β-D-glucan, while the microfibrillar material was probably composed of cellulose fibrils with tightly bound arabinoxylan and mannan.

The water-extractable endosperm arabinoxylan from barley flour (probably cell-wall-derived but not necessarily representative of the total wall arabinoxylan) was characterised by Aspinall and Ferrier (1958). Its composition approximated to 60% xylose, 40% arabinose, and the polymer consisted of a β-(1→4)-linked D-xylan backbone with terminal arabinofuranosyl residues attached to C-2 and/or C-3 of some xylose residues.

B. Structure

The xylans of higher plants have been well characterised as polymers possessing linear, β-(1→4)-linked D-xylan backbones (Aspinall, 1970a, 1980, 1982). A low degree of (1→3) branching has been observed in some samples (Aspinall and Stephen, 1973; Wilkie, 1979), but homoxylans are relatively uncommon. Two xylans isolated from endocarp of mango seeds (Amin and El-Sayed, 1973) possessed both these unusual features, being homoxylans containing (1→3)-linked branches.

Xylans from most sources contain short side chains of other sugar residues; these side chains are of three types.

1. Single (4-O-methyl)-α-D-glucopyranosyluronic acid residues, most frequently attached by (1→2) linkage to C-2 of backbone xylose units.
2. Single, α-L-arabinofuranosyl residues, most frequently attached by (1→3) linkage to xylose but with double branching [(1→3) and (1→2)-linked] on xylose units in more highly substituted arabinosylans.
3. More complex side chains in which (1,3)-linked α-L-arabinfuranose residues carry additional glycosyl substituents (Wilkie and Woo, 1977).

A general structure for higher plant xylans is shown in Fig. 4.

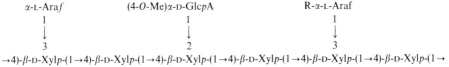

Fig. 4. General structure of higher plant xylans showing commonly encountered side chains. Ara*f*, arabinofuranose; Xyl*p*, xylopyranose; (4-*O*-Me)α-D-Glc*p*A, 4-*O*-methyl-α-D-glucopyranosiduronic acid; R, variously linked galactopyranose and/or xylopyranose residues in those higher plant xylans that have been characterised.

C. Biosynthesis and Mobilisation

There is limited information available on the biosynthesis of xylans. Bailey and Hassid (1966) reported that particulate preparations from immature corn cobs incorporated UDP-xylose and UDP-arabinose into a polymer similar to natural corn arabinoxylan, and a similar preparation from mung bean shoots incorporated UDP-xylose into a homoxylan (Odzuck and Kauss, 1972). More recently, Bolwell and Northcote (1983a) showed that xylan and arabinan synthase activities, utilising the nucleotide sugars as substrates were located largely in Golgi-derived membranes, prepared from *Phaseolus vulgaris* hypocotyls and suspension cultures. They concluded that these activities might be due to simple transglycosylases since no proteinaceous or lipid intermediate acceptors were found in either case. The transglycosylases were localised in both endoplasmic reticulum and Golgi apparatus but appeared to be subject to posttranslational control mechanisms giving rise to increased glycosylation in the Golgi *in vivo*.

It has long been established that barley endosperm cell-wall poly-saccharides are mobilised during germination (Brown and Morris, 1890), and Morrall and Briggs (1978) have studied this mobilisation in barley endosperm and embryo. Arabinoxylans (along with accompanying β-D-glucan) were extensively catabolised over 6 days, these polymers constituting almost 20% of the total endospermic polysaccharides mobilised, contributing significantly to carbohydrate supply to the developing embryo. Morrall and Briggs (1978) confirmed the observation of McNeil *et al.* (1975) that the cell walls of the aleurone layer are particularly rich in arabinoxylan.

With regard to the mobilisation of this polymer, it is probably significant that the activities of α-L-arabinofuranosidase, endo-β-D-xylanase, and β-D-xylosidase in barley endosperm all increase during germination, production of these enzymes in the aleurone layer being apparently stimulated by gibberellin released from the embryo (Taiz and Honigman, 1976). It is known that gibberellin also stimulates the aleurone layer to produce (1→3)-β-D-glucanase (Taiz and Jones, 1970) and β-amylase and other starch-degrading enzymes (Halmer and Bewley, 1982). Cell-wall degradation in the aleurone layer precedes the release of the hydrolytic enzymes into the starchy endosperm (Ashford and Jacobson, 1974); this wall dissolution is probably necessary to enable access of the amylolytic enzymes to their substrate. Morrall and Briggs (1978) have shown that wall degradation and starch hydrolysis proceed simultaneously.

The water-soluble, nonstarch polysaccharides of other grains are also mobilised following germination (MacLeod and McCorquodale, 1958; MacLeod and Sandie, 1961). Arabinoxylan is the major component of wheat endosperm cell walls, small amounts of cellulose, β-D-glucomannan, and galactan also being present (Mares and Stone, 1973a,b,c).

IV. GALACTANS AND ARABINANS

A. Occurrence

These neutral polysaccharides are generally considered to form part of the pectic polymer fraction of higher plant cell walls, being present as side chains linked to the acidic pectic backbone, which is believed to be a rhamnogalacturonan (Northcote, 1972; Keegstra et al., 1973; Albersheim, 1978; Darvill et al., 1980; Aspinall, 1980, 1982). In the primary cell wall of suspension-cultured sycamore cells, these neutral polysaccharides are believed to form covalently bonded bridges between the pectic rhamnogalacturonan and hemicellulosic xyloglucans hydrogen-bonded to the cellulose microfibrils (Keegstra et al., 1973; Albersheim, 1978; Darvill et al., 1980). They thus probably constitute important components of the matrix material of the plant cell wall.

B. Structure

Galactans have been isolated from citrus pectin (Labavitch et al., 1976), white willow (Toman et al., 1972), and beech (Meier, 1962). These pectic galactans are primarily β-(1→4)-linked, although galactans have been isolated that contain β-(1→6)-linked branching points (Meier, 1962; Toman et al., 1972; M. McNeil and P. Albersheim, unpublished results, 1978, cited in Darvil et al., 1980). The existence of such homogalactans has been questioned because the isolation procedures employed might cause acid hydrolysis of arabinofuranosidic linkages in arabinogalactans (Aspinall, 1980).

Polysaccharides containing β-(1→4)-linked D-galactan chains with up to 25% of L-arabinofuranose residues in side chains form a family of structurally related polymers. They are usually minor components of plants, but in a few cases—for example Centrosema plumari seeds and soybean cotyledons

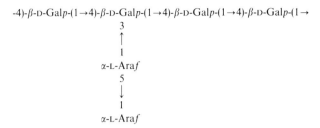

Fig. 5. Structure of soybean cotyledon arabinogalactan. Gal*p*, galactopyranose; Ara*f*, arabinofuranose.

(Aspinall, 1970b)—they are relatively abundant. The soybean cotyledon arabinogalactan has been partly characterised and has the structure shown in Fig. 5 (Aspinall and Cottrell, 1971). This polysaccharide occurs in association with a complex rhamnogalacturonan, and the mild isolation procedures employed for the neutral arabinogalactan render it unlikely to be an experimental artifact. The associated rhamnogalacturonan was shown to contain β-D-galactan units in side chains (Aspinall and Cottrell, 1971).

β-(1→4)-Linked D-galactans devoid of arabinose residues but containing 5–10% of uronic acid residues are found in compression wood of gymnosperms, for example, red spruce (Schreuder *et al.*, 1966) and tamarack (Jiang and Timell, 1972). In these polysaccharides it is likely that the uronic acid residues are end groups.

No homogalactan has been isolated directly from a primary cell wall, although analysis of the primary walls of suspension-cultured sycamore cells (Talmadge *et al.*, 1973) has indicated the probable presence of β-(1→4)-linked D-galactan chains in neutral pectic side chains. Those galactans that have been studied have degrees of polymerisation ranging from 33 (Toman *et al.*, 1972) to 50 (McNeil and Albersheim, 1978).

Arabinans have been isolated from the cell walls of many dicotyledonous plants (Hirst and Jones, 1947; Rees and Richardson, 1966; Aspinall and Cottrell, 1971; Siddiqui and Wood, 1974; Karacsonyi *et al.*, 1975; Joseleau *et al.*, 1977). However, most of these arabinans have contained small proportions of other glycosyl residues of uncertain significance and pure arabinans have been obtained only from mustard seeds (Rees and Richardson, 1966), the inner bark of *Rosa glauca* (Joseleau *et al.*, 1977), and a methylated primary cell-wall polysaccharide fraction of suspension-cultured sycamore cells (Darvill *et al.*, 1980). Methylation analysis of the primary walls of suspension-cultured pea cells suggested the presence of a similar polymer.

All of the arabinans that have been investigated appear to have similar structures, being highly branched polymers containing α-(1→5)-linked and α-(1→3)-linked L-arabinofuranosyl residues, generally represented by the

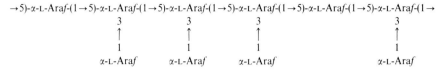

Fig. 6.   General structure of higher plant arabinans. Ara*f*, arabinofuranose .

partial structure shown in Fig. 6. Several arabinans contain linkages in addition to those shown in Fig. 6, notably side chains attached by α-(1→2)-L-arabinofuranosyl linkages to branch points in the backbone, which may or may not also carry (1→3)-linked units.

Arabinans are regarded as forming part of the neutral pectic fraction of the cell wall, and the so-called sugar beet arabinan preparation, extracted under strongly basic conditions (Hough and Powell, 1960; Hullar, 1965), is probably a degradation product, the galacturonic acid residues it contains originating from the rhamnogalacturonan chains of a pectin backbone. Estimation of the degree of polymerisation of pectic arabinans ranges from 34 (Joseleau *et al.*, 1977) to 90 (Karacsonyi *et al.*, 1975).

C.   Biosynthesis and Mobilisation

Particulate enzyme preparations from *Phaseolus aureus* seedlings in-corporated UDP-galactose into a galactan (McNab *et al.*, 1968) that was later characterised as a β-(1→4)-linked D-galactan (Panayotatos and Villemez, 1973). Bolwell and Northcote (1983a) found that an arabinan synthase that incorporated UDP-arabinose into arabinan was located in endoplasmic reticulum and Golgi-derived membranes from *P. vulgaris* hypocotyls and suspension cultures. Like the xylan synthase from the same tissues, this synthase activity may have been due to a single transglycosylase, since no lipid or proteinaceous intermediate acceptor was found. Posttranslational control of this activity appeared to be operating within the Golgi fraction. Like xylan synthase, the development of this activity was related to the stage of differentiation of the tissue (Bolwell and Northcote, 1981; Dalessandro and Northcote, 1981a,b). The cessation of galactan (neutral pectin) synthesis is followed by the onset of xylan synthesis in both hypocotyl and callus of *P. vulgaris* (Bolwell and Northcote, 1981). It was concluded that the control for the induction of the synthases by plant growth regulators incorporated in the culture medium involves transcription and possibly translation (Bolwell and Northcote, 1983b).

Seed cotyledon cells of *Lupinus* species possess thickened walls (Schulze and Steiger, 1889, 1892; Nadelmann, 1890; Elfert, 1894). Controversy exists as to whether these thickenings constitute reserves. According to Nadelman (1890), the thickenings were mobilised following germination, and Schulze and Steiger (1889), who demonstrated that 30% of the dry weight of the seed of *Lupinus luteus* consisted of galactose-rich polysaccharide material, reached similar conclusions, which they later (Schulze, 1895–1896) extended to *L. angustifolius*. In contrast, Elfert (1894) contended that the wall thickenings were merely "metamorphosed" in the course of cotyledon expansion.

This controversy has carried over into more recent studies. Matheson and Saini (1977) conducted a study of the water- and oxalate/EDTA-solubilised "pectic" polysaccharides of *L. luteus* cotyledons following germination. They noted a net loss of galactose- and arabinose-containing polysaccharides and attributed these findings to selective hydrolysis of certain wall polymers during the later stages of cell-wall expansion. The alternative possibility, that these polysaccharides might be primarily storage reserves, was not considered. Parker (1976), utilizing microscopic techniques, established that the cell-wall thickenings of *L. albus* and *L. angustifolius* are mobilised following germination. Furthermore, the recent work of A. Crawshaw and J. S. G. Reid (unpublished results, 1982, cited in Meier and Reid, 1982) has firmly established that the bulk of the cell wall in *L. angustifolius* cotyledons consists of galactose- and arabinose-rich polysaccharides that are degraded following germination.

Although the chemical structure of these polymers, apart from that of a water-soluble β-(1→4)-linked D-galactan (Hirst *et al.*, 1947) that is probably not fully representative of this class of polysaccharides, has not been established, it seems clear that pectic arabinans/galactans and/or arabino-galactans act as storage reserves in the cotyledons of *Lupinus* species. Perhaps the true answer to the controversy concerning their role is that the cell-wall "storage" polysaccharides may also have other functions. These polymers may fulfill a structural role within the pectic network of the cotyledon cell wall and be implicated in cell wall expansion in addition to their reserve role following germination.

Although the galactose- and arabinose-containing polysaccharides of the parenchymatous tissues of fleshy fruits have not been characterised, there is strong circumstantial evidence to suggest that primary wall galactans and arabinans are degraded in fruit tissues during ripening. Galactose and/or arabinose is lost from the primary wall during the ripening of apple (Knee, 1973; Bartley, 1976), strawberry (Knee *et al.*, 1977), pear (Ahmed and Labavitch, 1980), tomato (Gross and Wallner, 1979; Lackey *et al.*, 1980), and mango (P. M. Dey, K. Brinson, M. A. John, and J. B. Pridham, unpublished results). In all of these cases, loss of these sugars preceded or was accompanied by solubilisation of polygalacturonan from the wall. Ahmed and Labavitch (1980) showed that nearly all the arabinose liberated from the wall during

ripening of pear could be recovered in polymeric form from tissue homogenates. The synthesis of wall components in ripening tomato fruit was
measured by Lackey *et al.* (1980), and galactan turnover was demonstrated.
No doubt, the liberated sugars are also used for respiration in ripening fruits.
It is believed that degradation of cross-linking pectic galactans/arabinans
may lead to dissolution of the wall matrix and detachment of cellulose
microfibrils from the pectic rhamnogalacturonan backbone, contributing to
tissue softening. Although this is essentially a senescent process, it may share
common features with the auxin-stimulated cell-wall extension in coleoptiles
(Loescher and Nevins, 1972; Labavitch and Ray, 1974a,b; Albersheim, 1976),
where turnover of wall components is known to occur. Elucidation of the
enzymology of these two processes may throw considerable light on the roles
of galactans and arabinans within the cell wall.

V. ARABINO-3,6-GALACTANS

A. Occurrence

The so-called type II arabinogalactans, characterised by possessing galactan backbones containing β-D-galactosyl residues linked by both $(1\rightarrow3)$ and
$(1\rightarrow6)$ linkages plus a significant proportion of $(3\rightarrow6)$-linked galactose residues and L-arabinofuranosyl units attached as side chains, are widely
distributed in higher plant tissues. Such polysaccharides containing only
galactose and arabinose have been isolated from rapeseed cotyledons
(Siddiqui and Wood, 1972) and rapeseed flour (Larm *et al.*, 1976), larch
heartwood (Aspinall *et al.*, 1968), *Coffea arabica* (coffee) seeds (Courtois *et al.*,
1963; Wolfrom and Patin, 1965; Wolfrom and Anderson, 1967), and soybean
seeds (Morita, 1965; Kikuchi, 1972). Similar polysaccharides containing
minor amounts of other glycosyl residues have also been isolated from
monocot and dicot tissues including seeds, leaves, roots, and fruits (Clarke
et al., 1979a,b).

Glycoproteins in which the carbohydrate moieties are type II arabinogalactans (usually referred to as β-lectins) are also widely distributed in both dicots
and monocots (Clarke *et al.*, 1979a,b; Fincher *et al.*, 1983). These polymers are
typified by the arabinogalactan protein of ryegrass endosperm (Anderson
et al., 1977). Type II arabinogalactans are also constituents of maple xylem sap
(Lamport, 1977) and of many exudate gums of great commercial significance,
both of angiosperms (e.g. the *Acacia*s) and gymnosperms (most notably in the
genus *Larix*). The sources and structures of gum exudate arabinogalactans
have been reviewed periodically (Jones and Smith, 1949; Smith and
Montgomery, 1959; Aspinall, 1969; Glinksman and Sand, 1973; Clarke *et al.*,
1979a,b).

Although no arabinogalactan has been isolated from a source known to contain only primary cell walls, these polysaccharides have been obtained from the culture medium of suspension-cultured sycamore (Aspinall *et al.*, 1969) and tobacco (Kato *et al.*, 1977) cells. Further, glycosyl linkage analysis of a pectic fraction from suspension-cultured sycamore cells suggested the presence of the type of linkages characteristic of type II arabinogalactans (Talmadge *et al.*, 1973). Components of pectic complexes with rhamnogalacturonan backbones to which arabino-3,6-galactan side chains are attached have been described from *Acer* callus cell walls (Barrett and Northcote, 1965; Stoddart *et al.*, 1967; Cook and Stoddart, 1973), *Panax* (Solov'eva *et al.*, 1969; Ovodov, 1975), *Zostera* (Ovodova *et al.*, 1968; Ovodov, 1975; Ovodov *et al.*, 1975), and *Hibiscus* (Bajpai and Mukherjee, 1971). The side chain may be attached either to the rhamnose residues, as in the *Panax* pectin, or to the galacturonic acid residues, as in the *Zostera* pectin. The available evidence thus strongly suggests that type II arabinogalactans are constituents of the pectic polymers of cell walls in at least some higher plants.

B. Structure

It is believed that all type II arabinogalactans are similar in possessing a galactan backbone containing both $(1 \rightarrow 3)$- and $(1 \rightarrow 6)$-linked β-D-galactosyl residues, some of the galactose units being $(3 \rightarrow 6)$-linked, with L-arabinofuranosyl residues present as side chains. However, beyond this similarity, the available evidence suggests wide structural variation between arabino-3,6-galactans from various plant sources. There is considerable variation in their glycosyl compositions: the rapeseed arabinogalactan (Larm *et al.*, 1976), for example, contains 90% arabinose, whereas the larch arabinogalactan (Aspinall *et al.*, 1968) contains 88% galactose. The significance of other glycosyl residues, which include uronic acids, rhamnose, xylose, glucose, and mannose, present in or extracted with arabinogalactans, is not clear. Whether these minor sugars are constituents of the arabinogalactan polymers, residual stubs of other polysaccharides to which arabinogalactans are bonded in the intact plant tissue, or merely impurities coextracted with the arabinogalactans is not known.

Although it is not possible at present to depict the full structure of any arabino-3,6-galactan, the larch polymer (Haq and Adams, 1961) has been characterised to a greater degree than most, and a partial structure for this polysaccharide is presented in Fig. 7. The ryegrass endosperm arabinogalactan protein has also been partly characterised (Anderson *et al.*, 1977), and the tentative structure of the carbohydrate moiety is shown in Fig. 8. The linkage between the polysaccharide and protein in these glycoproteins has received

→6)-β-D-Gal*p*-(1→6)-β-D-Gal*p*-(1→3)-β-D-Gal*p*-(1→3)-β-D-Gal*p*-(1→6)-β-D-Gal*p*-(1→6)-β-D-Gal*p*-(1→3)-β-D-Gal*p*-(1→

```
                6             6             3              3
                ↑             ↑             ↑              ↑
                1             1             1              1
            β-D-Gal p      β-D-Gal p     β-D-Gal p       L-Ara f
                6                                          3
                ↑                                          ↑
                1                                          1
            β-D-Gal p                                   β-L-Ara p
```

Fig. 7. Partial structure of larch (*Larix laricina*) arabinogalactan. Gal*p*, galactopyranose; Ara*p*, arabinopyranose; Ara*f*, arabinofuranose. After Haq and Adams (1961).

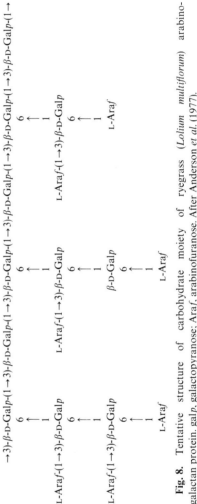

→3)-β-D-Gal*p*-(1→3)-β-D-Gal*p*-(1→3)-β-D-Gal*p*-(1→3)-β-D-Gal*p*-(1→3)-β-D-Gal*p*-(1→3)-β-D-Gal*p*-(1→

```
          6                    6                    6                      6
          ↑                    ↑                    ↑                      ↑
          1                    1                    1                      1
   L-Ara f-(1→3)-β-D-Gal p   L-Ara f-(1→3)-β-D-Gal p   β-D-Gal p      L-Ara f-(1→3)-β-D-Gal p
          6                    6                    6
          ↑                    ↑                    ↑
          1                    1                    1
   L-Ara f-(1→3)-β-D-Gal p   β-D-Gal p            L-Ara f
          6                    6
          ↑                    ↑
          1                    1
       L-Ara f               L-Ara f
```

Fig. 8. Tentative structure of carbohydrate moiety of ryegrass (*Lolium multiflorum*) arabino-galactan protein. gal*p*, galactopyranose; Ara*f*, arabinofuranose. After Anderson *et al.* (1977).

relatively little attention, but linkages via galactose to both hydroxyproline (Fincher *et al.*, 1974) and serine (Hillstead *et al.*, 1977a,b) have been suggested. Further details of the structure and properties of galactose- and arabinose-containing protoglycans and glycoproteins are available in a recent review (Fincher *et al.*, 1983).

C. Biosynthesis

There is little specific information available on the biosynthesis of arabinogalactans or arabinogalactan proteins. With regard to the glycoproteins, by analogy with the synthesis of the cell-wall protein and animal glycoproteins, it might be expected that the protein component would be synthesised on the rough endoplasmic reticulum, with proline being incorporated into the nascent protein and enzymatically hydroxylated as a posttranslational event (Savada and Chrispeels, 1971, 1973; Clarke *et al.*, 1979a). In suspension-cultured sycamore cells, a membranous fraction originating from either the endoplasmic reticulum or the Golgi complex appears to be associated with the synthesis of both hydroxyproline-rich protein and carbohydrate moieties of cell-wall glycoprotein, although the mechanism of formation of the protein–carbohydrate linkage is not established (Pope, 1977).

The metabolic pathways for formation of the monosaccharides that are incorporated into the carbohydrate chains—namely, galactose, arabinose, rhamnose, and the uronic acids—are well established (Nikaido and Hassid, 1971; Feingold and Fan, 1973; Dalessandro and Northcote, 1977). However, the mechanism and control of the addition of monosaccharides to the polysaccharide chain is not understood. Elongation of the carbohydrate chains could occur by sequential addition of monosaccharides to either the glycosylated protein or a carbohydrate primer, or possibly the monosaccharides may be assembled on some kind of intermediate and added in a block. This possibility of block addition of a preformed fragment during arabinogalactan biosynthesis is suggested by the presence in some *Acacia* gums of discrete components possessing molecular weights that are multiples of 6000 (Churms and Stephen, 1972, 1975; Adam *et al.*, 1977). Smith degradation of these and certain other arabinogalactans yields similar polymer homologues (Bouveng, 1961; Churms and Stephen, 1972, 1975; Adam *et al.*, 1977).

Whatever the method of chain elongation, it is likely that glycosyltransferases specific for the sugar transferred, the acceptor, and the position and anomeric configuration of the linkage formed and acting in accordance with the "one enzyme–one linkage" concept (Hagopian and Eylar, 1968) are involved (Roden and Swartz, 1975). The means of termination of chain growth and the signal for secretion of the completed molecule are obscure. It is presumed that glycosylation proceeds as the molecules pass from the endoplasmic reticulum to the Golgi apparatus, after which they are transported to the plasma membrane and secreted. Secretion of polysaccharides and hydroxyproline-rich protein from suspension-cultured sycamore cells is calcium-dependent (Morris and Northcote, 1977). These generalities probably apply equally to the biosynthesis of the other complex carbohydrates discussed earlier in this chapter, for example the arabinogalactans.

D. Physiological Role

No specific function can be ascribed to any of the arabinogalactans or arabinogalactan proteins discussed. It is presently a matter of speculation as to whether their ubiquitous distribution reflects a fundamental role in plant tissues or whether they fulfill specific functions in different situations. Even their subcellular location has not been unquestionably defined. Although arabinogalactans are knwon to be present in the pectic fractions extracted from plant tissues (Barrett and Northcote, 1965; Stephen and de Bruyn, 1967; Stoddart et al., 1967; Solov'eva et al., 1969; Ovodova et al., 1968; Cook and Stoddart, 1973; Ovodov, 1975; Ovodov et al., 1975), there has been no reported cytochemical examination of the middle lamella for arabinogalactan proteins.

However, the specific characteristics of the arabinogalactans suggest certain roles. Unlike most other noncellulosic cell-wall polysaccharides, which are essentially linear or only sparsely branched, arabinogalactans are multibranched, "bush-like" molecules. Their most marked physical properties are their water-solubility, adhesiveness, and ability to associate with other macromolecules (Clarke et al., 1979a) Arabinogalactans in the middle lamella may function as adhesives to cement cell–cell contact. Specialised adhesive functions for arabinogalactans or arabinogalactan proteins may occur in certain situations, for example, the adhesion of callus cell clumps (Bauer et al., 1973) or capture and adhesion of pollen to stigma surfaces (Clarke et al., 1979b). In view of their widespread distribution, it is perhaps surprising that little interest has been expressed in a possible storage role for these polysaccharides, especially in seeds and cotyledons where they occur. For example, arabinogalactan protein has been associated with the aleurone layer of cereal seeds, especially at the cytoplasm–wall interface (Anderson et al., 1977). In experiments using Vinca rosea cell suspension cultures, [^{14}C]arabinose and galactose were shown to be lost from cell walls, suggesting turnover of arabinogalactan protein (Takeuchi et al., 1980; Fincher and Stone, 1982). Enzymes capable of degrading arabinogalactans are present in plant tissues. In at least one situation, a nutritive role has been assigned to an arbinogalactan: in the style canal of *Lilium*, an arabinogalactan mucilage within the canal acts as a source of carbohydrate precursors for the growing pollen tube cell-wall (Loewus and Labarca, 1973).

The gum-exudate arabinogalactans may be secretions (Ghosh and Purkayastha, 1962; Seita *et al.*, 1977) or products of tissue necrosis and are considered to be produced in some situations in response to infection or other pathological conditions (Ghosh and Purkayastha, 1962). The pronounced water-holding properties of the gums may mean that they act as a seal to isolate the infected or damaged area of the plant (Smith and Montgomery, 1959; Anderson and Dea, 1971; Rosik, 1971). However, some trees secrete gum while in an apparently healthy condition (Anderson and Dea, 1971). In the most extreme case, *Larix* species secrete immense quantities of arabinogalactan gum into the inner, nonfunctioning part of the xylem. Perhaps the gums are products of normal metabolic processes that are influenced or modified by environmental conditions such as water stress, temperature (Howes, 1949; Glinksman and Sand, 1973), infection, or damage (Ghosh and Purkayastha, 1962).

The localisation of arabinogalactan proteins at the protoplast surface (Larkin, 1977, 1978; Keller and Stone, 1978) and the membrane–wall interface in intact cells (Anderson *et al.*, 1977; Clarke *et al.*, 1978) suggests that these polymers may have specific membrane-related functions. Speculations as to the functions of these molecules include protection of membranes from dessication, presence as membrane-bound proteoglycolipids analagous to blood group glycolipids influencing cell–cell contact and recognition in plant cells (Anderson and Dea, 1969), and a role as lectin-type substances, possessing specific carbohydrate-binding properties (Jermyn and Yeow, 1975; Anderson *et al.*, 1977; Clarke *et al.*, 1978), possibly affecting cell–cell interaction.

Although such speculations are highly interesting and may ultimately prove of great importance, it must be borne in mind that, at present, they are based on fragments of information drawn from diverse sources. Arabinogalactans, currently among the most obscure of plant polysaccharides with regard to biological function, may prove to be of great physiological importance.

REFERENCES

Adam, J. W. H., Churms, S. C., Stephen, A. M., Streefkerk, D. G. and Williams, E. H. (1977). *Carbohydr. Res.* **54**, 304–307.

Ahmed, A. El-Rayah (1979). Ph.D. Thesis, University of California, Davis.

Ahmed, A. El-Rayah and Labavitch, J. M. (1980). *Plant Physiol.* **65**, 1009–1013.

Albersheim, P. (1976). *In* "Plant Biochemistry" (J. Bonner and J. E. Varner, eds.), 3rd ed., pp. 225–274. Academic Press, New York.

Albersheim, P. (1978). *In. Rev. Biochem.* 127–150.

Amin, E. S. and El-Sayed, M. M. (1973). *Carbohydr. Res.* **27**, 39–46.

Anderson, D. M. W. and Dea, I. C. M. (1969). *Phytochemistry* **8**, 167–176.

Anderson, D. M. W. and Dea, I. C. M. (1971). *J. Soc. Cosmet. Chem.* **22**, 61–76.

Anderson, R. L., Clarke, A. E., Jermyn, M. A., Knox, R. B. and Stone, B. A. (1977). *Aust. J. Plant Physiol.* **4**, 143–158.

Ashford, A. E. and Jacobson, J. V. (1974). *Planta* **120**, 81–105.

Aspinall, G. O. (1969). *Adv. Carbohydr. Chem. Biochem.* **24**, 333–379.

Aspinall, G. O. (1970a). "Polysaccharides," pp. 104–114. Pergamon, Oxford.

Aspinall, G. O. (1970b). *In* "The Carbohydrates" (W. Pigman and D. Horton, eds.), Vol. 2B, pp. 515–536. Academic Press, New York.

Aspinall, G. O. (1980). *In* "The Biochemistry of Plants" (J. Preiss, ed.), Vol. 3, pp. 473–500. Academic Press, New York.

Aspinall, G. O. (1982). *In* "Encyclopedia of Plant Physiology: New Series" (W. Tanner and F. A. Loewus, eds.), Vol. 13B, pp. 3–8. Springer-Verlag, Berlin and New York.

Aspinall, G. O. and Cottrell, I. W. (1971). *Can. J. Chem.* **49**, 1019–1022.

Aspinall, G. O. and Ferrier, R. J. (1958). *J. Chem. Soc.* pp. 638–642.

Aspinall, G. O. and Stephen, A. M. (1973). *MTP Int. Rev. Sci.: Org. Chem., Ser. One* **7**, 285–313.

Aspinall, G. O., Fairweather, R. M. and Wood, T. M. (1968). *J. Chem. Soc. C* pp. 2174–2179.

Aspinall, G. O., Molloy, J. A. and Craig, J. W. T. (1969). *Can. J. Biochem.* **47**, 1063–1070.

Aspinall, G. O., Krishnamurthy, T. N. and Rosell, K. G. (1977). *Carbohydr. Res.* **55**, 11–19.

Bailey, R. W. and Hassid, W. Z. (1966). *Biochemistry* **56**, 1586–1593.

Bajpai, K. S. and Mukherjee, S. (1971). *Indian J. Chem.* **9**, 33–35.

Barnoud, F., Mollard, A. and Dutton, G. G. S. (1977). *Physiol. Veg.* **15**, 153–161.

Barrett, A. J. and Northcote, D. H. (1965). *Biochem. J.* **94**, 617–627.

Bartley, I. M. (1976). *Phytochemistry* **15**, 625–626.

Bauer, W. D., Talmadge, K. W., Keegstra, K. and Albersheim, P. (1973). *Plant Physiol.* **51**, 174–187.

Bolwell, G. P. and Northcote, D. H. (1981). *Planta* **152**, 225–233.

Bolwell, G. P. and Northcote, D. H. (1983a). *Biochem. J.* **210**, 497–507.

Bolwell, G. P. and Northcote, D. H. (1983b). *Biochem. J.* **210**, 509–515.

Bouveng, H. O. (1961). *Sven. Kem. Tidskr.* **73**, 115–131.

Bowles, D. J. and Northcote, D. H. (1976). *Planta* **128**, 101–106.

Brown, H. T. and Morris, G. H. (1890). *J. Chem. Soc.* **57**, 458–528.

Burke, D., Kaufman, P., McNeil, M. and Albersheim, P. (1974). *Plant Physiol.* **54**, 109–115.

Churms, S. C. and Stephen, A. M. (1972). *Carbohydr. Res.* **21**, 91–98.

Churms, S. C. and Stephen, A. M. (1975). *Carbohydr. Res.* **45**, 291–298.

Clarke, A. E., Gleeson, P. A., Jermyn, M. A. and Knox, R. B. (1978). *Aust. J. Biol. Sci.* **5**, 707–722.

Clarke, A. E., Anderson, R. L. and Stone, B. A. (1979a). *Phytochemistry* **18**, 521–540.

Clarke, A. E., Gleeson, P. A., Harrison, S. and Knox, R. B. (1979b). *Proc. Natl. Acad. Sci. U.S.A.* **76**, 3358–3362.

Cook, G. M. W. and Stoddart, R. W. (1973). "Surface Carbohydrates of the Eukaryotic Cell." Academic Press, London.

Courtois, J. E. and Le Dizet, P. (1974a). *An. Quim.* **70**, 1067–1072.

Courtois, J. E. and Le Dizet, P. (1974b). *C.R. Hebd. Seances Acad. Sci., Ser. C* **278**, 81–83.

Courtois, J. E., Percheron, F. and Glomaud, J. C. (1963). *Cafe, Cacao, The* **7**, 231–236.

Courtois, J. E., Le Dizet, P. and Robic, D. (1976). *Carbohydr. Res.* **49**, 439–449.

Crawshaw, A. and Reid, J. S. G. (1982). Unpublished results cited in Meier, H. and Reid, J. S. G. *In* "Encyclopedia of Plant Physiology: New Series" (F. A. Loewus and W. Tanner, eds.). Vol. 13A, pp. 418–471. Springer-Verlag, Berlin.

Dalessandro, G. and Northcote, D. H. (1977). *Planta* **134**, 39–44.

Dalessandro, G. and Northcote, D. H. (1981a). *Planta* **151**, 53–60.

Dalessandro, G. and Northcote, D. H. (1981b). *Planta* **151**, 61–67.

Darvill, A. G., Smith, C. J. and Hall, M. A. (1977). *In* "Regulation of Cell Membrane Activities in Plants" (E. Marre and O. Ciferri, eds.), pp. 275–282. North-Holland Publ., Amsterdam.

Darvill, A. G., Smith, C. J. and Hall, M. A. (1978). *New Phytol.* **80**, 503–516.

Darvill, A. G., McNeil, M. and Albersheim, P. (1980). *In* "The Plant Cell" (N. E. Tolbert, ed.), pp. 91–162. Academic Press, New York.

Elfert, T. (1894). *Bibl. Bot.* **30**, 1–25.

Feingold, D. S. and Fan, D. (1973). *In* "Biogenesis of Plant Cell-Wall Polysaccharides" (F. Loewus, ed.), pp. 69–84. Academic Press, New York.

Fincher, G. B. (1975). *J. Inst. Brew.* **81**, 116–122.

Fincher, G. B., and Stone, B. A. (1982). *In* "Encyclopedia of Plant Physiology: New Series" (W. Tanner and F. A. Loewus, eds.), Vol. 13B, pp. 68–132. Springer-Verlag, Berlin and New York.

Fincher, G. B., Sawyer, W. H. and Stone, B. A. (1974). *Biochem. J.* **139**, 535–545.

Fincher, G. B., Stone, B. A. and Clarke, A. E. (1983). *Annu. Rev. Plant Physiol.* **34**, 47–70.

Ghosh, S. S. and Purkayastha, S. K. (1962). *Indian For.* **88**, 92–98.

Glinksman, M. and Sand, R. E. (1973). *In* "Industrial Gums" (R. L. Whistler and J. N. Be Miller, eds.), 2nd ed., pp. 197–210. Academic Press, New York.

Gould, S. E. B., Rees, D. A. and Wight, N. J. (1971). *Biochem. J.* **124**, 47–53.

Gross, K. C. and Wallner, S. J. (1979). *Plant Physiol.* **63**, 117–120.

Hagopian, A. and Eylar, E. H. (1968). *Arch. Biochem. Biophys.* **126**, 785–794.

Halmer, P. and Bewley, J. D. (1982). *In* "Encyclopedia of Plant Physiology: New Series" (F. A. Loewus and W. Tanner, eds.), Vol. 13A, pp. 748–793. Springer-Verlag, Berlin and New York.

Haq, C. and Adams, G. A. (1961). *Can. J. Chem.* **39**, 1563–1573.

Heinricher, E. (1888). *Flora (Jena)* **71**, 163–185.

Hillstead, A., Wold, J. K. and Engen, T. (1977a). *Phytochemistry* **16**, 1953–1956.

Hillstead, A., Wold, J. K. and Paulsen, B. S. (1977b). *Carbohydr. Res.* **57**, 135–144.

Hirst, E. L. and Jones, J. K. N. (1947). *J. Chem. Soc.* pp. 1221–1225.

Hirst, E. L., Jones, J. K. N. and Walder, W. O. (1947). *J. Chem. Soc.* pp. 1225–1229.

Hough, L. and Powell, D. B. (1960). *J. Chem. Soc.* pp. 16–22.

Howes, F. N. (1949). "Vegetable Gums and Resins." Chronica Botanica, Waltham, Massachusetts.

Hsu, D. S. and Reeves, R. E. (1967). *Carbohydr. Res.* **5**, 202–209.

Hullar, T. L. (1965). *Diss. Abstr.* **26**, 671.

Jermyn, M. A. and Yeow, Y. M. (1975). *Aust. J. Plant Physiol.* **2**, 501–531.

Jiang, K. S. and Timell, T. E. (1972). *Sven. Papperstidn.* **75**, 592–594.

Jones, J. K. N. and Smith, F. (1949). *Adv. Carbohydr. Chem.* **4**, 243–291.

Joseleau, J. P. and Barnoud, F. (1974). *Phytochemistry* **13**, 1155–1158.

Joseleau, J. P., Chambat, G., Vignon, M. and Barnoud, F. (1977). *Carbohydr. Res.* **58**, 165–175.

Karacsonyi, S., Toman, R., Janecek, F. and Kubackova, M. (1975). *Carbohydr. Res.* **44**, 285–290.

Kato, K., Watanabe, F. and Eda, S. (1977). *Agric. Biol. Chem.* **41**, 533–538.

Kawasaki, S. (1981). *Plant Cell Physiol.* **22**, 431–442.

Keegstra, K., Talmade, K. W., Bauer, W. D. and Albersheim, P. (1973). *Plant Physiol.* **51**, 188–196.

Keller, F. and Stone, B. A. (1978). *Z. Pflanzenphysiol.* **87**, 167–172.

Kikuchi, T. (1972). *J. Agric. Chem. Soc. Jpn.* **46**, 405–409.
Knee, M. (1973). *Phytochemistry* **12**, 637–653.
Knee, M., Sargent, J. A. and Osborne, D. J. (1977). *J. Exp. Bot.* **28**, 377–396.
Kooiman, P. (1960). *Acta Bot. Neerl.* **9**, 208–219.
Kooiman, P. (1961). *Recl. Trav. Chim, Pays-Bas* **80**, 849–865.
Kooiman, P. (1967). *Phytochemistry* **6**, 1665–1673.
Labavitch, J. M. (1981). *Annu. Rev. Plant Physiol.* **32**, 385–406.
Labavitch, J. M. and Ray, P. M. (1974a). *Plant Physiol.* **53**, 669–673.
Labavitch, J. M. and Ray, P. M. (1974b). *Plant Physiol.* **54**, 449–502.
Labavitch, J. M. and Ray, P. M. (1978). *Phytochemistry* **17**, 933–938.
Labavitch, J. M., Freeman, L. E. and Albersheim, P. (1976). *J. Biol. Chem.* **251**, 5904–5910.
Lackey, G. D., Gross, K. C. and Wallner, S. J. (1980). *Plant Physiol.* **66**, 532–533.
Lamport, D. T. A. (1977). *In* "The Structure, Biosynthesis and Degradation of Wood" (F. A. Loewus and V. C. Runeckles, eds.), pp. 79–115. Plenum, New York.
Larkin, P. (1977). *J. Cell. Sci.* **26**, 31–46.
Larkin, P. (1978). *J. Cell. Sci.* **30**, 283–292.
Larm, O., Theander, O. and Aman, P. (1976). *Acta Chem. Scand.* **30**, 627–630.
Loescher, W. and Nevins, D. J. (1972). *Plant Physiol.* **50**, 556–563.
Loewus, F. and Labarca, C. (1973). *In* "Biogenesis of Plant Cell Wall Polysaccharides" (F. Loewus, ed.), pp. 175–193. Academic Press, New York.
MacLeod, A. M. and McCorquodale, H. (1958). *New Phytol.* **57**, 168–182.
MacLeod, A. M. and Sandie, R. (1961). *New Phytol.* **60**, 117–128.
McNab, J. M., Villemez, C. L. and Albersheim, P. (1968). *Biochem. J.* **106**, 355–360.
McNeil, M., Albersheim, P., Taiz, L. and Jones, R. L. (1975). *Plant Physiol.* **55**, 64–68.
Mares, D. J. and Stone, B. A. (1973a). *Aust. J. Biol. Sci.* **26**, 793–812.
Mares, D. J. and Stone, B. A. (1973b). *Aust. J. Biol. Sci.* **26**, 813–830.
Mares, D. J. and Stone, B. A. (1973c). *Aust. J. Biol. Sci.* **26**, 1005–1007.
Matheson, N. K. and Saini, H. D. (1977). *Phytochemistry* **16**, 59–66.
Meier, H. (1962). *Acta Chem. Scand.* **16**, 2275–2283.
Meier, H. and Reid, J. S. G. (1982). *In* "Encyclopedia of Plant Physiology: New Series" (F. A. Loewus and W. Tanner, eds.), Vol. 13A, pp. 418–471. Springer-Verlag, Berlin and New York.
Morita, M. (1965). *Agric. Biol. Chem.* **29**, 564–573.
Morrall, P. and Briggs, D. E. (1978). *Phytochemistry* **17**, 1495–1502.
Morris, M. R. and Northcote, D. H. (1977). *Biochem. J.* **166**, 603–618.
Nadelmann, H. (1890). *Jahrb. Wiss. Bot.* **21**, 1–83.
Nikaido, H. and Hassid, W. Z. (1971). *Adv. Carbohydr. Chem. Biochem.* **26**, 351–483.
Northcote, D. H. (1972). *Annu. Rev. Plant Physiol.* **23**, 113–132.
Northcote, D. H. (1974). *In* "Plant Carbohydrate Biochemistry" (J. B. Pridham, ed.), pp. 165–181. Academic Press, New York.
Odzuck, W. and Kauss, H. (1972). *Phytochemistry* **11**, 2489–2494.
Ovodov, Y. S. (1975). *Proc. Int. Symp. Carbohydr. Chem.*, 7TH, 1976 pp. 351–363.
Ovodov, Y. S., Ovodova, R. G., Shibaeva, V. I. and Mikheyskaya, L. V. (1975). *Carbohydr. Res.* **42**, 197–199.
Ovodova, R. G., Vashousky, V. E. and Ovodov, Y. S. (1968). *Carbohydr. Res.* **6**, 328–332.
Panayotatos, N. and Villemez, C. L. (1973). *Biochem. J.* **133**, 263–271.
Parker, M. L. (1976). Ph.D. Thesis, University of Wales, Bangor.
Pope, D. G. (1977). *Plant Physiol.* **59**, 894–900.
Ray, P. M. (1973a). *Plant Physiol.* **51**, 601–608.
Ray, P. M. (1973b). *Plant Physiol.* **51**, 609–614.
Ray, P. M. (1975). *Plant Physiol.* **56**, Suppl., 16, Abstr. 84.

Ray, P. M. (1980). *Biochim. Biophys. Acta* **629**, 431–444.

Rees, D. A. and Richardson, N. G. (1966). *Biochemistry* **5**, 3009–3107.

Reiss, R. (1889). *Landwirtsch. Jahrb.* **18**, 711–765.

Robinson, D. G. (1977). *Adv. Bot. Res.* **5**, 89–151.

Robinson, D. G., Eisinger, W. R. and Ray, P. M. (1976). *Ber. Dtsch. Bot. Ges.* **89**, 147–161.

Roden, L. and Swartz, N. B. (1975). *MTP Int. Rev. Sci., Biochem., Ser. One* **5**, 95–122.

Rosik, J. (1971). *Biologia (Bratislava)* **26**, 13–18.

Sarkar, K. P. (1963). *Sci. Cult.* **29**, 51.

Savada, D. and Chrispeels, M. J. (1971). *Biochim. Biophys. Acta* **227**, 278–287.

Savada, D. and Chrispeels, M. J. (1973). *Dev. Biol.* **30**, 49–53.

Schreuder, H. R., Cote, W. A. and Timell, T. E. (1966). *Sven. Papperstidn.* **60**, 641–657.

Schulze, E. (1895–1896). *Hoppe-Seyler's Z. Physiol. Chem.* **21**, 392–411.

Schulze, E. and Steiger, E. (1889). *Landwirtsch. Vers.-Stn.* **36**, 391–476.

Schulze, E. and Steiger, E. (1892). *Landwirtsch. Vers.-Stn.* **41**, 223–229.

Seita, R. C., Parthasarathy, M. V. and Shah, J. J. (1977). *Ann. Bot. (London)* [N.S.] **41**, 999–1007.

Siddiqui, I. R. and Wood, P. J. (1971). *Carbohydr. Res.* **17**, 97–108.

Siddiqui, I. R. and Wood, P. J. (1972). *Carbohydr. Res.* **24**, 1–9.

Siddiqui, I. R. and Wood, P. J. (1974). *Carbohydr. Res.* **36**, 35–44.

Siddiqui, I. R. and Wood, P. J. (1977). *Carbohydr. Res.* **53**, 85–94.

Smith, D. L. (1974). *Protoplasma* **79**, 41–57.

Smith, F. and Montgomery, R. (1959). *ACS Monogr.* Series No. 141.

Solov'eva, T. P., Arsenyuk, L. V. and Ovodov, Y. S. (1969). *Carbohydr. Res.* **10**, 13–18.

Stephen, A. M. and de Bruyn, D. C. (1967). *Carbohydr. Res.* **5**, 256–265.

Stoddart, R. W., Barrett, A. J. and Northcote, D. H. (1967). *Biochem. J.* **102**, 194–204.

Taiz, L. and Honigman, W. A. (1976). *Plant Physiol.* **58**, 380–386.

Taiz, L. and Jones, R. L. (1970). *Planta* **92**, 73–84.

Takeuchi, Y., Komamine, A., Saito, T., Watanabe, K. and Morikawa, N. (1980). *Physiol. Plant.* **48**, 536–541.

Talmadge, K. W., Keegstra, K., Bauer, W. D. and Albersheim, P. (1973). *Plant Physiol.* **51**, 158–173.

Toman, R., Karacsonyi, S. and Kovacik, V. (1972). *Carbohydr. Res.* **25**, 371–378.

Villemez, C. L. and Hinman, M. (1975). *Plant Physiol.* **56**, Suppl., 15, Abstr. 79.

Vogel, T. and Schleiden, M. J. (1839). *Ann. Phys. Chem.* pp. 327–330.

Wada, S. and Ray, P. M. (1978). *Phytochemistry* **17**, 923–931.

Wilder, B. M. and Albersheim, P. (1973). *Plant Physiol.* **51**, 889–893.

Wilkie, K. C. B. (1979). *Adv. Carbohydr. Chem. Biochem.* **36**, 215–264.

Wilkie, K. C. B. and Woo, S. L. (1977). *Carbohydr. Res.* **57**, 145–162.

Wolfrom, M. L. and Anderson, L. E. (1967). *J. Agric. Food Chem.* **15**, 685–687.

Wolfrom, M. L. and Patin, D. L. (1965). *J. Org. Chem.* **30**, 4060–4063.

Index